Springer Series in Synergetics   Editor: Hermann Haken

Synergetics, an interdisciplinary field of research, is concerned with the cooperation of individual parts of a system that produces macroscopic spatial, temporal or functional structures. It deals with deterministic as well as stochastic processes.

Instabilities and Chaos in Quantum Optics

Editors: F. T. Arecchi and R. G. Harrison

With 135 Figures

Springer-Verlag Berlin Heidelberg New York
London Paris Tokyo

Professor Dr. F. Tito Arecchi
Dipartimento di Fisica dell' Università and
Istituto Nazionale di Ottica, Largo E. Fermi 6,
I-50125 Firenze

Professor Dr. Robert G. Harrison
Department of Physics, Heriot-Watt University, Riccarton,
Edinburgh EH14 4AS, United Kingdom

Series Editor:
Professor Dr. Dr. h. c. Hermann Haken
Institut für Theoretische Physik der Universität Stuttgart, Pfaffenwaldring 57/IV,
D-7000 Stuttgart 80, Fed. Rep. of Germany and
Center for Complex Systems, Florida Atlantic University,
Boca Raton, FL 33431, USA

ISBN-13:978-3-642-71710-9   e-ISBN-13:978-3-642-71708-6
DOI: 10.1007/978-3-642-71708-6

Library of Congress Cataloging-in-Publication Data. Instabilities and chaos in quantum optics. (Springer series in synergetics ; v. 34) 1. Quantum optics–Congresses. 2. Lasers–Congresses. 3. Frequency stability–Congresses. 4. Chaotic behavior in systems–Congresses. I. Arecchi, F.T. II. Harrison, R.G. (Robert G.), 1944-. III. Series. QC446.15.I58 1987   535   86-31653

2153/3150-543210

Foreword

Since its advent more than a quarter of a century ago, the laser has turned out to be a magnificent device for investigations and applications in science and technology. Again and again we are surprised by the discovery of new types of lasers or by important new applications. From a more fundamental point of view the laser has become a beautiful system which allows us to study in great detail and with extreme precision phenomena which are typical of complex systems. In this way the laser has become an extremely useful tool in the investigation of general properties such as instabilities and deterministic chaos.

The laser has thus allowed the verification of new concepts in fields such as dynamic systems theory, bifurcation theory, catastrophe theory or, more generally, synergetics. At the same time we realize that the laser is a device exhibiting complex behavior extending far beyond the simple production of coherent light, which was the original reason for its construction. In this way we may expect important new applications of this kind of behavior, e.g., of deterministic chaos or other kinds of instabilities. Indeed, the study of these phenomena will be of great importance for the construction of all-optical computers, to mention just one example.

I am sure that this book, which has been edited and written by leading scientists in the field of quantum optics, will find a broad readership, ranging from students to research workers and professors, and will appeal to both experimental and theoretical physicists. I am convinced that engineers working in laser applications will also find very useful hints in this book.

Stuttgart, April 1987 *Hermann Haken*

Contents

Part II Instabilities and Chaos in Passive Systems

List of Contributors

Abraham, N.B.
Department of Physics, Bryn Mawr College, Bryn Mawr, PA 19010, USA

Adams, S.P.
Department of Physics, Widener University, Chester, PA 19013, USA

Albano, A.M.
Department of Physics, Bryn Mawr College, Bryn Mawr, PA 19010, USA

Al-Saidi, I.A.
Department of Physics, Heriot-Watt University, Riccarton,
Edinburgh EH14 4AS, United Kingdom
Permanent address: Department of Physics, College of Education,
Basrah University, Basrah, Iraq

Arecchi, F.T.
Dipartimento di Fisica dell'Università and Istituto Nazionale di Ottica,
Largo E. Fermi 6, I-50125 Firenze, Italy

Bandy, D.K.
Physics Department, Bryn Mawr College, Bryn Mawr, PA 19010, USA

Biswas, D.J.
M.D.R.S. (Physics Group), Bhabha Atomic Research Centre,
Bombay, 400 085, India

Bonifacio, R.
Dipartimento di Fisica dell'Università, Via Celoria 16,
I-20133 Milano, Italy

Casagrande, F.
Dipartimento di Fisica dell'Università, Via Celoria 16,
I-20133 Milano, Italy

Cecchi, S.
Istituto Nazionale di Ottica, Largo E. Fermi 6, I-50125 Firenze, Italy

Chyba, T.H.
Department of Physics and Astronomy, University of Rochester,
Rochester, NY 14627, USA

Dangoisse, D.
Laboratoire Spectroscopie Hertzienne, Université de Lille,
F-59655 Villeneuve d'Ascq, France

Derstine, M.W.
Honeywell Inc., Physical Sciences Center, 10701 Lyndale Avenue South,
Bloomington, MN 55420, USA

Firth, W.J.
Department of Physics and Applied Physics,
University of Strathclyde, John Anderson Building, 107 Rottenrow,
Glasgow G4 0NG, United Kingdom

Gibbs, H.M.
Optical Sciences Center, University of Arizona, Tucson, AZ 85721, USA

Gioggia, R.S.
Department of Physics, Widener University, Chester, PA 19013, USA

Giusfredi, G.
Istituto Nazionale di Ottica, Largo E. Fermi 6, I-50125 Firenze, Italy

Glorieux, P.
Laboratoire Spectroscopie Hertzienne, Université de Lille I,
F-59655 Villeneuve d'Ascq, France

Harrison, R.G.
Department of Physics, Heriot-Watt University, Riccarton,
Edinburgh EH14 4AS, United Kingdom

Hoffer, L.M.
Department of Physics, Bryn Mawr College, Bryn Mawr, PA 19010, USA

Hopf, F.A.
Optical Sciences Center, University of Arizona, Tucson, AZ 85721, USA

Jewell, J.L.
Optical Sciences Center, University of Arizona, Tucson, AZ 85721, USA
Present address: AT & T Bell Labs., Holmdel, NJ 07733, USA

Klische, W.
Physikalisch-Technische Bundesanstalt,
D-3300 Braunschweig, Fed. Rep. of Germany

Lawandy, N.M.
Division of Engineering and Department of Physics, Brown University,
Rhode Island, RI, USA

Lugiato, L.A.
Dipartimento di Fisica, Universitá di Milano, I-20133 Milano, Italy

Moloney, J.V.
Department of Physics, Heriot-Watt University, Riccarton,
Edinburgh EH14 4AS, United Kingdom

Narducci, L.M.
Department of Physics and Atmospheric Science, Drexel University,
Philadelphia, PA 19104, USA

Petriella, E.
Istituto Nazionale di Ottica, Largo E. Fermi 6, I-50125 Firenze, Italy and
CEILAP (CITEFA-CONICET)-Zufriategui, Vareca,
Villa Martelli, Argentina

Rushford, M.C.
Optical Sciences Center, University of Arizona, Tucson, AZ 85721, USA

Salieri, P.
Istituto Nazionale di Ottica, Largo E. Fermi 6, I-50125 Firenze, Italy

Sanders, L.D.
Optical Sciences Center, University of Arizona, Tucson, AZ 85721, USA

Tai, K.
Optical Sciences Center, University of Arizona, Tucson, AZ 85721, USA

Tarroja, M.F.H.
Department of Physics, Bryn Mawr College, Bryn Mawr, PA 19010, USA

Weiss, C.O.
Physikalisch-Technische Bundesanstalt,
D-3300 Braunschweig, Fed. Rep. of Germany

1. Introduction

F.T. Arecchi and R.G. Harrison

1.1 A Heuristic Approach to Instabilities

Let us see in a heuristic way, why instabilities and chaos are relevant to Quantum Optics, which was once considered as the realm of coherent, or ordered, phenomena. Already for $N = 3$ degrees of freedom a dynamic system is very different from the $N = 2$ problem since, in general, there may be asymptotic instabilities. This means a divergence, exponential in time, of phase space trajectories stemming from nearby initial points. The uniqueness theorem for solution of differential systems seems to offer an escape way: be more and more precise in localising the coordinates of the initial point. However, a fundamental difficulty arises. Only rational numbers can be assigned by a finite number of digits. A "precise" assignment of a real number requires an infinite acquisition time and an infinite memory capacity to store it, and neither of these two infinities is available to the physicist. Hence any initial condition implies a truncation. A whole range of initial conditions, even if small, is usually given and from within it trajectories may arise whose difference becomes sizeable after a given time, if there is an exponential divergence. This means that predicitions are, in general, limited in time and that motions are complex, starting already from the three-body case. In fact, we know nowadays from very elementary topological considerations that the nonlinear coupling of three degrees of freedom is already sufficient to yield a positive Lyapunov exponent, and accordingly an expanding phase space direction. Thus deterministic chaos is not due to coupling with a noise source as a thermal reservoir, but to unavoidable inaccuracy in setting the initial conditions.

When we consider a system made of many interacting components, three cases are conceptually possible [1.1]:

i) The mutual interactions can be linearised. A linear system can be diagonalised, that is, transformed into a set of uncoupled degrees of freedom or normal modes, thus we face just a one-body dynamics (compare, e.g., the phonon theory).

ii) The nonlinearities are essential, and we are in the presence of many coupled degrees of freedom. There may be however a heuristic rule to distinguish between a "system", which is the part we are interested in, and a "thermal bath", which is the rest of the world weakly perturbing the system. We have to use a statistical approach, and the fluctuation-dissipation theorem tells us that the effect of the bath is to provide a deterministic dissipation plus noise. The system may have a reasonable, but still large, number of degrees of freedom (3 in a single mode laser, 10^6 in a cubic centimeter of stirred water undergoing a turbulent motion). If all of them are equally "fast" (that is, have comparable damping times) then one has to expect deterministic chaos, and here only the recent developments in nonlinear dynamics are able to tackle some qualitative aspects of their motion.

iii) in the case (ii) it may happen that one (or a few) degrees of freedom are particularly slow (as for instance the amplitude of the tuned mode in a high Q electromagnetic cavity). On these long time scales the fast variables can be considered through their equilibrium values (adiabatic elimination) and the whole dynamics becomes low dimensional, sometimes even one- or two-dimensional. At variance with case (i) however, the resulting dynamics is highly nonlinear and multiple bifurcations may occur depending on the values of the "control parameters" which summarise the effect of the fast variables. Some of these bifurcations are equal to those already considered in thermodynamic phase transitions and may lead to highly ordered states starting from a disordered phase. This way, we recover the characteristic of order which is of one- and two-body problems.

Let us then see how disorder, order and deterministic chaos play a role in "open systems". We use this term for a many-body system far away from thermal equilibrium so that the linearisation around equilibrium no longer holds and nonlinear interactions play an essential role. Open systems transfer energy from a source to a sink. Source and sink can be large reservoirs at constant temperature, however the body can not be described in terms of a uniform temperature. This is, for instance, the case of a living being which receives nutrients and disposes of metabolic residuals. Depending on the values of the control parameters, the same open system can go from disorder to order and then to turbulence. Hence, the three levels appear no longer as three possible conceptual abstractions, but as three different behaviours of the same physical system.

Let us refer to a laser. As the energy supplied to the radiators is increased, there are three successive situations, namely:

a) **Disorder.** For low excitation, we have a regular lamp which ejects light in many directions and with many frequencies. The photoelectric output from a fast detector appears as a noisy signal without apparent correlations.

b) **Order.** Increasing the excitation above a threshold value the system undergoes a "symmetry breaking". Only one direction and one frequency are privileged and all the others are quenched. The photoelectric signal is highly ordered.

2

How did order spring from randomness: We sketch here a heuristic answer. In thermal equilibrium, any component of a large system executes a small motion described by linear dynamics (like the harmonic oscillator or the linearised pendulum). Such small motions are necessary to balance what each component gets from elsewhere and what it has to give back, in order to keep a constant temperature. In such small motion the component explores just its immediate surroundings and no long-range features can build up in the system. For larger motions, the components have necessarily to interfere with one another's behaviour, and beyond the threshold the motion becomes "cooperative", that is, ordered over a long range. In the laser case, as the field increases beyond a critical value, the excited molecules contribute by "stimulated emission" to build up the same "coherent" field, rather than emitting uncorrelated contributions.

c) Deterministic Chaos. The onset of order was explained in terms of all radiators driven by a single field which imposes its phase. Technically, this corresponds to a dynamics with a single relevant degree of freedom (the field) all the others being "slaved" to adapt instantly to the slowly varying main variable. We have already called this slaving as adiabatic elimination of the fast variables. When however the excitation increases above a second threshold, all the dynamical variables have to be considered on equal footing. The single-mode homogeneous-line laser happens to be described by just three collective variables: the electromagnetic field, plus the induced polarisation and the amount of population. But three is a sufficient number for having a positive Lyapunov exponent, and hence an expanding direction in phase space. Thus deterministic chaos may arise.

Similarly, we can describe the convective hydrodynamic instability. The physical system consists of a cell with rigid walls filled with a liquid and heated from below. As the temperature difference between bottom and top increases, we have three distinct regimes, namely:

a) Disorder. Initially the molecules are heated at the bottom, that is, they receive an excess of kinetic energy with respect to the average, and they transfer their excitation by individual molecular collisions. This process is heat conduction and no particular correlations appear with respect to the isothermal fluid.

b) Order. Above a critical temperature difference, convective motions start. Small regions of fluid near the bottom acquire a lower density and rather than rapidly thermalising with the surrounding, act as miniscule "bubbles" pushed up by buoyancy. This motion is correlated over the whole cell showing a "roll" feature, since the upward motion has to turn away at the top where the constraint of the wall stops the upward tendency.

c) Deterministic Chaos. For larger heating more degrees of freedom play a relevant role, thus expanding directions in phase space arise, with associated irregularities in the motion. The overall motion appears as turbulent.

Whatever has been said for active (laser) systems can be rephrased for passive systems, that is radiators in their ground states strongly perturbed by an external field. In both situations a crucial parameter is the adimensional "cooperation number" [1.2,3] that is the total number of radiators within a "cooperation length" [1.4]

$$C = \frac{g^2 N}{\gamma_\perp k} \tag{1.1}$$

where

$$g^2 = \frac{\omega \mu^2}{2\hbar \varepsilon_0 V} \tag{1.2}$$

is the coupling constant (ω is the transition frequency, μ the dipole moment, ε_0 the dielectric constant and V the radiating volume), $\nu_\perp$ is the decay rate of the induced dipoles, k the decay rate of photons (that is, k^{-1} is one transit time across volume V if there are no mirrors, and $1/\theta$ transit times for a cavity with a loss factor θ), and N the number of radiators. For active systems, C may be readily modified through N, by the pump strength whereas for passive systems, N corresponds to ground state population.

Here C plays the same role as the adimensional numbers which characterise hydrodynamic instabilities, viz. Reynolds, Rayleigh, Marangoni, Taylor numbers [1.1,5]. In fact $C = 1$ is the first laser threshold, $C = 9$ is the second Haken threshold [1.6], $C = 4$ is the threshold for absorptive optical bistability [1.7], and so on.

However, the crucial difference between quantum optics and hydrodynamical instabilities is that in quantum optics it is much easier to isolate a single domain (as the single-mode laser) whereas in hydrodynamics, even when working with cells with a small "aspect ratio", spatial features still play a role. From a theoretical point of view this means that in quantum optics a small set of coupled ordinary differential equations, in general, are a much better approximation than any sensible truncation of partial differential equations to a small number of ordinary equations, as done in hydrodynamics starting from Navier-Stokes equations and arriving, e.g., at the three Lorenz equations.

1.2 About This Book

This book consists of nine topics, five on active systems and four on passive ones.

The first part reviews the work done at Istituto Nazionale di Ottica (I.N.O.) Florence, on the onset of oscillatory instabilities and chaos in single-mode homogeneous-line lasers. Here a laser classification based on time scale considerations is introduced. Precisely, single-mode lasers are called class A, B, and C depending on whether, after suitable adiabatic elimination of fast variables, the laser dynamics is ruled by one, two or three equations, respectively. An example of a class A laser is the usual He-Ne 632.8 nm transition; an example of class B is any of the CO_2 transitions around 9 to 10 μm as, e.g., the $10P(20)$ line; an example of class C is the 81.5 μm rotational line of $^{14}NH_3$, even though the optical pumping makes the two level approximation somewhat questionable. Chapter 2 is devoted to class B lasers and it considers several means to increase the number of dynamical variables from two to three or more, in order to have a positive Lyapunov exponents, as: (i) loss modulation, (ii) injection of an external frequency, (iii) feedback of the output intensity upon the loss rate, and (iv) a bidirectional ring cavity. While (i) to (iii) represent an external manipulation of the laser dynamics, (iv) is an intrinsic chaos, based on the decoupling of forward and backward field waves which act as independent variables.

The limitation to low-dimensional chaos allows a close correspondence between experiments and theoretical models. Such a correspondence is by no means obvious. While in a demonstration kit, such as a mechanical or an eletronic device (e.g., two coupled pendulums or a driven nonlinear oscillator) the number of coupled variables is imposed by the design of the experiment, in a single-mode laser the small number of relevant degrees of freedom is due to the "slaving" of a large number of radiators by the same field [1.1].

Chapter 3 summarises recent results of detailed and ongoing experimental investigations of a single-mode laser with an inhomogeneously broadened medium. Recognised as particularly attractive experimental systems for the generation of instabilities, the threshold for such operation being at or near that for lasing, they have nevertheless posed formidable barriers to theoretical analysis which are only now being overcome [1.8,9]. The high gain 3.51 μm transition in xenon is selected for study, the laser cavity comprising a unidirectional ring-laser configuration popularly used in theoretical studies. Careful measurements of the output intensity and the electric field spectrum by heterodyne techniques identify many details of the development of pulsations that have hitherto been overlooked. At variance with the homogeneous case, it is here hard to assign a definite route to chaos. When chaos appears however, it is again low dimensional, thus showing again a drastic shrinking of the available phase space.

Chapter 4 is a theory of single and multimode operation of a laser with an injected signal. It is well known how difficult it still is to carry out global analyses of chaotic behaviour [1.8] at variance with the well established theories on two-dimensional nonlinear oscillators [1.9]. Thus the present approach is based on an accurate steady-state analysis, followed by numerical solutions and evaluations of the Lyapunov exponents, which allows detailed reconstruction of a route to chaos for a relevant range of injected amplitudes. The work reports several new results on the steady-state and linear-stability properties of this laser system. Notably off-resonant modes display wide domains of instabilitiy and suggest that earlier single mode dynamical studies should be re-examined.

Chapter 5 is a collaborative survey of the very recent experimental investigations by several groups devoted to far-infrared and mid-infrared optically-pumped lasers which work at a sufficiently low pressure to have a collisional broadening comparable with the cavity damping rate. Being class C, the three crucial degrees of freedom are present in single-mode operation so providing under resonant excitation one of the few practical systems identifiable under certain conditions with the Lorenz theoretical model [1.6]. Alternatively, off-resonant excititation leads to Raman lasing and results here identify these lasers as a further rich source of instability phenomenon.

Chapter 6 is a first-principle theory of the onset of collective instabilities in a free-electron laser (FEL). The expanding availability of FEL facilities across the world makes these devices among the most promising sources of tunable coherent radiation in the coming years. Thus the physics of their instabilities will receive great attention, and this theory, besides its intrinsic beauty, is going to be extremely valuable.

Part II consists of four chapters, one theoretical and three experimental, on instabilities and passive systems.

Chapter 7 is a theoretical analysis of the transverse structure of optical instabilities in passive media. Compared to familiar plane-wave analysis the more realistic Gaussian-beam problem extends significantly the allowable class of bifurcations and indeed period doubling cascades are shown to be an unlikely scenario to chaos at least in high-finesse resonators. Furthermore, self-focussing nonlinearities cause the beam profile to break up into transverse soliton or solitary wave structures that may exist as steady state of the field envelope or as spatially coherent structures that undergo regular or chaotic motion, the latter being analogous to experimentally measured low-level turbulence in fluids.

Chapter 8 is a review of recent work by the Tucson group on instabilities in both intrinsic and hybrid optical bistable systems. Besides characterising regenerative pulsations arising in intrinsic bistable semiconductor devices, e.g., GaAs and multiple quantum-well structures, potentially useful as square wave generations in optical logic systems, Ikeda instabilities [1.10] arising from time retardation effects inherent to nonlinear optical systems with feedback, are comprehensively analysed using a hybrid bistable optical device extending their earlier investigations in this area in quantifying the effect of noise and the observed paths to chaos. Preliminary observations of apparent instabilities and chaos in a novel intrinsic cavity-less bistable device are also described comprising sodium vapour.

Chapter 9 reports the Edingburgh research line on passive multistability and instabilities in molecular gases. Dispersive optical bistability and instabilities leading to period doubling, higher harmonics and chaos in all-optical passive quantum systems are discussed theoretically and experimentally. These phenomena are observed in both ring and Fabry-Perot resonators containing molecular gases as discrete level nonlinear media. The results are shown to be well modelled by a generalisation of standard two-level system theory to include standing wave, reservoir and transverse effects.

Chapter 10 reviews the recent Florence experiments on multistabilities and instabilities in a sodium filled Fabry-Perot. By eliminating the buffer gas and controlling accurately the magnetic fields, the Florence group was able to isolate the role of narrow velocity packets, localising various types of bifurcations at different detuning values between the impinging light and the $D1$ transition. While the set of data is rich, the theoretical interpretation is still incomplete. The interesting phenomenon of polarisation switch on which optical tristability was based is here associated with oscillatory changes of power between the two circular components. The coexistence of Zeeman pumping and hyperfine pumping seems to lead to a codimension-two bifurcation, as guessed from the experimental evidence of a line of Hopf bifurcations with smoothly changing frequencies.

As conclusion, we are here offering some samples of a new exciting field of radiation physics, which upsets many established ideas on the coherence and stability of quantum optical devices.

References

1.1   H. Haken: *Synergetics*, 3rd ed., Springer Ser. Syn., Vol. 1 (Springer, Berlin, Heidelberg 1983)
1.2   R. Bonifacio, P. Schwendimann, F. Haake: Phys. Rev. A4, 382, 854 (1971)
1.3   F.T. Arecchi: in *Order and Fluctuations in Equilibrium and Nonequilibrium Statistical Mechanics*, Proc. XVII Solvay Conf. on Physics, ed. by G. Nicolis et al. (Wiley, New York 1981)
1.4   F.T. Arecchi, E. Courtens: Phys. Rev. A2, 1720 (1970)
1.5   C. Normand, Y. Pomeau, M.G. Velarde: Rev. Mod. Phys. **49**, 581 (1977)
1.6   H. Haken: Phys. Lett. **53A**, 77 (1975)
1.7   R. Bonifacio, L. Lugiato: Opt. Commun. **19**, 172 (1976)
1.8   J. Guckenheimer, P. Holmes: *Nonlinear Oscillations, Dynamical Systems & Bifurcations of Vector Fields* (Springer, Berlin, Heidelberg 1983)
1.9   N. Minorslki: *Nonlinear Oscillations* (Van Nostrand, New York 1962)
1.10   K. Ikeda: Opt. Commun. **30**, 257 (1979)

2. Instabilities and Chaos
in Single-Mode Homogeneous Line Lasers

F.T. Arecchi

With 26 Figures

This chapter presents a review of the experimental investigation done in Florence on dynamical instabilities and deterministic chaos in Quantum Optics. In a dissipative system such as a laser we distinguish between a transient regime, strongly dependent on the initial conditions, and an asymptotic one, where the motion is confined on an attractor independent of the initial conditions. First, laser transients carry relevant information on the birth or death of a coherent state. Thus, a set of experiments is reported on the characterization of nonlinear transients and of their statistical features. Second, the onset of deterministic chaos is studied by referring to the invariant properties of low-dimensional attractors, in order to isolate the characteristics of chaos from the random fluctuations due to the coupling with a thermal reservoir. For this purpose, attention is focused on single-mode homogeneous-line lasers, whose dynamics is ruled by a low number of coupled variables. In the examined cases, experiments and theoretical model are in close agreement. In particular, when many attractors co-exist for the same parameter values (generalized multistability) the presence of random noise induces long lived transients with $1/f$ like low frequency spectra.

2.1 Background

Quantum optics from its beginning in 1960 with the first laser was considered as the physics of coherent and intrinsically stable radiation sources. Lamb's semiclassical theory [2.1] showed the role of the EM field in the cavity in ordering the phases of the induced atomic dipoles, thus giving rise to a macroscopic polarization and making possible a description in terms of very few collective variables. In the case of a single-mode laser and a homogeneous-gain line this meant just five coupled degrees of freedom, namely, a complex field amplitude E, a complex polarization P, and a population inversion ΔN. A corresponding quantum theory, even for the simplest model laser (the so called Dicke model, that is, a discrete collection of modes interacting with a finite number of two-level atoms) does not lead to a closed set of equations. However the interaction with other degrees of freedom acting as a thermal bath (atomic collisions, thermal radiation) provides truncation of high-order terms in the atom-field interation [2.2–4]. The problem may be reduced to five coupled

9

equations (the so-called Maxwell-Bloch equations) but now they are affected by noise sources to account for the coupling with the thermal bath [2.5]. Being stochastic, or Langevin equations, the corresponding solution in closed form refers to a suitable weight function or phase-space density. Anyway the average motion matches the semiclassical one, and fluctuations play a negligible role if one excludes the bifurcation points where there are changes of stability in the stationary branches. Leaving out the peculiar statistical phenomena which characterize the threshold points and which suggest a formal analogy with thermodynamic phase transitions [2.6] the main point of interest is that a single-mode laser provides a highly stable or coherent radiation field.

From the point of view of the associated information, the standard interferometric or spectroscopic measurements of classical optics, relying on average field values or on their first-order correlation functions, are insufficient. In order to characterize the statistical features of Quantum Optics it was necessary to make extensive use of photon statistics [2.7,8].

As discussed in detail in Sect. 2.3, coherence is equivalent to having a stable fixed point attractor and this does not depend on details of the nonlinear coupling, but on the number of relevant degrees of freedom. Since such a number depends on the time scales on which the output field is observed, coherence becomes a question of time scales. This is the reason why for some lasers coherence is a robust quality, persistent even in the presence of strong perturbations, whereas in other cases coherence is easily destroyed by the manipulations common in the laboratory use of lasers, such as modulation, feedback or injection from another laser.

This chapter is a review of the simplest variety of instabilities and chaos on active Quantum Optics. Precisely, Sect. 2.2 explores transient instabilities, related to the decay of an initial unstable state, Sect. 2.3 is a general presentation of low-dimensional chaos in lasers, including the description of the relevant measurements upon which any assessment on chaos has to rely. Sections 2.4–7 describe the investigations developed at Istituto Nazionale di Ottica, Firenze, and are respectively devoted to lasers with modulated losses, lasers with injected signals, lasers with feedback and bidirectional ring lasers. For convenience, mathematical derivations are collected in an appendix.

2.2 Transient Decay Toward a Stable State

As will be discussed in the next section, the notion of chaotic dynamics is based on the geometric properties of the attractor. This is defined heuristically as the locus of the phase space of a dissipative system where any initial condition asymptotically merges. For a more formal definition, see [2.9]. The simplest dynamical systems are the stable ones, where the attractor is just one point (as in a coherent laser) or two points (as in a bistable device) or a finite number of points (as in a multistable system). In such a case it is interesting to study the transient instability as the system precipitates from a generic initial condition toward the attracting fixed point.

A nonequilibrium system, under the action of external parameters, may undergo a transition in the sense that one (or a set) of its macroscopic observables have a sizeable change. Usually these changes are studied by a slow variation of the external parameter, in order to measure the stationary fluctuations and their associated spectra around each equilibrium point.

More dramatic evidence on the decay of an unstable state can be obtained by applying sudden jumps to the driving parameter and observing the statistical transients. The decay is initiated by microscopic fluctuations. In the first linear part of the decay process the fluctuations are amplified, hence during the transient, and until nonlinear saturation near the new stable point reduces them, fluctuations do not scale with the reciprocal of the systems size, as it is at equilibrium.

An experiment on the photon statistics of the laser field during its switch initiated this investigation [2.10]. Figures 2.1–3 give the transient photon statistics during a laser build-up and the associated average photon number and variance.

Similar experiments were afterward done on gas-liquid or magnetic transitions, and received the name of spinodal decomposition [2.11].

Limiting to the case of one stochastic amplitude x, the most natural approach was to measure the probability density $P(x,t)$ at a given time t after the sudden jump of the driving parameter. A time-dependent solution in terms of an eigenfunction expansion is unsuitable for the large number of terms involved, with the exception of small jumps near threshold [2.12] or the asymptotic behavior for long times.

Solving for the moments $\langle x^k(t) \rangle$ leads to an open hierarchy of coupled equations. A two-piece approximation [2.13] consists in first letting the system decay from the unstable point under the linearized part of the deterministic force, diffusing simultaneously because of the stochastic forces. This yields a

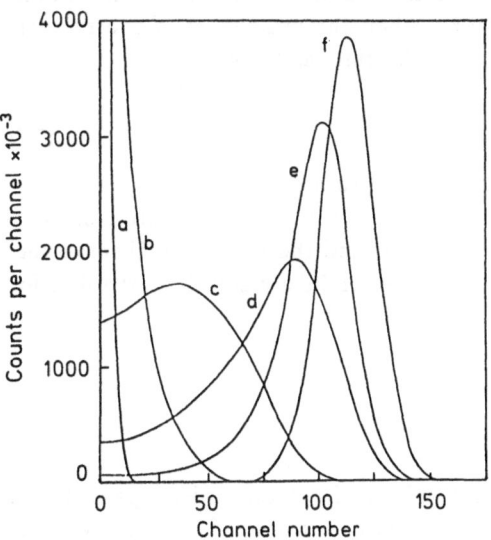

Fig. 2.1. Experimental statistical distributions with different time delays obtained on a laser transient. All distributions are normalized to the same area (a) 2.6 μs, (b) 3.7 μs, (c) 4.3 μs, (d) 5 μs, (e) 5.6 μs, (f) 8.8 μs

11

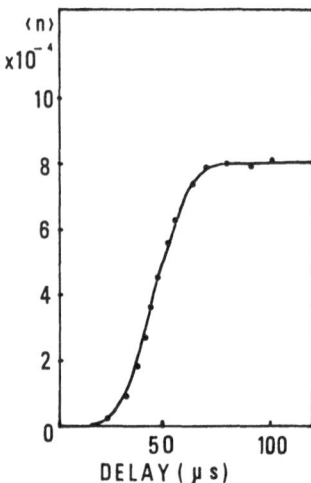

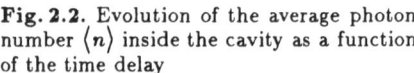

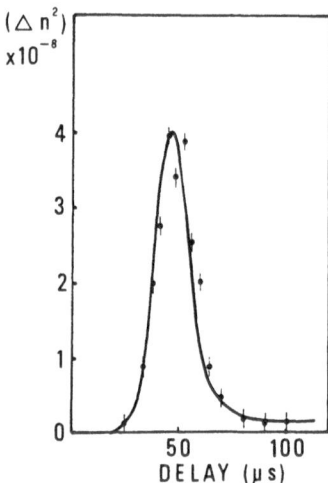

Fig. 2.2. Evolution of the average photon number $\langle n \rangle$ inside the cavity as a function of the time delay

Fig. 2.3. Evolution of the variance $\langle \Delta n^2 \rangle$ of the statistical distribution of photons inside the cavity, as a function of the time delay

short-time probability distribution of easy evaluation. Then we solve for the nonlinear deterministic path and spread it over the ensemble of initial conditions previously evaluated in the linear regime.

A different approach leads instead to closed moment equations of easy solution [2.14,15]. We consider the time t, at which a given threshold z_F is crossed, as the stochastic parameter, whose distribution $Q(t, z, z_F)$ in terms of the interval between the initial position z and z_F must be assigned. Here the time is no longer an ordering parameter but an interval limited by a start-stop operation. Let the position z get unstable under a force $F(z)$ and a noise delta-correlated with a correlation $D(z)$. Then the evaluation of the moment $T_m \equiv \langle t^m \rangle$ is given by a simple recurrence formula as [2.14]

$$F(z)T'_m + D(z)T''_m = -mT_{m-1} \qquad (2.1)$$

(the apex denoting differentiation with respect to z).

When we apply this formalism to the decay of unstable states, since D scales with the inverse system size, we can expand the results in D series and display the first relevant correction to the deterministic solution. We find for the mean time T_1

$$T_1(z) = \int_z^{z_F} \frac{dy}{F(y)} + \int_z^{z_F} dy \frac{dF}{dy} \frac{D}{F^3} \ , \qquad (2.2)$$

where the first term on the right-hand side is the deterministic part. For a spread in the initial position z, $T_1(z)$ should still be averaged over the set of z.

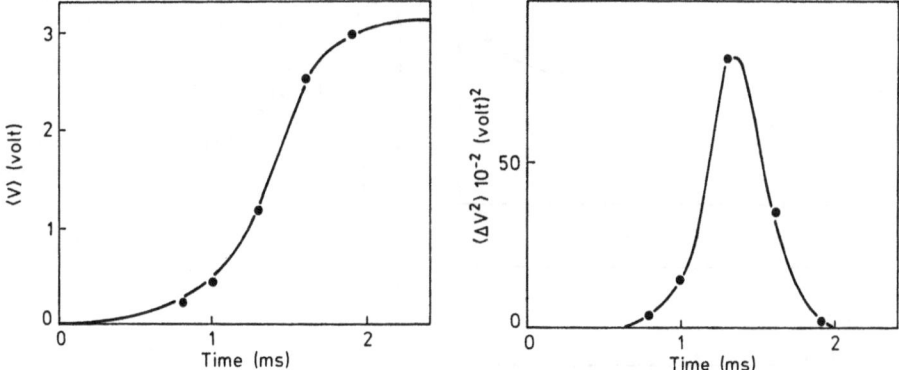

Fig. 2.4. Transient statistical evolution of an electron oscillator driven from below to above threshold by a sudden jump. No external noise added. Average amplitude V and variance. *Dots* denote experiment and *solid line* denotes theory

Similarly, performing the same approximation for T_2 we obtain for the variance $\Delta T \equiv T_2 - T_1^2$ the following relation

$$\Delta T = 2 \int\limits_{z}^{z_F} dy \, \frac{D(y)}{F^3(y)} \quad . \tag{2.3}$$

In order to show the power of this approach, we have measured the crossing time probability distributions for an electronic oscillator driven from below to above threshold [2.15].

Figure 2.4 gives the mean oscillator amplitude and its variance versus time as in the usual stochastic treatment of transients, Fig. 2.5 gives the variance of crossing times for increasing threshold as defined here.

The following comments convey some of the relevant physics: (i) the first term of (2.2) yields an average decay time which scales as $T_1 \approx \ln(N)$, that is, a logarithmic divergence with the system size N; (ii) a constant variance for increasing threshold means that the various trajectories are shifted versions of the same deterministic curve, and the noise scaling as $1/N$ plays a role only in spreading the initial condition; (iii) introduction of an external noise adds a fluctuation peculiar for each path, giving a ΔT dependent on z_F.

In conclusion, this new experimental characterization of a statistical transient shows a clear separation between the role of the initial spread and the noise along each path. In the case of superfluorescence, a delay time statistics in the pulse onset with respect to the preparation time has offered the most detailed tools for a comparison between measurements [2.16] and the corresponding theory [2.17].

Similar considerations apply also to bistable or multistable situations [2.15] and have been recently re-considered with specific reference to optical bistability [2.18].

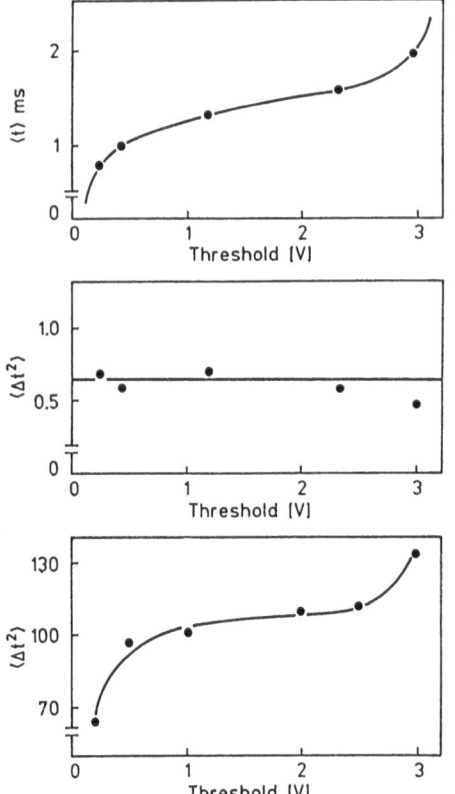

Fig. 2.5a–c. Transient oscillator as in Fig. 2.4. Statistical distribution of the time intervals between the initial condition and the crossing time (**a**), variance under the action of the internal noise (**b**), variance for an added external noise (**c**). In (**b**) and (**c**) the scale is in units of $10^{-9}\,s^2$. *Dots* denote experimental and *solid line* denotes theory

2.3 Deterministic Chaos

2.3.1 Historical Aspects

Until recently the current point of view as that of few-body dynamics was fully predictable, and that only addition of noise sources, due to coupling with a thermal reservoir, could provide statistical fluctuations. Lack of long-time predictability, or turbulence, was considered as resulting from the interaction of a large number of degrees of freedom, as in a fluid above the critical Reynolds number (Landau-Hopf model of turbulence).

On the contrary, it is now known that even in systems with few degrees of freedom nonlinearities may give rise to expanding directions in phase space and this, together with the lack of precision in assigning initial conditions, is sufficient to induce a loss of predictability over long times.

This level of dynamical description was born with the three-body problem in celestial mechanics (Poincare). Already a three-body dynamic system is very different from the two-body problem since, in general, there are asymptotic instabilities. This means a divergence, exponential in time, of two phase space trajectories stemming from nearby initial points. The uniqueness theorem for solutions of differential systems seems to offer an escape way: be more and more precise in localizing the coordinates of the initial point. However a fundamental difficulty arises. Only rational numbers can be assigned by a finite number of digits. A "precise" assignment of a real number requires an infinite acquisition time and an infinite memory capacity to store it, and neither of these two infinities is available to the physicist. Hence any initial condition implies a truncation. A whole range of initial conditions, even if small, is usually given and from within it trajectories may arise whose difference becomes sizeable after a given time, if there is an exponential divergence. This means that predictions are in general limited in time and that motions are complex, starting already from the three-body case. In fact, we know nowadays from very elementary topological considerations that a three-dimensional phase space corresponding to three coupled degrees of freedom is already sufficient to yield a positive Lyapunov exponent, and accordingly an expanding phase-space direction. This complexity is not due to coupling with a noise source as a thermal reservoir, but to sensitive dependence on initial conditions. It is called determinstic chaos.

The birth of this new dynamics was motivated by practical problems, such as fixing the orbit of a satellite or forecasting meteorology [2.19], and it was strongly helped by the introduction of powerful computers. The mathematics of multiple bifurcations leading from a simple behavior to a complex one is under current investigation. Some regularities, such as the "scenarios" [2.10] or routes to deterministic chaos, have already been explored. We are hopefully on the verge of a new formalism, which will describe in a unified way the passage from order to complicated behaviors such as developed turbulence in a fluid.

2.3.2 Dynamical Aspects

A dissipative system (i.e., with damping terms) does not conserve the phase-space volume. If we start with initial conditions confined in a hypersphere of radius ε, that is, with an initial phase volume

$$V_0 = \varepsilon^N$$

as time goes on, the sphere transforms into an ellipsoid with each axis modified by a time dependent factor. Its volume is

$$V_t = \varepsilon^N \exp\left(\Sigma_i \lambda_i t\right)$$

(λ_i: Lyapunov exponents). Since the volume has to contract, $V_t < V_0$, then

$$\Sigma_i \lambda_i < 0 \ . \tag{2.4}$$

We denote the sequence of λ exponents, starting from the smallest up to the highest as the Lyapunov spectrum. Let us consider for simplicity just the signs of nonzero λ_i, keeping the zero for $\lambda_i = 0$. We then describe a sequence of, negative, zero and positive λ_i as, e.g., $(--0+)$. For $N = 1$, we have $(-)$ and a segment $V_0 = \varepsilon^1$ of initial conditions shrinks to a single point for $T \to \infty$, that is, the attractor is a fixed point. For $N = 2$ the system goes either to a fixed point $(--)$, or to a limit cycle (-0). Chaotic motion $[(-+)$ with $\lambda_+ < |\lambda_-|]$ is forbidden in two dimensions by the Poincaré-Bendixon theorem. For $N = 3$, besides fixed point $(---)$, and limit cycle $(--0)$, we can have motion on a torus with two incommensurate frequencies (-00), but we can also have $(-0+)$, that is, a positive λ which gives an expanding direction along which we rapidly get uncertainty.

An example of chaotic motion is offered by the Lorenz model of hydrodynamic instabilities [2.19] which corresponds to the following equations where the parameter values have been chosen in order to yield one positive Lyapunov exponent:

$$\dot{x} = -10x + 10y \ ,$$
$$\dot{y} = -y + 28x - xz \ ,$$
$$\dot{z} = -(8/3)z + xy \ . \tag{2.5}$$

The above considerations suggest the system will exhibit low-dimensional chaos, with the simplest phase-space topology allowing for the appearence of a positive Lyapunov exponent.

Focusing on these situations in quantum optics permits close comparison between experiments and theory. By purpose, I do not tackle the vast class of inhomogeneously broadened lasers, where it is extremely difficult to drive close correspondences between experiments [2.21] and theory [2.22] because of the large number of coupled degrees of freedom (Sect. 2.2.2).

If we couple Maxwell equations with Schrödinger equations for N atoms confined in a cavity, and expand the field in cavity modes, keeping only the first mode E which goes unstable, this is coupled with the collective variables P and Δ describing the atomic polarization and population inversion as follows, see (2.A.5)

$$\dot{E} = -kE + gP \ ,$$
$$\dot{P} = -\gamma_\perp P + gE\Delta \ ,$$
$$\dot{\Delta} = -\gamma_{||}(\Delta - \Delta_0) - 4g\,PE \ . \tag{2.6}$$

For simplicity, we consider the cavity frequency at resonance with the atomic resonance, so that we can take E and P as real variables and we have three coupled equations. Here, k, $\gamma_\perp$, $\gamma_{||}$ are the loss rates for field, polarization and population, respectively, g is a coupling constant and Δ_0 is the population inversion which would be established by the pump mechanism in the atomic medium, in the absence of coupling. While the first equation comes from Maxwell equations, the two others imply the reduction of each atom to a two-level atom resonantly coupled with the field, that is, a description of each atom in a isospin space of spin $1/2$. The last two equations are like Bloch equations which describe the spin precession in presence of a magnetic field. For such a reason (2.6) are called Maxwell-Bloch equations.

The presence of loss rates means that the three relevant degrees of freedom are in contact with a "sea" of other degrees of freedom. In principle, (2.6) could be deduced from microscopic equations by statistical reduction techniques [2.5]. A simple derivation is shown in the Appendix 2.A.

The similarity of Maxwell-Bloch equations (2.6) with Lorenz equations (2.5) would suggest the easy appearence of chaotic instabilities in single-mode, homogeneous-line lasers. However, time-scale considerations rule out the full dynamics of (2.6) for most of the available lasers. Equations (2.5) have damping rates which lie within one order of magnitude of each other. On the contrary, in most lasers the three damping rates are wildly different from one another.

The following classification has been introduced [2.23]

Class A (e.g., He-Ne, Ar, Kr, dye): $\gamma_\perp \simeq \gamma_{||} \gg k$. The two last equations of (2.6) can be solved at equilibrium (adiabatic elimination procedure) and one single nonlinear field equation describes the laser. $N = 1$ means fixed point attractor, hence coherent emission.

Class B (e.g., ruby, Nd, CO): $\gamma_\perp \gg k \gtrsim \gamma_{||}$. Only polarization is adiabatically eliminated (middle equation of (2.6)) and the dynamics is ruled by two rate equations for field and population. $N = 2$ allows also for period oscillations.

Class C (e.g., FIR lasers) $\gamma_{||} \approx \gamma_\perp \approx k$. The complete set of (2.6) has to be used, hence Lorenz like chaos is feasible (Chap. 5).

We have carried a series of experiments on the birth of deterministic chaos in CO_2 lasers (class B). In order to increase, by at least 1, the number of degrees of freedom, we have tested the following configurations:

i) Introduction of a time dependent parameter to make the system non autonomous [2.24]. Precisely, an electro-optical modulator modulates the cavity losses at a frequency near the proper oscillation frequency Ω provided by a linear stability analysis, which for a CO_2 laser happens to lie in the 50–100 KHz range, providing easy and accurate sets of measurements.

ii) Injection of a signal from an external laser detuned with respect to main one, choosing the frequency difference near the above mentioned Ω. With respect to the external reference the laser field has two quadrature components which represent two dynamical variables. Hence we reach $N = 3$ and observe chaos [2.23].

iii) Use a bidirectional ring, rather than a Fabry-Perot cavity [2.25]. In the latter case the boundary conditions constrain the forward and backward waves, by phase relations on the mirror, to act as a single standing wave. In the former case forward and backward waves have just to fill the total ring length with an integer number of wavelengths but there are no mutual phase constrains, hence they act as two separate variables. Furthermore, when the field frequency is detuned with respect to the center of the gain line, a complex population grating arises from interference of the two counter-going waves, and as a result the dynamics becomes rather complex, requiring $N>3$ dimensions.

iv) Add an overall feedback, besides that provided by the cavity mirrors, by modulating the losses with a signal provided by the output intensity [2.26]. If the feedback has a time constant comparable with the population decay time, it provides a third equation sufficient to yield chaos.

Notice that while methods (i), (ii) and (iv) require an external device, (iii) provides intrinsic chaos. In any case, since feedback, injection or modulation are currently used in laser applications, the evidence of chaotic regions puts a caution on the optimistic trust in the laser coherence.

Of course, the requirement of three coupled nonlinear equations does not necessarily restrict attention to just the Lorenz equations. In fact, none of the explored case (i to iv) corresponds to Lorenz chaos.

2.3.3 Information Aspects

Here, we discuss what we measure to assess chaos. We plot two of the three (or more) variables on a plane phase-space projection. This way, we build projections of phase space trajectories on an $x - y$ oscilloscope. Simultaneously we can measure the power spectrum. In Sect. 2.4 a sequence of subharmonic bifurcations are shown, which eventually leads to an intricated trajectory (strange attractor) and to a continuous power spectrum. But how can we discriminate between deterministic chaos and noise? After all, noise also would give a continuous spectrum, and the phase space point would fill ergodically part of the plane, thus covering a two-dimensional set.

In order to discriminate deterministic chaos from order as well as from random noise, we introduce two invariants of the motion, one static the other dynamic.

We partition the phase space into small boxes of linear size ε and give ith box a probability $p_i = M_i/M$ equal to fractional number of times it has been visited by the trajectory. This way, we build a Shannon information $I(\varepsilon)$, and with it an "information dimension" $D_1(\varepsilon)$ [2.27] which is, in general, a fractional number, or a "fractal":

$$I(\varepsilon) = - \sum_i p_i \log p_i \ , \tag{2.7}$$

$$D_1(\varepsilon) = - \lim_{\varepsilon \to 0} \frac{I(\varepsilon)}{\log \varepsilon} \ . \tag{2.8}$$

To understand the meaning of a fractal, look up an operational definition of dimension [2.28]. Let us compare three sets: (i) a segment of unit length; (ii) the Cantor set, built by taking out the middle one-third of the unit segment and repeating the operation on each fragment; (iii) the Koch curve, built by replacing the middle third with the other two sides of an equilateral triangle and repeating the operation add infinitum. At each stage of the partition, we cover each set with beads of suitable size not to lose in resolution (e.g., diameter $1/3$ at the first partition) and count the number N for each set (at the first partition, we need 2 for the Cantor set, 3 for the segment, 4 for the Koch curve). We define the fractal dimension as the ratio

$$D_0(\varepsilon) = \frac{\log N(\varepsilon)}{\log 1/\varepsilon} \ .$$
(2.9)

This definition is independent of the partition. Indeed, for the Cantor set and the Koch curve we have $N = 2$, $\varepsilon = 1/3$ and $N = 4$, $\varepsilon = 1/3$, respectively, at the first partition, yielding

$$D_0(\text{Cantor}) = \frac{\log 2}{\log 3} \simeq 0.63\ldots\ , \quad \text{and}$$

$$D_0(\text{Koch}) = \frac{\log 4}{\log 3} \simeq 1.2618\ldots\ .$$

At the second partition the number of necessary beads goes as N^2 and the diameter of each as ε^2, hence D_0 remains invariant.

Going back to the information dimension $D_1(\varepsilon)$ we see that we have replaced $\log N$ with $I(\varepsilon)$ which is an *average* [for p_i all equal, we recover $I(\varepsilon) = \log N$]. Hence D_1 generalizes D_0 whenever the density of points is not uniform along the trajectory.

As D_0 was independent of the partition stage, similarly D_1 is an invariant, but static (time does not enter). It can be shown that $D_0 \geq D_1$, however for non pathological sets the difference is irrelevant. Let us refer for simplicity to an $N = 3$-dimensional phase space. If $D = 0$ (fixed point) or 1 (limit cycle) or 2 (torus) we have an ordered, or coherent, motion. In the other limit of random noise, fluctuations fill ergodically an N dimensional region of the space, hence $D = 3$. Deterministic chaos has to be in between, that is

$$2 < D < 3 \ .$$

Hence, a fractal dimension is an indicator of chaos. As we show later, this indicator is expressed in term of correlation functions thus, it requires the same measuring techniques introduced in photon statistics.

These features related to the topology of the attractor have a temporal counterpart in another invariant, which measures how information is dissipated in a motion to maintain knowledge on the system. To build this dynamic invariant, we partition both space and time in boxes of sizes ε and τ that we name $i_1, i_2, \ldots i_d$ at each of the discrete times $\tau, 2\tau, \ldots d\tau$, and introduce the joint probability over the d time intervals,

$$p_{i_1 i_2 \ldots i_d} \equiv \{x(t = \tau) \subset i_1; \ldots; x(t = d\tau) \subset i_d\} \ .$$

Correspondingly, we define a joint information

$$I_d(\varepsilon) = - \sum_{\{i_1...i_d\}} p_{i_1...i_d} \log p_{i_1...i_d} \; . \tag{2.10}$$

Then, by a limit operation, define the Kolmogorov entropy as the rate of information loss per unit time

$$K \equiv \lim_{\tau \to 0} \lim_{\varepsilon \to 0} \lim_{d \to \infty} \frac{1}{d\tau} \sum_{n=1}^{d} (I_{n+1} - I_n) = \lim \frac{1}{d\tau} I_d \; . \tag{2.11}$$

Now we have two indicators to gauge the difference among *order, random noise* (Brownian motion) and *deterministic chaos*. Referring to K, it is easily seen that

$K = 0 \quad$ for order (no information loss) ;

$K = \infty \quad$ for random noise (total information loss) ;

$0 < K < \infty \quad$ for deterministic chaos .

The box counting method desribed above is impractical. It may require 10^6 points for a convergent numerical result. On the contrary, the following method introduced by *Grassberger* and *Procaccia* [2.29] is applicable to only 10^3–10^4 independent data points. We generalize Shannon information defining the order-f information as

$$I_f(\varepsilon) = \frac{1}{1-f} \ln \sum_i p_i^f \; . \tag{2.12}$$

For $f \to 1$ we recover the usual definition. Associated with I_f, there is an order-f dimension of the attractor

$$D_f = \lim_{\varepsilon \to 0} \frac{I_f(\varepsilon)}{\ln 1/\varepsilon} \; . \tag{2.13}$$

For $f = 0$ and 1 we recover D_0 and D_1. Consider $f = 2$. The sum $\sum p_i^2$ is just the probability that a pair of random points on the attractor fall into the same box, that is, that two arbitrary points will have a distance less than ε. Calling this probability $C(\varepsilon)$, we expect thus

$$C(\varepsilon) = \lim_{N \to \infty} \frac{1}{N^2} \sum_{ij} \theta(\varepsilon - |\mathbf{x}_i - \mathbf{x}_j|) \simeq \varepsilon^{D_2} \tag{2.14}$$

where $C(\varepsilon)$ is measured as the number of pairs (i, j) with a distance $|\mathbf{x}_i - \mathbf{x}_j| < \varepsilon$. In (2.14), θ is the Heaviside step function.

Experimentally, we do not measure at each time the vector $\mathbf{x}(t)$ of phase space, but just one component $x_i(t)$ (for instance, just the light out of a laser).

However, in a nonlinear system, any component $x_k(t)$ will influence x_i at a later time (no normal mode transformation !). Hence, we can build an m-dimensional phase space $\boldsymbol{\xi}(t)$ by just measuring one single component x_i at successive times and considering the m-fold as a single point in m space:

$$\boldsymbol{\xi}(t) \equiv [x_i(t), x_i(t + \tau), \ldots x_i(t + (m - 1)\tau)] \ . \tag{2.15}$$

As we evaluate the slope $\log C$ vs $\log \varepsilon$ from our data, we can stop from increasing m when the slope shows saturation. The saturated slope is D_2.

2.3.4 Role of Transients: The Hyperchaos

Nonlinear dissipative systems can have many simultaneously coexisting basins of attraction (generalized multistability – GM). This situation can be destabilized by changes of the control parameters, merging two independent attractors into a single one via an intermediate region which is only sporadically visited near the transition. The associated dynamics implies a low frequency tail (deterministic diffusion) [2.30]. Vice versa, when the above coexistence is stable, application of external noise may induce jumps between two otherwise disjoint regions of phase space.

The simultaneous presence of deterministic chaos and noise should not introduce new features within one attractor, since trajectories are already irregular. When however many attractors coexist for the same parameters, addition of noise makes it possible to leave a basin and go to another (which would be otherwise forbidden by the uniqueness theorem). This "hyperchaos" gives rise to low frequency spectra of $1/f$ type.

Clear evidence of generalized multistability was first shown in an electronic oscillator [2.31] and then in the modulated laser [2.24]. In both cases, besides the qualitative appearance of different attractors in phase space, there was a low-frequency spectral component due to noise-induced jumps among different attractors. Both measurements, however, might be considered as experimental artifacts. In fact, there is evidence of single attractors made of two sub-regions with infrequent passages from one to the other (see, e.g., the Lorenz attractor). In such a case, the low-frequency tail corresponds to the sporadic passages, and does not require added noise, (deterministic diffusion). As a matter of fact, power spectra do not permit discrimination between the two phenomena.

An analysis of the role of noise in GM was given for a cubic iteration map, allowing for two simultaneous attractors [2.32a], and in the numerical studies of a forced Duffing oscillator with a double potential well, in a parameter region allowing for the simultaneous existence of more than one attractor [2.32b].

A recent solution of the $1/f$ spectral problem [2.33] is based on the double randomness due to both the irregular deterministic motion with long lived transients, where the trajectory wanders near a fractal basin boundary, and the presence of stochastic noise. In this case, $1/f$ behavior can be accurately traced over more than three decades in frequency.

2.4 The Modulated Laser

For a single-mode (class B) laser tuned at the center line, the phase space becomes two-dimensional, see (2.A.17). However introduction of a time-dependent parameter makes the system non autonomous adding a third equation, see (2.A.16), thus making possible the appearence of a positive Lyapunov exponent. It is then a pratical matter to localize the values of the control parameters (pump, modulation frequency and amplitude) for which this will occur.

When we apply a time-dependent loss the evolution follows (2.A.16,17) given as

$$\dot{I} = 2kI(z-1) \ ,$$
$$\dot{z} = \gamma_{||}(z_0 - z - zI) \ ,$$
$$\dot{k} = mk_1\Omega \sin \Omega t \ . \tag{2.16}$$

For $m \to 0$, we have small deviations from the equilibrium values

$$\bar{I} = z_0 - 1 \ ,$$
$$\bar{z} = 1 \ . \tag{2.17}$$

These deviations are linear in m and synchronous with the external frequency Ω. Destabilization of this limit cycle has to be dealt with by the Floquet theory [2.34]. It may be shown that even for $m \to 0$ a nonlinear resonance yields a positive Lyapunov exponent for Ω around the characteristic frequency given by the imaginary part of (2.A.21), that is

$$\Omega \simeq \sqrt{2k_1\gamma_{||}(z_0 - 1)} \ . \tag{2.18}$$

For a CO_2 laboratory laser near threshold $k \sim 3 \times 10^7 \mathrm{s}^{-1}$, $\gamma_{||} \sim 10^4 \mathrm{s}^{-1}$, and $z_0 - 1 \sim 0.1$ (10 % above threshold), the corresponding frequency $f = \Omega/2\pi$ is in the 50 kHz range, easily accessible.

Thus we have made two series of experimental observations, the first [2.24] devoted to an experimental assessment of chaotic instabilities by phase-space portraits and power spectra, the second [2.35] to fractal dimensions and Kolmogorov entropy.

The driving frequency f was chosen to vary in the region from $\Omega/2\pi$ to its third harmonic, that is from 60 to 190 kHz. We have explored modulation values between 1 % and 5 %. A complete state diagram would yield the dynamical features for all possible values of the modulation parameters m and Ω. However, the strip $m = 1 \% - 5 \%$ does not display m dependence; therefore we limit ourselves to giving experimental results at $m = 1 \%$ for various Ω values.

The experimental setup consists of a CO_2 laser carefully stabilized against thermal and acoustic disturbances, with the discharge current stabilized to better than $1/10^4$. No long-term stabilization was necessary. The electro-optical modulator was a CdTe, antireflex-coated, 6-cm-long crystal, with an absorption less than 0.2 %. The laser cavity includes also a $\lambda/4$ plate and a beam expander, both coated to limit the total losses per pass to 20 %. The laser output is detected on a fast (2.5-ns rise time) pyroelectric detector whose current, proportional to the photon number $n(t)$, is sent together with its time derivative $\dot{n}(t)$ to an $x-y$ scope, in order to have the phase-space portrait $(n, \dot{n})$. The detector is also sent to a Rockland spectrum analyzer to measure the power spectra. The limited range (up to 100 kHz) of the spectrum analyzer limited the frequency range explored in this first run. We show later that interesting bifurcations are also expected in 180-kHz domain.

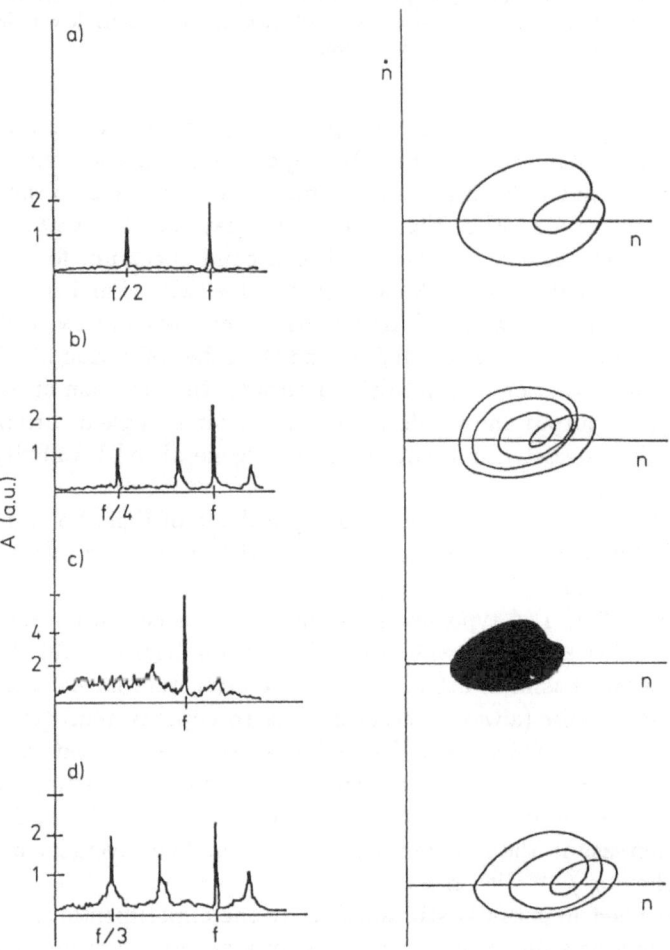

Fig. 2.6. Experimental phase-space portraits $(n - n')$ *(right side)* and the corresponding frequency spectra *(left side)* for different modulation frequencies f. *(a)* $f = 62.75$ kHz. Period-two limit cycle and corresponding $f/2$ subharmonic. *(b)* $F = 63.80$ kHz. Period-four limit cycle and $f/4$ subharmonic. *(c)* $f = 64.00$ kHz. The phase-space portrait shows a strange attractor (the oscilloscope spot could not resolve single windings). The power spectrum is a quasicontinuous one with a small peak at the modulation frequency (see the scale change with respect to previous figures). *(d)* $f = 64.13$ kHz. Period-three limit cycla and $f/3$ subharmonic

23

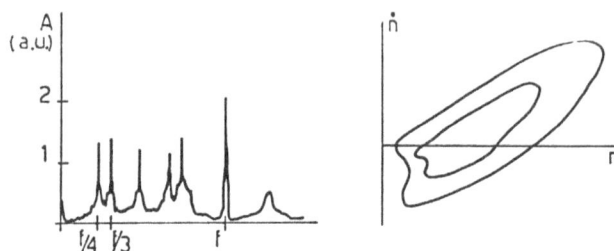

Fig. 2.7. $f = 63.85\,\mathrm{kHz}$. Experimental evidence of generalized multistability (coextistence of two independent attractors). The power spectrum shows that those attractors correspond to $f/3$ and $f/4$ subharmonic bifurcations, respectively; in phase space, the multiple windings merged within the thickness of the phase portrait contour

In Fig. 2.6 we show experimental data in a narrow region between 62.7 and 64.25 kHz where various bifurcations occur. This region is limited above and below by wide intervals with stable single-period limit cycles. Figure 2.6a shows the $f/2$ bifurcation at $f = 62.7\,\mathrm{kHz}$, Fig. 2.6b the $f/4$ case for $f = 63.8\,\mathrm{kHz}$; Fig. 2.6c exhibits the strange attractor and a broad-band spectrum for $f = 64.0\,\mathrm{kHz}$; and Fig. 2.6d displays the $f/3$ case for $f = 64.2\,\mathrm{kHz}$. Furthermore, at $f = 63.85\,\mathrm{kHz}$ a new feature appears, namely the coexistence of two independent stable attractors, one of period $4(f/4)$ and the other of period $3(f/3)$ (Fig. 2.7). This bistable situation has nothing to do with the common optical bistability where two dc output amplitude values appear for a single dc driving amplitude. We call this coexistence of two attractors "generalized bistability" (Sect. 2.3.4).

In Figs. 2.8 and 9 we report the theoretical equivalents of Figs. 2.6 and 7, respectively, obtained by computer solution of (2.16) with parameter values in the range of the experiment.

As stated in Sect. 2.3, $1/f$ type low-frequency divergences, with power spectra as $f^{-\alpha}$ ($\alpha = 0.6$–1.2), appear when the following conditions are fulfilled: (i) There are at least two basins of attraction; (ii) the attractors have become strange and any random noise (always present in a macroscopic system) acts as a bridge, triggering jumps between them. These jumps have the $f^{-\alpha}$ feature. In the region of bistability (Fig. 2.7) we have increased the modulator amplitude m up to the point where the two attractors have become strange. Figure 2.10 shows the sudden increase in the low-frequency spectrum. The divergent part has a power-law behavior $f^{-\alpha}$ with $\alpha \approx 0.6$.

The above described first run is still affected by the experimental uncertainties which characterize a phase space projection or a power spectrum. Does the first one show a self-similar structure beyond the chaotic threshold, as theoretically expected for a strange attractor, or does it just fill ergodically a two-dimensional region of $(n, \dot{n})$ plane, thus being trivial random noise? After all the latter test (continuous frequency spectrum) is also a common property of random noise and it is not a sufficient characterization of deterministic chaos.

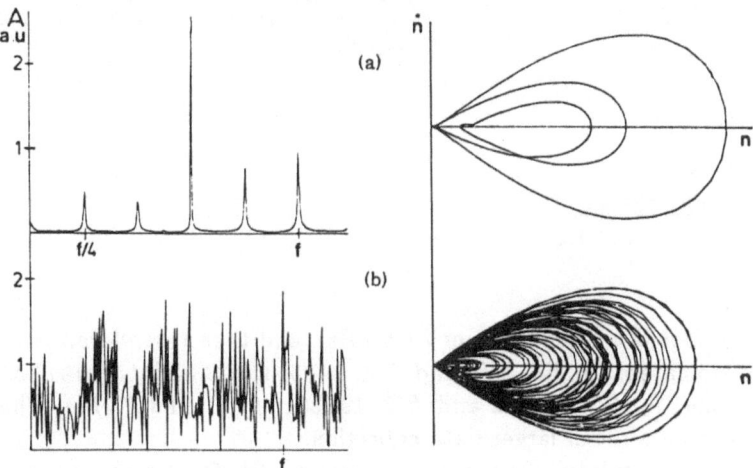

Fig. 2.8. Computer plots for the parameter values $\gamma_\| = 10^3\,s^{-1}$, $k = 7 \times 10^7\,s^{-1}$, $m = 2.0 \times 10^{-2}$, $\Delta_0 = 2.0 \times 10^{11}$. *(a)* $f = 64.33\,kHz$. Subharmonic bifurcation $f/4$, as in the experiment of Fig. 2.6b. *(b)* $f = 78.8\,kHz$, $m = 3 \times 10^{-2}$. Strange attractor and broad spectrum corresponding to a chaotic solution

In order to set a more reliable distinction between chaos and random noise, and also to specify the route to chaos (Fig. 2.6 is only a preliminary evidence of a Feigenbaum, n subharmonic, route), we have improved the stability of our system. This time, a stable $f/8$ subharmonic frequency and even an $f/10$ periodic window inside the chaotic region have been observed. To give an idea of the reliability of the apparatus we report here a series of behaviors observed at 2% modulation depth for slight changes of the modulation frequency, controlled via a programmable synthesizer driven by a microprocessor. In the following

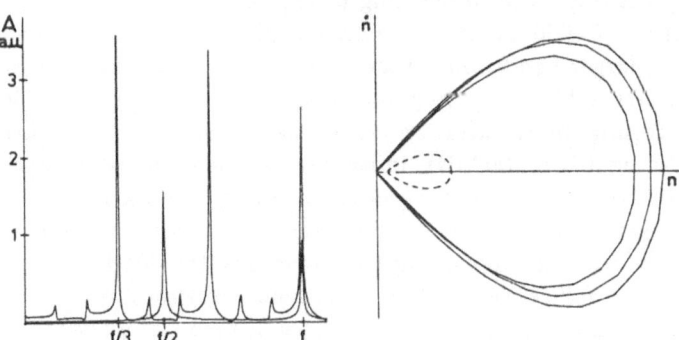

Fig. 2.9. Theoretical generalized bistability; $f = 119.0$ kHz, $m = 2.0 \times 10^{-2}$. The phase-space portrait shows the existence of two independent attractors, corresponding to the subharmonic frequencies $f/2$ (- - -) and $f/3$ (——); relative spectra are superimposed. It must be noted that one attractor remains inside the other as in the experiment of Fig. 2.7. If initial conditions are properly changed, a third attractor is found with a subharmonic frequency $f/10$ (not plotted for the sake of simplicity). Initial conditions: $n_0 = 4 \times 10^8$, $\dot{n}_0 = 0$ (- - -), $n_0 = 2 \times 10^5$, $\dot{n}_0 = -2 \times 10^6$ (——)

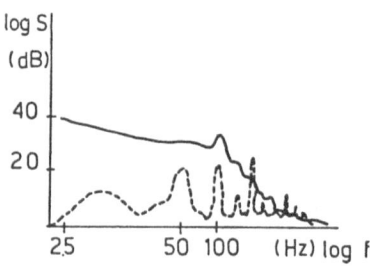

Fig. 2.10. Experimental power spectra in the case of two attractors, stable (- - -), and strange (——)

sequence the number is the set frequency (in kHz), and then the relevant sub-harmonics are indicated: 191 290, $f/5$ and $f/4$; 191 313, $f/3$ and $f/4$; 191 320, $f/2$ and f; 191 324, f; 191 327, $f/2$ and $f/3$; 191 331, $f/3$; 191 337, $f/4$. This is just a sample from a much larger data collection.

As we keep the modulation frequency constant at 191 000 kHz and increase the modulation depth from 1 % to 20 %, the system passes through a period-doubling cascade up to the accumulation point and enters a fully chaotic region. The chosen frequency is close to the third harmonic of the nonlinear laser resonance Ω. As we scanned the frequency we found a narrow tongue of maximum sensitivity around 3Ω, where the laser destabilizes with the least amount of modulation.

The signal was digitized by a LeCroy transient recorder with 32 000 samples in memory. Setting the internal clock at 320 ns, we obtained approximately 16 points for each period of the fundamental frequency with light-bit resolution. By synchronizing the sampling time to the external drive period we obtained a projection of the Poincaré section. The projection is onto a one-dimensional space (we measure only the intensity) independent of the other variables. In Fig. 2.11 we present the sections and the corresponding time series, respectively. The advantage of this signal processing is that we are able to analyze a high number of perods (32 000 maximum) with a single acquisition. Furthermore, it allows a much larger-bandwidth processing of narrow pulse sequences, which otherwise requires a high sampling rate with the related problems in data storing and processing. In Fig. 2.11, on the left-hand side, the band width is 300 kHz, and on the right it is 100 MHz; indeed we can notice already in the $f/8$ plot a loss of resolution in the smaller peaks on the left-hand side.

We analyze digitized time sequences of the laser output intensity and reconstruct the attractors with an embedding technique. For the determination of the fratal dimension we follow the method of Grassberger-Procaccia.

If we define $N_n(\varepsilon)$ as the number of vectors whose distance is smaller than ε, andif the embedding dimension n is large enough, then $N_n(\varepsilon) \sim \varepsilon^\nu$, where ν is the D_2 dimension of the attractor. In Figs. 2.12a to f we plot $\log N_n(\varepsilon)$ as a function of $\log \varepsilon$ for a sequence of bifurcations $f/4$, $f/8$, and chaos. We limit our analysis to the regions where the slope remains constant over a wide region of $\log \varepsilon$ and where it is independent of n, as it must be from theoretical predictions.

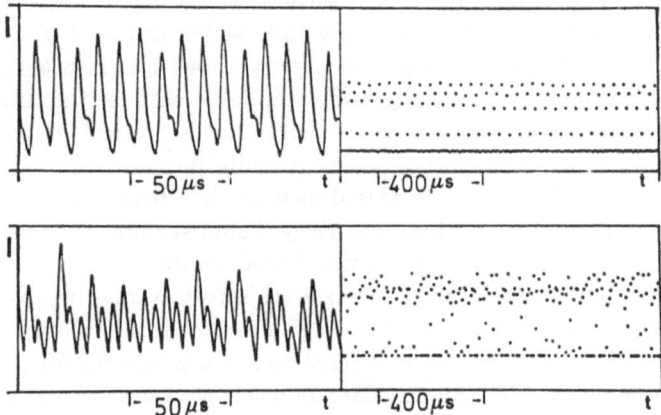

Fig. 2.11. *Top:* Laser intensity vs time for an $f/8$ subharmonic frequency, and corresponding stroboscopic intensity plot with the time interval between successive pints equal to the period of the modulation frequency (191 000 kHz). *Bottom:* Laser intensity vs time and stroboscopic plot for chaotic behavior. The period of the modulation is 5.2 μs. We note on the left-hand side the loss of resolution due to the limited acquisition bandwidth. This drawback is absent on the right-hand side because of the huge increase in band width

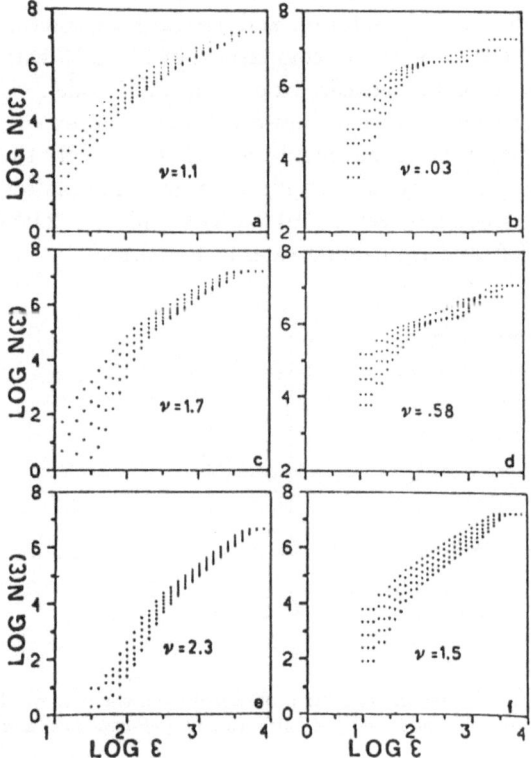

Fig. 2.12. Plots of log $N_n(\varepsilon)$ vs log ε for different values of n calculated from the time series (*left-hand panels*) and from the stroboscopic sections (*right-hand panels*) for different subharmonic frequencies *(a), (b)* $f/4$; *(c), (d)* $f/8$; and *(e), (f)* chaotic behavior. All best-fit values of the slope ν are assimed to have an overall estimated error of ± 0.1. 6000 points were used. The embedding dimensions for all reported plots run from 5 to 9. Dimensions from 1 to 15 were tested

27

From inspection of Figs. 2.12a and b it is clear that the slope obtained for the $f/4$ subharmonic saturates at $\nu \sim 1$ in the time series and $\nu \sim 0$ in the Poincaré section. For the $f/8$ subharmonic ν is slightly above 1.5 (Figs. 2.12c and d). This result, even though not readily understandable because the time signal still appears periodic, nevertheless agrees with the theoretical prediction for the dimension at the accumulation point (infinite periodicity) of the logistic map $(1.5376 < \nu < 1.53385)$. Indeed this dimension has been proven to be universial for those mappings for which the Feigenbaum scaling law holds [2.36]. We present here a heuristic interpretation based on our data. In our expermental system, the unavoidable noise yields a trajectory wandering over a nonzero range of parameter values, thus "testing" near by periodic attractors of the subharmonic sequence. For infinite resolution, we would see for the stroboscopic data a staircase of horizontal plateaus each with zero slope, as it appears at higher embedding dimensions in Figs. 2.12b and d. However, the finite resolution of the correlation measurements averages over adjacent steps, and thus provides the 0.58 slope, as it appears in Fig. 2.12d. This is the first time that the dimension at the accumulation point of a Feigenbaum cascade has been measured in an experimental system.

When the system enters the chaotic region, the fractal dimension suddenly jumps to a higher value $(\nu = 2.4)$.

The time behavior of the intensity obtained by numerical integration of (2.16) was processed in the same manner as the experimental signal. Figure 2.13 shows the results obtained for an $f/8$ solution and a strange attractor. Again near the accumulation point $\omega \sim 1.5$. Direct comparison of Fig. 2.13 with Figs. 2.12c and d shows a good agreement between experiment and model.

It is important to stress that this agreement between theory and experiment is obtained with no floating parameters, but just by feeding (2.16) with the values, for our CO_2 laser and the frequency and amplitude of loss modulation as in the experiment, that is, the frequency set at $191\,000\,\mathrm{kHz}$ and $m = 2.0\,\%$ and $2.85\,\%$, respectively, for the left and right-hand sides of Fig. 2.13.

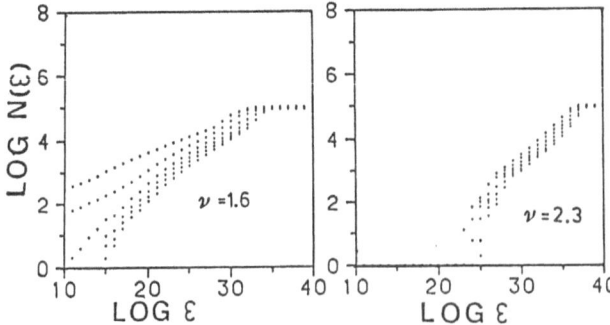

Fig. 2.13. Plots of log $N(\varepsilon)$ vs log ε for different dimensions n obtained from the numerical integration of themodel equations for two different cases, $f/8$ subharmonic (*left-hand side*) and chaos (*right-hand side*). 6000 points were used

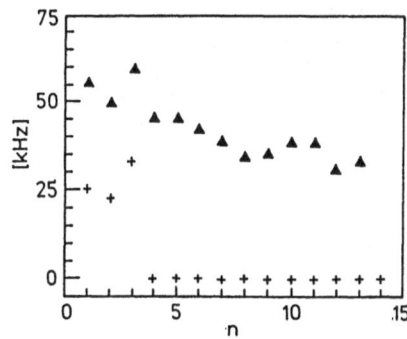

Fig. 2.14. Correlation entropy K_2 (in kHz) vs the embedding dimension for the $f/4$ ($+++$) and the chaotic ($\triangle\triangle\triangle$) attractor

The high regularity of the stroboscopic $N(\varepsilon)$ plots for increasing the embedding dimension suggests evaluation of an approximation to the Kolmogorov entropy. In Fig. 2.14 we report the correlation entropy K_2 versus the embedding dimension for the $f/4$ and for the chaotic attractor, versus the embedding dimension, from the data of Fig. 2.12. We see that while $K_2 = 0$ for $f/4$, $K_2 \simeq 35$ kHz for the chaotic attractor. As we have a single positive Lyapunov exponent and as the embedding time is $5.2\,\mu s$, we estimate that the half-loss of information corresponds to 3.8 periods of the modulation frequency.

2.5 The Laser with Injected Signal (LIS)

Injecting an external signal into a single-mode laser provides an extra degree of freedom. Indeed, in general, the field amplitude x has to be decomposed into two dynamical variables, that is, the two quadrature components $x_1 + ix_2 = x$ with respect to the external phase reference. In class C lasers this provides a fourth equation [2.37,38] which is more than the necessary requirement for deterministic chaos. A simpler situation is that of a class B LIS, which is ruled by (2.A.24) that we repeat here for convenience

$$
\begin{aligned}
\dot{I}/2k &= Iz(1+\delta^2)^{-1} - I + \sqrt{I}x_0 \cos\varphi \;, \\
\dot{\varphi}/k &= -\vartheta - \delta z(1+\delta^2)^{-1} - x_0 \sin\varphi/\sqrt{I} \;, \\
\dot{z}/\gamma_\| &= z_0 - z - zI(1+\delta^2)^{-1} \;.
\end{aligned}
\tag{2.19}
$$

The frequency relations among gain line, cavity, external field and internal lasers are given in Fig. 2.15.

Equation (2.A.25) yields a steady state relation between output and input intensity that shows bistability (Fig. 2.16). For each $\bar{I}$ value in Fig. 2.16, the other two steady-state variables are given by

$$
\bar{z} = z_0 \frac{1+\delta^2}{1+\delta^2+\bar{I}} \;,
\tag{2.20}
$$

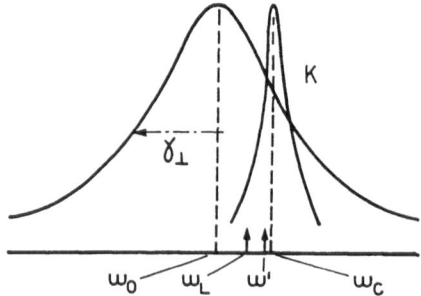

Fig. 2.15. Qualitative plot of the frequency relations among atomic resonance (homogeneous width $\gamma_\perp$) centered at ω_0, cavity resonance (width k) centered at ω_c, and injected field at ω_1

$$\bar{\varphi} = \arccos \frac{\sqrt{\bar{I}}}{x_0} \left(\frac{z_0}{1 + \delta^2 + \bar{I}} - 1 \right). \tag{2.21}$$

The hysteresis amount depends on both detunings (Fig. 2.16). The lower part of the bistable curve is always unstable. The upper part is wholly stable for zero detuning ($\theta - \delta = 0$). For nonzero detuning, it has a stable locked region and an unstable one where the laser oscillates either regularly or irregularly. We have observed two different ways to reach the locked regime, either by decreasing the oscillation frequency (tangent bifurcation), or by decreasing the amplitude of oscillation (Hopf bifurcation), and two different routes to chaos, either by intermittency or by period doubling.

An extensive linear stability analysis was reported in [2.39], together with numerical solutions of (2.19) for different values of the injected amplitude $A = x_0$ and mutual detuning $\theta - \delta$. Some results are summarized below.

In Fig. 2.17 we show regular oscillations in the output intensity; the higher frequency is that predicted by the linear stability analysis, while the lower one is related to a spiking – with amplitudes which can also be ten or more times higher than the steady state – due to field injection. As a matter of fact we have found that in this region, when the injecting amplitude is too low to lock the system steadily, the laser operates for most part of the time at ω_1 (external frequency) but it regularly unlocks going to ω_L (internal frequency). During

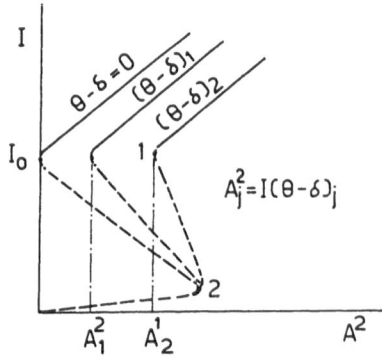

Fig. 2.16. Steady-state solution for LIS equations. Output intensity versus intensity of the injected field for constant value of the pump parameter (z_0) and different values of the detuning ($\omega_1 - \omega_L$). The *dashed line* shows the unstable region and the *solid line* shows the region where steady-state solutions are stable. For each curve, the critical value of the intensity of the injected field is marked with A_j^2. For A^2 larger than A_j^2, there is no instability

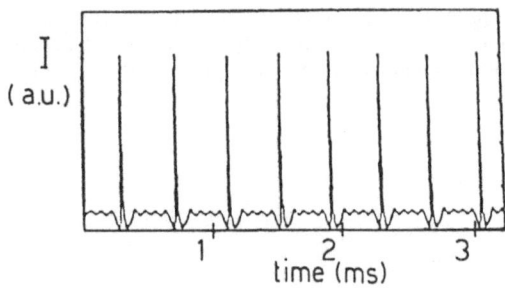

Fig. 2.17. LIS: Spiking and regular oscillations in the output intensity vs time

the oscillation at ω_L the energy of the injecting field enhances the population inversion so that it gives rise, with a delay related to the injecting intensity, to a giant pulse. Increasing the injecting amplitude, the frequency of these pulses goes to zero, because the system remains locked for longer times.

In Fig. 2.18 we show the temporal sequence which leads to chaos, on changing $x_0 = A$, by intermittency. Each dot representing the peak of an oscillation, we see, from bottom to top, how the laminar period becomes shorter and shorter and eventually dies in a wholly developed chaos.

The bifurcation sequence is shown in Fig. 2.19. The bifurcation occurs in the higher spikes while between two near-lying higher peaks we find oscillations at Ω.

Preliminary experimental data have been obtained by a three-laser setup [2.40], the first laser being a ring laser where the dynamics develops, the second one the external injecting laser and the third one a master oscillator with

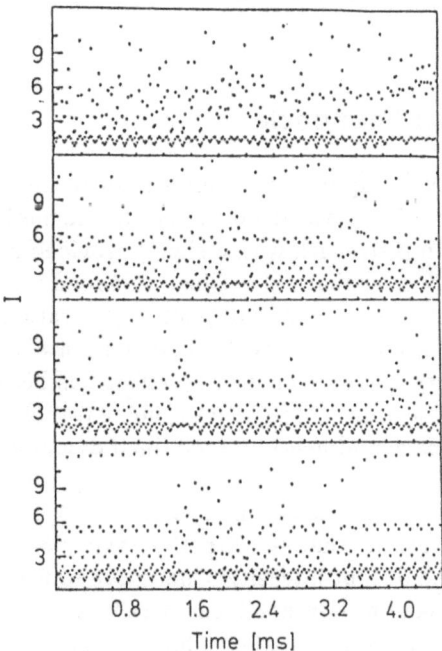

Fig. 2.18. LIS: Temporal sequence leading to chaos by intermittency increasing A (from *bottom* to *top*), each dot representing a maximum

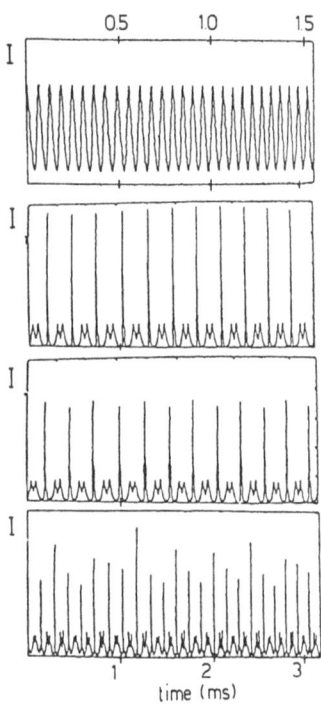

Fig. 2.19. LIS: Output intensity vs time. A is increased from top to bottom and chaos is reached by frequency doubling in the higher peaks

reference to which first and second laser are stabilized. The parameter region explored in this experiment was sufficient to yield oscillatory instabilities but not enough to reach chaos.

2.6 Instabilities in a Laser with Feedback

In laser applications where high stability is required, an overall negative feedback is currently used, besides that already provided by the electromagnetic cavity, for instance by controlling the pump strength with a signal provided by the detected output intensity [2.41]. Such a feedback is not just an added artifact, but it affects, in a fundamental way, the dynamics of photon generation; indeed, it has been proposed [2.42a] as a mean to provide squeezed states of the electromagnetic field, and preliminary evidence of such an effect has been given [2.42b]. However, a fundamental objection to a feedback scheme is that it provides one extra dimension to phase space and hence the modified dynamics can be affected by irregular behavior.

Indeed, self-pulsing and deterministic chaos has been reported [2.26] in a single-mode laser fed back by its own output, that is, with the cavity losses modulated by a signal proportional to the output intensity. Chaos due to feedback had already been observed in connection with nonlinear passive systems (either a KDP crystal between crossed polarizers [2.43a] or a long optical fiber in a ring cavity [2.43b]). In both cases the passive system was studied per se, being outside the laser cavity, and thus the laser dynamics was not affected by the feedback configuration.

In Sects. 2.4 and 5 we have shown the onset of chaos in low-dimensional optical systems, in controlled conditions displaying a one-to-one correspondence between experiment results and the predictions of their theoretical model.

Here we show how feeding the laser output back on an intracavity modulator introduces a third degree of freedom leading to chaotic instability. When the feedback loop is so fast that it practically provides an "instantaneously" adapted loss coefficient, it does not modify the phase space topology, which in the case of a class B laser remains two-dimensional. If however the time scale of the feedback loop is of the same order as that of the other relevant variables, the system becomes three dimensional. Such a system is ruled by three first-order equations for the intensity x, population difference z and modulation voltage v. With suitable normalizations (see Appendix 2.A, and notice that here x is the intensity, *not* the field) the equations are

$$\dot{x} = -k_0 x (1 + \alpha \sin^2 v - z) \ ,$$
$$\dot{z} = -\gamma_{||}(z - A + xz) \ ,$$
$$\dot{v} = -\beta(v - B + fx) \ , \tag{2.22}$$

where $k(v) = k_0(1 + \alpha \sin^2 v)$ is the loss rate modulated by the voltage v, k_0 is the nonmodulated cavity loss parameter, α is a coupling constant, $\gamma_{||}$ is the population decay rate, and β the damping constant of the feedback loop. Furthermore B is the voltage bias applied to the second input of the modulator amplifier, A is the pump parameter, f is a coupling coefficient between intensity x detected on D and voltage v. Notice that x is normalized to the saturation intensity, z and A to the threshold population (without feedback) and v is given in angular units, that is, if we call V the voltage applied to the modulator and V_0 the $\lambda/2$ modulator voltage, then $v = V/V_0$.

The experimental system of [2.26] has $k_0 = 1.17 \times 10^7 \ (\mathrm{s}^{-1})$, $\gamma_{||} = 0.98 \times 10^4 \ (\mathrm{s}^{-1})$, $\beta = 3.0 \times 10^4 \ (\mathrm{s}^{-1})$ and a normalized pump $A = 4.2$. The stationary solutions $(\bar{x}, \bar{z}, \bar{v})$ of (2.22) imply the condition

$$B = f\bar{x} + \arcsin \left(\frac{A/\alpha}{1 + \bar{x}} - \frac{1}{\alpha} \right)^{1/2} . \tag{2.23}$$

Depending on the feedback coupling f, for different bias values B we can have mono or bistability (Fig. 2.20). In particular, around $f = 0.1$ we expect an ambiguity, since (2.23) provides a quasi-vertical curve. Indeed as we show later, this is the region where we observe chaos.

By a linear stability analysis around the stationary solution, we evaluate the points where the system starts self pulsing (Hopf bifurcations). The lines of Hopf bifurcations are drawn in Fig. 2.20 (dashed) for three different β values. In fact, we have a slight uncertainty in the assignement of the open loop

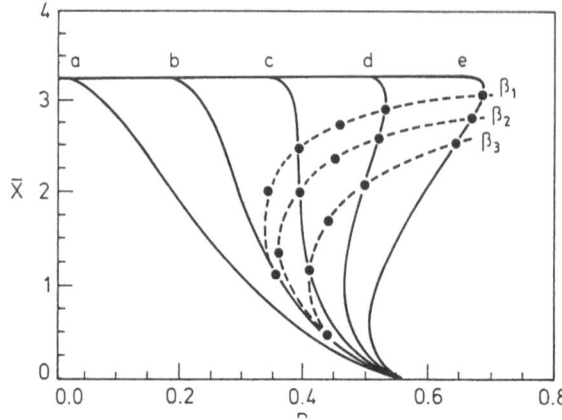

Fig. 2.20. Laser with feedback: Plots of normalized stationary intensity x vs B (the bias-voltage B is expressed in angular units) for different values of the feedback coupling constant f. The curves (a to e) refer to $f = 0$, 0.052, 0.102, 0.152 and 0.202, respectively. (- - -) correspond to the loci of the first Hopf bifurcations for three different values of the damping constant (s^{-1}) of the feedback loop, namely: $\beta_1 = 3.5 \times 10^4$, $\beta_2 = 3.0 \times 10^4$ and $\beta_3 = 2.5 \times 10^4$.

damping constant. At the intersection points of the stationary solutions (lines with fixed f) with the lines of onset of Hopf bifurcations (lines with fixed β) the corresponding pulsing frequency in kHz is given in Table 2.1.

Table 2.1. Laser with feedback: Values [kHz] of the first Hopf bifurcation

$10^{-4} \times \beta$	$f = 0.052$	$f = 0.102$	$f = 0.152$	$f = 0.202$
2.5		31	39	57
3.0	45	16	39	49
3.5	51	15	29	36

In Fig. 2.21 we present the power spectra of the intensity detected in the experiment. Figure 2.21a shows the first Hopf bifurcation, Fig. 2.21b the appearence of a subharmonic $f/2$, and Fig. 2.21c corresponds to the appearence of chaos. Beyond chaos, there are periodic time windows. In order to get full assurance of the chaotic nature of the time plot of Fig. 2.21c, the correlation dimension was measured along the lines already outlined in Sect. 2.3.

Figure 2.22 shows clear evidence of a fractal exponent $D_2 = 2.6 \pm 0.1$. While Fig. 2.22 comes from the experiment, the same D_2 value is obtained by solving numerically (2.22) for $\beta = \beta_2 = 3.0 \times 10^4$ and $B = 0.383$. The theoretical plots, shown in Fig. 2.22, closely follow the experimental ones with uncertainties smaller than the dot sizes. Narrow regions with higher-order subharmonics ($f/4$ and $f/8$) plus $f/3$ windows were observed beyond chaos. In order to have a better understanding of the chaotic scenario, we have solved nu-

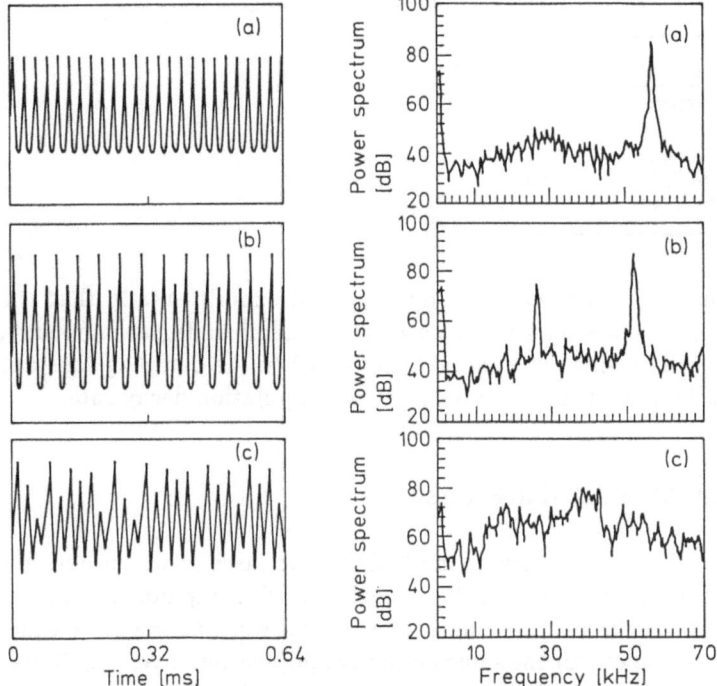

Fig. 2.21a–c. Laser with feedback: Digitizer time plots of the experimental laser intensity (*left*) and the corresponding power spectra (*right*) for increasing values of the control parameter B. (a) corresponds to the onset of the first Hopf bifurcation, at a frequency $\nu = 57.3\,\text{kHz}$, $B = 0.364$; (b) shows the appearence of subharmonic bifurctation $f/2$ where the fundamental frequency is $\nu = 52.0\,\text{kHz}$, $B = 0.378$ and (c) shows the appearence of chaos, $B = 0.383$

merically (2.22). An accurate localization of the bifurcation points was done by studying the stability of the phase space orbits in term of their Floquet multipliers. More specifically, the multipliers were evaluated by determining the Poincaré sections with the Henon method [2.26], and finding the zero of the associated recursive relation by the Newton method. In Table 2.2 we give the bifurcation diagram, which shows clear evidence of a Feigenbaum scenario with a Feigenbaum converging ratio in fair agreement with the asymptotic value.

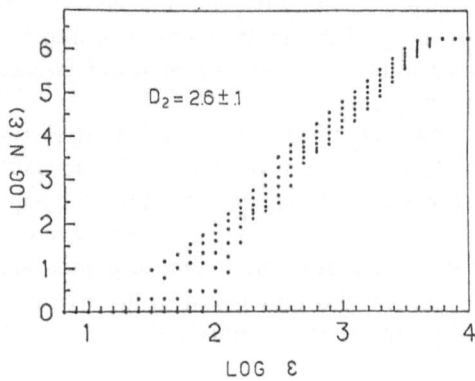

Fig. 2.22. Laser with feedback: Plots of log $N_n(\varepsilon)$ vs log ε for different values of embedding dimension n ($n = 10$–15). Square dots come from experiment. Theoretical plots coincide with the experimental ones within the dot sizes

Table 2.2. Laser with feedback: Bifurcation parameters and Feigenbaum ratio

Bifurcation	B	δ
f	0.394907	
$f/2$	0.395355	
$f/4$	0.395662	...5.11
$f/8$	0.395722	...5.00
chaos	0.395734	

In conclusion, we have shown theoretically and experimentally that the introduction of negative feedback provides a third dynamical degree of freedom sufficient to yield self pulsing and chaotic behavior when the damping constant β of the feedback loop is of the same order as the population decay rate.

2.7 The Bidirectional Ring Laser

Last we consider a longitudinal single mode CO_2 ring laser in which both directions of propagation are allowed [2.25]. The line width being homogeneously broadened, the two counterpropagating beams cannot work at the same time, because they must compete for the same amount of population inversion. Moreover they are slightly detuned between each other − and with respect to line centre − because, for intrinsic asymmetries, cavity losses are different in the two propagation directions (k_1 and k_2); this results in a different mode pulling and then different lasing frequency. Indeed, having a gas flow in the laser tube, this already induces a small amount of Doppler shift in the interaction with one or the other of the two counter-running modes. The detuning has been shown experimentally as well as in the numerical solution to be essential for breaking the symmetry between the two directions. A forbidden gap around the center of the molecular line, as well as the interchange of role of forward and backward fields at right and left of the line center are evidence of such a detuning. If $k_{1,2}$ were the "cold" damping rates, they could not differ for reciprocity (in a passive medium thermodynamics forbids such a symmetry breaking). However the $k_{1,2}$ in the active medium differ for the above-mentioned gas flow effect.

Through the grating induced in the population inversion by the interference of the two waves we have an interchange of energy from one field into the other by backscattering, so that we may consider the system as a LIS (where the injection comes from the counterpropagating-mode).

A modelling of this system has been given in Appendix 2.A.5. Here together with the detuning from the cavity mode we have to take into account two complex running waves and the induces time-dependent grating in the population inversion (truncated at the first order in the expansion). Being x and y the two complex fields, z (real) the spatially uniform component of population inversion, and w the complex amplitude of the grating inchced in the inversion we have from (2.A.28) when losses and pumping are included

$$\dot{x} = \frac{1}{1+i\delta}(zx + w^*y) - x \;,$$

$$\dot{y} = \frac{1}{1+i\delta}(zy + wx) - \frac{k_2}{k_1}y \;,$$

$$\frac{k_1}{\gamma_\|}\dot{z} = (z_0 - z) + \frac{1}{1+\delta^2}[z(|x|^2 + |y|^2) + w^*x^*y + wxy^*] \;,$$

$$\frac{k_1}{\gamma_\|}\dot{w} = -w - \frac{1}{1+\delta^2}[zx^*y + w(|x|^2 + |y|^2)] \qquad\qquad (2.24)$$

with δ cavity detuning, z_0 pump parameter and normalized time $\tau = Kt$. Numerical solutions of this seven equation system closely matches all experimental results.

In our parameter space (Fig. 2.23) we can distinguish three main different regions showing completely different behavior. In the first one we observe a self-pulsing very similar to that of the laser with an injected signal. One mode is also running CW while in the other one we observe only spikes, in phase with the main mode, which occur at a repetition rate (ω_S) of the order of $\gamma_\|$. In fact, as in the LIS system, the CW working mode injects some energy into the other one letting population inversion increase up to a level at which a giant pulse takes place (the height may be 500 times greater than the stationary level). During the pulse both modes go above threshold and spike in phase. Superimposed to the decay we see relaxation oscillations typical of CO_2 lasers with a frequency (ω_0) very near to Ω : they are out of phase because of competition between the two modes.

For higher excitation currents we observe a deterministic switching due to competition between the two fields with low frequency (30 Hz). During interchange jumps we observe again the two frequencies of Fig. 2.24 but with the lower one increased because of a larger value of $\gamma_\|$ (higher current) while the higher one can be varied also by adjusting the cavity length and alignment by moving a mirror mounted on a piezoelectric crystal.

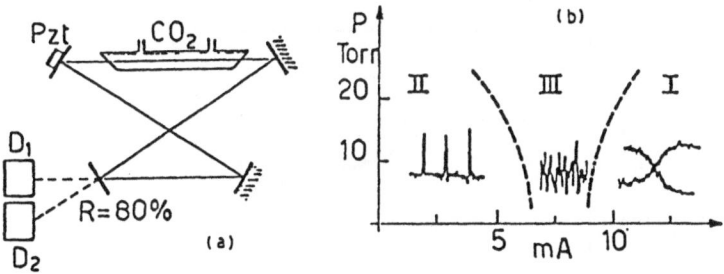

Fig. 2.23a,b. Bidirectional ring laser: (a) Experimental set-up: Gain cell with partial pressures: CO_2 1, N_2 1.5 Torr, He variable. Cavity length 4.2 m; PZT piezoelectric mirror translator; D_1 and D_2 detectors for forward and backward intensities. (b) Phase diagram for total pressure (P) and discharge current (mA). Regions are: (*I*) mode alternation, (*II*) self Q-switch, (*III*) irregular pulsation

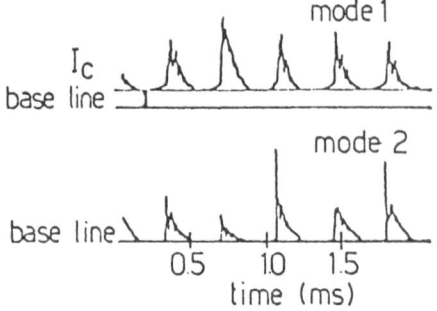

Fig. 2.24. Bidirectional ring laser – Region (*II*): Output intensity vs time. The upper signal has a CW baseline, while the lower oscillates over a zero level. Oscillations relax to the steady state before a new jump (interchange between CW action) takes place

The transition between these two regimes is not abrupt and it takes place through a region which shows chaotic behavior. Here both phenomena related to population inversion, spiking (lower currents) and oscillation (higher currents) take place; effective output frequency results also as a combination of the two others ($\omega_s + \omega_0$). At the same time if we adjust the cavity mirror position so that we brinig $\Omega \simeq (\omega_s + \omega_0)$ we obtain a competition of two different variables (population inversion and field) on the same time scale. The result is a fully developed chaos (Fig. 2.25).

If now we inject back one field into the laser with an external mirror (a fifth mirror in the configuration of Fig. 2.23) we obtain: stabilization of self-spiking, stable laser action instead of switching between the two modes and chaotic behavior. At the boundary between the spiking and the chaotic region we observe a phenomenology typical of a laser with an injected signal (Fig. 2.26). It means that in this situations we have parameters practically equal to those responsible of such behavior in the LIS case, although the system here is more complicated.

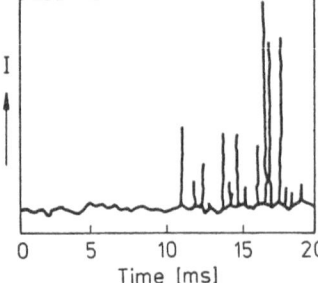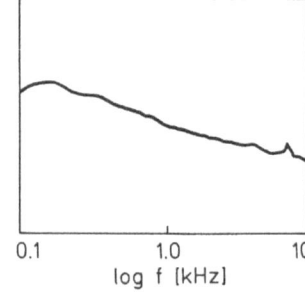

Fig. 2.25. Bidirectional ring laser – Region (*III*): Output intensity vs time for a wholly chaotic signal (*left*). Log-log power spectrum with low frequency divergence $f^{-\alpha}$, $\alpha \simeq 0.6$ (*right*)

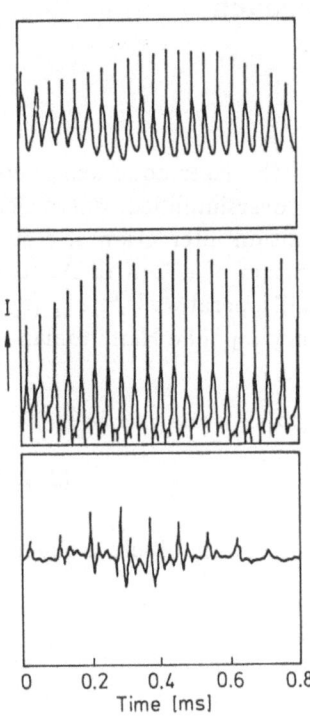

Fig. 2.26. Bidirectional ring with extra mirror to reinject the forward mode into the backward one: Output intensity vs time. Bifurcation sequence in analogy with a laser with injected signal

2.8 Conclusion

We have seen that the single-mode CO_2 laser has a rich phenomenology, which we can reproduce numerically with simple theoretical models.

In experiments involving a parameter modulation or feedback we obtain much more stable and noise-free output so that we can easily compute fractal dimensions and Kolmogorov entropies. At the same time the phenomenology is here not so rich as in experiments where interaction with another field takes place (LIS and bidirectional ring). This must be attributed to the higher complexity of such an interaction where not only an amplitude but also a phase coupling takes place, while in the parameter modulation case interaction is carried only through amplitude modulation.

Summarizing we have found:

i) in a laser with modulated parameters or feedback, a clear Feigenbaum route to chaos, with related δ_F and accumulation point evidence;

ii) in a laser with an injected signal, two different routes to chaos, by intermittency and period doubling;

iii) in a bidirectional ring laser, self-spiking and chaos.

In this last system we find also a surprising coincidence with a laser with an injected signal when we reflect back one mode into the laser.

2.A Appendix: A Simple-Minded Approach to Laser Equations

2.A.1 The Laser Equations

Leaving to textbooks [2.5] a detailed derivation of the laser equations, here we use a pedagogical approach which may appear oversimplified, but which contains the relevant physics. We consider the quantum interaction of a single mode of frequency ω_c described by Bose operators a, a^+, with N non-interacting two-level atoms, each described by Pauli operators $\sigma_i^\pm$, σ_{3i} (r_i : position of i-th atom) and with a transition frequency ω_0. The Hamiltonian is

$$\frac{H}{\hbar} = \omega_c a^+ a + \frac{\omega_0}{2} \sum_i \sigma_{3i} + ig \sum_i [a^+ \exp(-ikr_i)\sigma_i - a \exp(ikr_i)\sigma_i^+]$$

$$(2.A.1)$$

where $g^2 = (\omega_c \mu^2)/(2\hbar\varepsilon_0 V)$ (V : cavity volume, μ : atomic dipole moment). Using the commutation rules

$$[a, a^+] = 1$$
$$[\sigma^+, \sigma^-] = \sigma_3$$
$$[\sigma_3, \sigma^\pm] = \pm 2\sigma^\pm \qquad (2.A.2)$$

for the same atom, otherwise zero. It is easily seen that the collective operators

$$J_3 = \sum_i \sigma_{3i} \ ,$$
$$J^\pm = \sum_i \sigma_i^\pm \exp(\pm ikr_i) \qquad (2.A.3)$$

obey also Pauli commutation rules.

The associated Heisenberg equations of motion can then be written

$$\dot{a} = -i\omega_c a + gJ^- \ ,$$
$$\dot{J}^- = -i\omega_0 J^- + gaJ_3 \ ,$$
$$\dot{J}_3 = -2g(a^+ J^- + aJ^+) \ . \qquad (2.A.4)$$

Equations (2.A.4) look formally like the dissipation-less of (2.6). However, their operator means that the corresponding equations for expectation values imply an infinite hierarchy. We truncate it by factoring out expectation values (semi-classical approximation):

$$\langle aJ^+ \rangle \sim \langle a \rangle \langle J^+ \rangle \ .$$

Thus (2.A.4) can be taken as C-number equations, and decorated with the three phenomenological damping rates $-ka$, $-\gamma_\perp J^-$, $-\gamma_\parallel(J_3 - J_0)$, respectively, where J_0 is the inversion imposed by the pump in the absence of field. Recurring to a more familiar notation $(a \to E,\ J^- \to P,\ J_3 \to \Delta)$ and considering also an external field E_0 at frequency ω_1, the single-mode Maxwell-Bloch equations are

$$\dot{E} = -(i\omega_c + k)E + gP + kE_0 e^{-i\omega_1 t} \ ,$$
$$\dot{P} = -(i\omega_0 + \gamma_\perp)P + gE\Delta \ ,$$
$$\dot{\Delta} = -\gamma_\parallel(\Delta - \Delta_0) - 2g(E^*P + EP^*) \ . \tag{2.A.5}$$

Equilibrium solutions at resonance $(\omega_c = \omega_0)$ and in the absence of external field yield

$$\overline{\Delta} = \frac{\Delta_0}{1 + \overline{E}^2/E_s^2} = \frac{\gamma_\perp k}{g^2} \tag{2.A.6'}$$

where we have called E_s^2 the saturation photon number

$$E_s^2 = \frac{\gamma_\perp \gamma_\parallel}{4g^2} \ . \tag{2.A.6''}$$

It then follows

$$\overline{E} = E_s \sqrt{\frac{\Delta_0}{\overline{\Delta}} - 1} \quad \text{and}$$

$$\overline{P} = \frac{g}{\gamma_\perp} \overline{\Delta E} \ . \tag{2.A.6'''}$$

To get the order of magnitude, take an allowed visible transition in a dilute gas confined in $V = 1\,\text{cm}^3$ (as, e.g., in He-Ne or A^+ lasers). Then $\gamma_\perp \sim \gamma_\parallel \sim 10^8\,\text{s}^{-1}$ and $E_s^2 \sim 10^8$. Add a cavity 1 m long with 1 % losses, then $k \sim 3 \times 10^6\,\text{s}^{-1}$, and $\overline{\Delta} \sim 3 \times 10^6$ inverted atoms. It is convenient to scale all variables to these parameters, as follows

$$x = E/E_s \ ,$$
$$z = \Delta/\overline{\Delta} \quad (z_0 = \Delta_0/\overline{\Delta}) \ ,$$
$$y = P/(\overline{\Delta}E_s g/\gamma_\perp) \ . \tag{2.A.7}$$

So that the scaled equilibrium values at resonance are

$$\overline{x} = \overline{y} = \sqrt{z_0 - 1} \ ,$$
$$\overline{z} = 1 \ . \tag{2.A.8}$$

Let us scale the frequencies as follows

$$\vartheta = \frac{\omega_c - \omega_L}{k} \quad \text{(cavity mistuning)} ,$$

$$\delta = \frac{\omega_0 - \omega_L}{\gamma_\perp} \quad \text{(atomic detuning)} . \tag{2.A.9}$$

Here ω_L is either the frequency ω_1 of the external laser or, in the absence of an external reference, the frequency of the laser field (which does not coincide with the cold cavity ω_c and will be determined later). The scaled equations are

$$\dot{x}/k = -(i\vartheta + 1)x + y + x_0 ,$$
$$\dot{y}/\gamma_\perp = -(i\delta + 1)y + xz ,$$
$$\dot{z}/\gamma_\parallel = -z + z_0 - \tfrac{1}{2}(xy^* + x^*y) . \tag{2.A.10}$$

For $x_0 = 0$, at equilibrium the first and second yield, for $\bar{x} \neq 0$:

$$\bar{z} = (1 + i\theta)(1 + i\delta) . \tag{2.A.11}$$

Since $\bar{z}$ is real, this implies the pulling condition

$$\theta = -\delta . \tag{2.A.12}$$

This assignes the value of ω_L as we see rewriting it as

$$\frac{\omega_c - \omega_L}{k} = \frac{\omega_L - \omega_0}{\gamma_\perp} . \tag{2.A.12$'$}$$

Setting (2.12) into (2.11) we have the increase in threshold due to detuning

$$\bar{z} = 1 + \delta^2 \tag{2.A.13$'$}$$

and replacing it into the last of (2.10)

$$|\bar{x}|^2 = z_0 - (1 + \delta^2) . \tag{2.A.13$''$}$$

Equation (2.13) generalizes (2.8) off resonance.

2.A.2 Adiabatic Elimination of Polarization – Modulation and Injection

Solving (2.10) at equilibrium

$$y = \frac{xz}{1 + i\delta} .$$

Replacing this result into the field and population equations we obtain

$$\dot{x}/k = -(1 + i\vartheta)x + \frac{xz}{1 + i\delta} + x_0 \ ,$$

$$\dot{z}/\gamma_\| = -z + z_0 - z\frac{|x|^2}{1 + \delta^2} \ . \tag{2.A.14}$$

First, take $x_0 = 0$.

For $\theta = \delta = 0$ (resonance) we recover the equations of Sect. 2.3. If $k = k(t)$ we have loss modulation. If $k = $ const, $\delta = \delta(t)$ we have frequency modulation. In the first case, working at resonance, we have a *real* x, and just 3 equations. Indeed, because the first two are non-autonomous, a third equation must be introduced to account for the explicit time dependence

$$k(t) = k_1(1 + m \cos \Omega t) \tag{2.A.15}$$

The three equations are

$$\dot{x} = -k(t)x(1 + z) \ ,$$
$$\dot{z} = -\gamma_\|(z - z_0 + z|x|^2) \ ,$$
$$\dot{k} = -mk_1\Omega \sin \Omega t \ , \tag{2.A.16}$$

or equivalently if $I = |x|^2$ the first two (2.16a,b) can be written as

$$\dot{I} = -2kI(1 - z) \ ,$$
$$\dot{z} = -\gamma_\|(z - z_0 + zI) \ . \tag{2.A.17}$$

In the second case the field has two non-zero components, and can not be reduced to a single variable. Writing

$$x = x_1 + ix_2 = \sqrt{I}e^{i\varphi} \tag{2.A.18}$$

it is easily seen that in terms of intensity the equations shown in [2.44] are valid

$$\frac{\dot{I}}{k} = -2I\left(1 - \frac{2}{1 + \delta^2}\right) \ ,$$

$$\frac{\dot{z}}{\gamma_\|} = z_0 - z - \frac{zI}{1 + \delta^2} \ ,$$

$$\delta = \delta(t) \ . \tag{2.A.17'}$$

But this is *not* the full story. In fact, a fourth equation for $\varphi(t)$ should be added or, equivalently, the above set must be replaced by

$$\frac{\dot{x}}{k} = -x_1 + \theta x_2 + z\frac{x_1 + \delta x_2}{1 + \delta^2} \ ,$$

$$\frac{\dot{x}_2}{k} = -x_2 - \theta x_1 + z\frac{x_2 - \delta x_1}{1 + \delta^2} \ ,$$

$$\frac{\dot{z}}{\gamma_\|} = -z + z_0 - z\frac{(x_1^2 + x_2^2)}{1 + \delta^2} \ ,$$

$$\delta = \delta|x| \ . \tag{2.A.19}$$

This shows that, at variance with loss modulation, in case of phase modulation a class B laser seems to be ruled by four equations. However, in the absence of an external field frequency, thus the decomposition (2.18) should be immaterial, since $\varphi = 0$ always. We deal with this problem later when performing a linear stability analysis. For $x_0 \neq 0$, we must add it to (2.19), and redefine θ and δ with respect to ω_1, rather than ω_L. They were written in [2.39] in a slightly different form, because z was there defined as

$$z \rightarrow \frac{z}{1 + \delta^2} \quad .$$

Thus the factor $(1 + \delta^2)^{-1}$ present in (2.19) is missing in [Ref. 2.39, Eq. (2.10)].

Notice that keeping x_0 in the 1st of (2.14) makes the equilibrium version $(\dot{x} = 0)$ of that equation non homogeneous in $\bar{x}$, hence one can no longer derive the full relation (2.12′). For this, [Ref. 2.39, Eq. (2.14)] is wrong.

2.A.3 Linear Stability Analysis of Class B Solutions

At resonance $(\theta = \delta = 0)$ we have two coupled equations, and the deviations from the steady state yield two eigenvalues $\lambda_{1,2}$. For $x_0 = 0$, the steady-state solutions are given by

$$\bar{z} = 1, \quad \bar{I} = |\bar{x}|^2 = z_0 - 1 \quad .$$

Writing $I = \bar{I} + u$, $z = \bar{z} + v$, we have from (2.17)

$$\dot{u} = 2k(z_0 - 1)v \quad ,$$
$$\dot{v} = -\gamma_{||}u - \gamma_{||}z_0 v \quad . \tag{2.A.20}$$

Thus the eigenvalues λ correspond to

$$\lambda = -\frac{\gamma}{2}z_0 \pm i\sqrt{2k\gamma_{||}(z_0 - 1) - \frac{\gamma^2}{4}z_0^2}$$
$$\approx -\frac{\gamma}{2}z_0 \pm i\sqrt{2k\gamma_{||}(z_0 - 1)} \tag{2.A.21}$$

since $\gamma_{||} \ll k$.

Equation (2.21) shows that a disturbance in a class B laser decays over a time scale $1/\gamma_{||}$ (assume z_0 and $z_0 - 1$ of the order of 1, for simplicity) and it has a ringing with period $1/\sqrt{k\gamma_{||}}$. We remember that for a CO_2-laser the two time scales are, respectively, 10^{-3} s and 10^{-5} s.

If now we consider a non-zero detuning, but still $x_0 = 0$ the equations are given by (2.10) or equivalently by the first two of (2.17′) plus a third equation for phase φ. Indeed, writing x as in (2.18), we have for the phase $\varphi = \arctan x_2/x$,

$$\frac{\dot{\varphi}}{k} = \frac{1}{kI}(x_1\dot{x}_2 - x_2\dot{x}_1) = -\theta + \frac{\delta}{1 + \delta^2}z \quad . \tag{2.A.22}$$

At equilibrium $\varphi = 0$, since $\theta = -\delta$. Notice that having chosen a frame rotating at the frequency ω_L of the laser field, by definition $\varphi = 0$ always, hence the extended stability analysis should provide a third root $\lambda = 0$. Indeed, extending the analysis of (2.20) for $\delta \neq 0$ we obtain three linear equations (we use the matrix form)

$$
\begin{pmatrix} \dot{u}/k \\ \dot{v}/\gamma \\ \dot{\varphi}/k \end{pmatrix} = \begin{pmatrix} 0 & 2\left(\frac{z_0}{1+\delta^2} - 1\right) & 0 \\ -1 & -\frac{z_0}{1+\delta^2} & 0 \\ 0 & -\frac{\delta}{1+\delta^2} & 0 \end{pmatrix} \begin{pmatrix} u \\ v \\ \varphi \end{pmatrix}
\tag{2.A.23}
$$

and the corresponding roots of $\tilde{M} - \lambda \tilde{I} = 0$ are again the two given by (2.21) plus $\lambda_3 = 0$, thus showing that the phase is irrelevant.

In presence of modulation, for a small modulation index, the motion is a periodic orbit at the frequency of the external perturbation, around the stationary values $\bar{I}$, $\bar{z}$, and with a radius linear in the perturbation. The stability of this synchronous orbit has to be tested by Floquet theory [2.34].

2.A.4 Laser with Injected Signal (LIS)

The equations are (2.19a–c), with the addition of x_0 to the first one. These equations have been discussed in [2.39]. Alternatively, writing $x = \sqrt{I} \exp(i\varphi)$, we have two equations for I and φ, so that the full set of LIS equations can be written as

$$
\dot{I}/2k = \frac{Iz}{1+\delta^2} - I + \sqrt{I} x_0 \cos\varphi ,
$$

$$
\dot{\varphi}/k = -\theta - \frac{\delta}{1+\delta^2} z - \frac{x_0}{\sqrt{I}} \sin\varphi ,
$$

$$
\dot{z}/\gamma_\| = z_0 - z - \frac{zI}{1+\delta^2} ,
\tag{2.A.24}
$$

where, we recall that θ and δ are defined with reference to the external frequency ω_1. At equilibrium calling $I_0 = x_0^2$, we have the relation

$$
I_0 = \bar{I}\left[\left(\theta + \frac{\delta z_0}{1+\delta^2+\bar{I}}\right)^2 + \left(\frac{z_0}{1+\delta^2+\bar{I}} - 1\right)^2\right]
\tag{2.A.25}
$$

which, yields the curves of Fig. 2.16 showing bistability. In [2.39] one can find an extended stability analysis yielding the three roots of the linearized (2.24).

2.A.5 The Bidirectional Class B Ring Laser

When two counterpropagating fields $a \exp[-i(\omega t - kr)] + c.c.$ and $b \exp[-i(\omega t + kr)] + c.c.$ coexist as separate dynamical variables, then

$$
[a, b^+] = 0
$$

and we have a bidirectional ring. The corresponding Hamiltonian is

$$\frac{H}{\hbar} = \frac{\omega_0}{2}\sum \sigma_{3i} + \omega(a^+a + b^+b)$$
$$- ig\sum_i [\sigma_i^+(ae^{ikr_i} + be^{-ikr_i}) - \sigma_i^-(a^+e^{-ikr_i} + b^+e^{ikr_i})] \ . \quad (2.A.26)$$

We now introduce a whole class of k-dependent collective operators, which are Fourier transforms of the position-dependent single-atom operators, defined as follows

$$J^\pm(k) = \sum \sigma_i^\pm e^{ikr_i} \ ,$$
$$J_3(k) = \sum \sigma_{3i} e^{ikr_i} \ .$$

It is easily verified that

$$[J^+(k), J^-(-k)] = J_3(0) \ ,$$
$$[J_3(0), J^\pm(k)] = \pm 2J^\pm(k) \ ,$$

and furthermore

$$[J^+(k), J^-(k)] = J_3(2k) \ ,$$
$$[J_3(2k), J^+(k)] = 2J^+(3k) \ , \quad (2.A.27)$$

and so on.

One immediately sees that the Heisenberg equations for fields are

$$\dot{a} = -i\omega a + gJ^-(-k) \ ,$$
$$\dot{b} = -i\omega b + gJ^-(k) \ .$$

Adiabatic elimination of polarization yields for single atom

$$\sigma^- = \frac{g\sigma_3}{\gamma_\perp + i\Delta\omega}(ae^{-ikr_i})$$

where $\Delta\omega = \omega_0 - \omega$.

Multiplying both sides for $\exp(-ikr_i)$ one builds a relation for $J^-(-k)$. This way we build the following equations

$$\dot{a} = \frac{g^2}{\gamma_\perp + i\Delta\omega}[aJ_3(0) + bJ_3(-2k)] \ ,$$

$$\dot{b} = \frac{g^2}{\gamma_\perp + i\Delta\omega}[aJ_3(2k) + bJ_3(0)] \ ,$$

$$\dot{J}_3(0) = \frac{4g^2}{\gamma_\perp^2 + \Delta\omega^2}[J_3(0)(|a|^2 + |b|^2) + J_3(2k)ab^+ + J_3(-2k)a^+b] \ ,$$

$$\dot{J}_3(2k) = \frac{4g^2}{\gamma^2 + \Delta\omega^2}[J_3(2k)(|a|^2 + |b|^2) + J_3(4k)ab^+ + J_3(0)a^+b] \ ,$$

$$(2.A.28)$$

and similar for $\dot{J}_3(-2k)$. There is a hierarchy built from the third equation, scaling upward (or downward) the k arguments by $2k$ at each step. If we introduce a one-dimensional lattice, with site i corresponding to $2ki$, then we have $J_3(i)$ coupled with $J_3(i\pm 1)$. The above equations have to be considered as classical, and completed with phenomenological damping terms. Truncation problems are discussed in Sect. 2.7. If we set $J_3(4k) \approx 0$, already (2.28) is a set of 7 real closed equations: two each for a and b, one for $J_3(0)$ and two for $J_3(\pm 2k)$.

Acknowledgement. I must express my appreciation and gratitude to all colleagues and students with whom I have been collaborating over the past years, and with whom I have shared the excitement of discovering new "whats" and "hows" and understanding the "whys". The reason why I had to prepare this manuscript alone is that most of them are now scattered around the world in different Laboratories. Let me list them in alphabetic order (their individual contributions appear from the References): N.B. Abraham, R. Badii, A. Califano, W. Gadomski, G.L. Lippi, F. Lisi, R. Meucci, A. Poggi, A. Politi, G.P. Puccioni, N. Ridi, J.R. Tredicce and L. Ulivi.

Part of this research was supported by the EJOB project of European Economic Community.

References

2.1 W.E. Lamb Jr.: Phys. Rev. **134**, A1429 (1964)

2.2 H. Haken: *Laser Theory*, Corr. Printing (Springer, Berlin, Heidelberg 1984)

2.3 M. Scully, W.E. Lamb Jr.: Phys. Rev. Lett. **16**, 853 (1966), Phys. Rev. **159**, 208 (1967); and **166**, 246 (1968)

2.4 J.P. Gordon: Phys. Rev. **161**, 367 (1967)

2.5 H. Haken: *Synergetics*, 3rd ed., Springer Ser. Syn., Vol. 1 (Springer, Berlin, Heidelberg 1983)

2.6 F.T. Arecchi: In *Order and Fluctuations in Equilibrium and Nonequilibrium Statistical Mechanics* (Proc. XVII Solvay Conf. on Physics) ed. by G. Nicolis et al. (Wiley, New York 1981) p. 107

2.7 R.J. Glauber: In *Quantum Optics and Electronics*, ed. by D. De Witt et al. (Gordon and Breach, New York 1965)

2.8 F.T. Arecchi: In *Quantum Optics*, ed. by R.J. Glauber (Academic, New York 1969)

2.9 J.P. Eckmann: Rev. Mod. Phys. **53**, 643 (1981); J.P. Eckmann, D. Ruelle: Rev. Mod. Phys. **57**, 617 (1985)

2.10 F.T. Arecchi, V. Degiorgio, B. Querzola: Phys. Rev. Lett. **19**, 1168 (1967)

2.11 J.S. Langer: In *Flucutations, Instabilities and Phase Transitions*, ed. by T.Riste (Plenum, New York 1975)

2.12 H. Risken, H.D. Vollmer: Z. Physik **201**, 323 and **104**, 240 (1967)

2.13 F.T. Arecchi, V. Degiorgio: Phys. Rev. **A3**, 1108 (1971)

2.14 F.T. Arecchi, A. Politi: Phys. Rev. Lett. **45**, 1215 (1980)

2.15 F.T. Arecchi, A. Politi, L. Ulivi: Nuovo Cimento **71B**, 119 (1982)

2.16 Q.H.F. Vrehen, H.M. Gibbs: In *Dissipative Systems in Quantum Optics*, ed. by R. Bonifacio, Topics Curr. Phys., Vol. 27 (Springer, Berlin, Heidelberg 1982) p. 111

2.17 F. Haake, J. Haus, H. King, G. Schröder, R. Glauber: Phys. Rev. Lett. **45**, 558 (1980) and Phys. Rev. **A23**, 1322 (1981)

2.18a W. Lange, F. Mitschke, R. Deserno, J. Mlynek: Phys. Rev. **32A**, 1271 (1985)

2.18b E.Arimondo, C.Gabbanini, A. Gozzini, R. Longo, F. Maccarrone, F. Mango, E. Menchi: In *Optical Bistability III*, ec. by H.M. Gibbs et al., Springer Proc. Phys. 8, 256–259 (Springer, Berlin, Heidelberg 1986)

2.19 E.M. Lorenz: J. Atmos. Sci. **20**, 130 (1963)

2.20 P. Cvitanovich (ed.): *Universality in Chaos* (Hilger, Bristol 1984)

2.21 L.M. Hoffer, T.H. Chyba, N.B. Abraham: J. Opt. Soc. Am. B**2**, 102 (1985)

2.22 L.W. Casperson: J. Opt. Soc. Am. B**2**, 62 (1985)

2.23 F.T. Arecchi, G.L. Lippi, G.P. Puccioni, J.R. Tredicce: Opt. Commun. **51**, 308 (1984)

2.24 F.T. Arecchi, R. Meucci, G.P. Puccioni, J.R. Tredicce: Phys. Rev. Lett. **49**, 1217 (1982)

2.25 G.L. Lippi, J.R. Tredicce, N.B. Abraham, F.T. Arecchi: Opt. Commun. **53**, 129 (1985)

2.26 F.T. Arecchi, W. Gadomski, R. Meucci: Phys. Rev. A**34**, 1617 (1986)

2.27 J.D. Farmer: Physica **4D**, 366 (1982)

2.28 B.B. Mandelbrot: *The Fractal Geometry of Natur* (Freeman, San Francisco 1982)

2.29 P. Grassberger, I. Procaccia: Phys. Rev. Lett. **50**, 346 (1983)

2.30 T. Geisel, J. Niewetberger: Phys. Rev. Lett. **48**, 7 (1982)

2.31 F.T. Arecchi, F. Lisi: Phys. Rev. Lett. **49**, 94 (1982); and **50**, 1328 (1983)

2.32 F.T. Arecchi, R. Badii, A. Politi: Phys. Lett. **103A**, 3 (1984);
     F.T. Arecchi, R. Badii, A. Politi: Phys. Rev. A**32**, 402 (1985)

2.33 F.T. Arecchi, A. Califano: Europhysics Lett. **3**, 5 (1987)

2.34 G. Ioos, D.D. Joseph: *Elementary Stability and Bifurcation Theory* (Springer, Berlin, Heidelberg 1980)

2.35 G.P. Puccioni, A. Poggi, W. Gadomski, J.R. Tredicce, F.T. Arecchi: Phys. Rev.Lett. **55**, 339 (1985)

2.36 P. Grassberger: J. Stat. Phys. **26**, 173 (1981)

2.37 L. Lugiato, L.M. Narducci, D.K. Bandy, C.A. Pennise: Opt. Commun. **46**, 64 (1983)

2.38 D.K. Bandy, L.M. Narducci, L. Lugiato: J. Opt. Soc. Am. B**2**, 248 (1985)

2.39 J.R. Tredicce, F.T. Arecchi, G.L. Lippi, G.P. Puccioni: J. Opt. Soc. Am. B**2**, 173 (1985)

2.40 J.L. Boulnois, P. Cottin, A. Van Lenberghe, F.T. Arecchi, G.P. Puccioni: Opt. Commun. **58**, 124 (1986)

2.41 See, e.g., Instruction Manuals of current Argon or Kripton lasers commercially supplied by Coherent or Spectra Physics

2.42 H.A. Haus, Y. Yamamoto: MIT Workshop on Squeezed States of Light,Cambridge, Mass. (21 October 1985) unpublished;
     S. Machida, Y. Yamamoto: Opt. Commun. **57**, 290 (1986)

2.43 F.A. Hopf, B.L. Kaplan, H.M. Gibbs, R.L. Shoemaker: Phys. Rev. A**25**, 2172 (1982);
     H. Nakatsuka, S. Asaka, H. Itoh, K. Ikeda, M. Matsuoka: Phys. Rev. Lett. **50**, 109 (1983)

2.44 T. Midavaine, D. Dangoisse, P. Glorieux: Phys. Rev. Lett. **55**, 1989 (1985)

3. Experimental Measurements of Transitions to Pulsations and Chaos in a Single Mode, Unidirectional Ring Laser with an Inhomogeneously-Broadened Medium

N.B. Abraham, A.M. Albano, T.H. Chyba, L.M. Hoffer, M.F.H. Tarroja, S.P. Adams, and R.S. Gioggia

With 14 Figures

We review a series of measurements we have made on a single-mode laser designed and constructed to optimally match the unidirectional ring-laser configuration popularly assumed in theoretical studies. The active medium is made up of xenon atoms excited in a gas discharge in a mixture with helium. The $3.51\,\mu$m transition in xenon provides exceedingly high gain and is inhomogeneously broadened at typical working pressures, providing the optimum situation for studying the transition from stable CW operation to the emission of periodic and chaotic pulsations as the degree of excitation is increased. Careful measurements of both the output intensity and the electric field spectrum provide a clear picture of many details of the development of pulsations that have been overlooked heretofore. The presence of deterministic chaotic behavior of the laser is confirmed by quantitative studies of the strange attractor that underlies this form of dynamical evolution of the laser system. These results are placed in context by reviewing other forms of low-frequency pulsations which occur in single-mode and multimode inhomogeneously broadened lasers.

3.1 Background

Over the past five years, systematic experimental studies of inhomogeneously-broadened lasers have revealed a variety of pulsation patterns, which, when matched with the popular interest in nonlinear dynamics, turbulence and chaos, have caused wide-spread interest in finding deterministic chaos in laser systems [3.1–17]. We wish to distinguish several different types of laser systems and to present a review of some of our results for one particular type, the single-mode, unidirectional, inhomogeneously-broadened ring laser.

Time-dependent output from a laser system hardly seems to merit special attention as many lasers emit either irregular pulses (as in the spiking mode of solid state lasers [3.18]) or regular pulses (generated by periodic Q-switching [3.19] or the beating between modes of different optical frequencies [3.20]). While the sustained irregular spiking that occurs in some solid-state lasers is still not very well understood, the basic time scale is derived from a kind of relaxation oscillation involving the slowly changing population inversion and the more rapidly changing electric field in the laser cavity [3.21].

To focus our discussion and to place the present results in context we choose to distinguish several different kinds of laser systems which may have time-dependent output. The simplest of these are those systems which have some sort of modulation imposed as part of their operation, (Chap. 2) for example, a rotating or vibrating mirror, or modulated losses or excitation [3.22–24]. Among systems with time-independent parameters, that is, with the dynamical evolution determined by the interaction of the intracavity field and a medium which is excited in a steady fashion, we wish to separate those effects arising from the coexistence or interaction of different spatial field patterns from those arising from a single spatial pattern, which we will term multimode and single-mode, respectively. The multimode case (Chap. 4) may involve patterns having different optical frequencies or patterns having nearly the same optical frequency [3.25,26] (as may occur due to polarization degeneracy or in a bidirectional ring laser). Time-dependent output also may come from the beating between the optical carrier frequencies of modes with constant amplitudes or it may arise from mode-mode interactions that cause each "mode" to have a time dependent amplitude [3.27].

Intuitively, single-mode systems seem less likely to develop time-dependent output which must arise from the dynamical interaction of the field and material variables. Two distinct cases again merit mention, the standing-wave laser (in which, in a sense, the wave is forced to interact with itself as do the counter-propagating modes in a ring laser) and the unidirectional ring laser which seems to be a "clean" dynamical system, lacking any sort of geometrical stimulus to cause pulsations.

Multimode phenomena are easily expected to give pulsations in the total intensity because different resonant spatial patterns are likely to have different optical carrier frequencies. If two or more of these patterns coexist without interacting in the laser cavity, then the total output intensity involves the absolute square of the sum of oscillating terms with different frequencies with the result that difference frequencies modulate the total intensity. While optical detectors and other electronic measuring devices generally fail to respond at the high optical frequencies, their harmonics, or their sum frequencies, these devices can readily detect the difference frequencies for reasonably sized lasers.

At this point in a discussion, many laser physicists are so comfortable with the intuitive nature of the phenomena that there seems to be little room for further study. From a dynamical point of view it is worth remembering that laser systems are nonlinear and thus should not be expected to have solutions which are linear combinations of normal modes. Indeed, reading the early history of laser physics, we see that there was initially considerable doubt that a laser would have a mode structure [3.28]. Numerical calculations integrating Maxwell's equations for the field between two imperfect mirrors were required to show that in the presence of an amplifying medium there were certain field distribution patterns which were self-replicating after a single round trip [3.29].

As these patterns are generally in one-to-one correspondence with the modes of the lossless cavity of the same dimensions we now, almost instinctively, use the empty cavity mode nomenclature when discussing lasers with very nonideal cavities.

The "coexistence" of such modes is a subject of some considerable dispute. Perhaps in an extremely inhomogeneously-broadened medium one can picture each mode interacting with its own class of atoms and having vanishingly small interaction with the atoms which interact with the other modes. In almost all practical cases, however, the fields within the cavity interact to varying degrees with all of the material. Nevertheless the modes of the cavity often prove to be a reasonable basis set, though they may be separately perturbed by the medium (notably by the dispersive effects known as mode pulling [3.30,31]). When the modes coexist in this sense, one may find that the interaction leads only to small constant perturbations (e.g., mode-mode pushing) although some forms of mode interaction lead to strong competition (resulting in suppression or extinction of one mode). In general, however, we will not be interested in these forms of interaction or in the pulsations which result (which are the simple beating of the mode frequencies).

Dynamical interactions become more interesting when even the best set of perturbed eigenmodes is found to have time-dependent amplitudes. Two interesting classes of this type have recently been studied. Bidirectional ring lasers can show amplitude modulation effects on the time scale of the relaxation rates of the field and the population inversion which lead to new frequencies completely independent of the frequency differences between modes of different spatial patterns [3.32,33]. A second form of low-frequency modulation can be observed in the presence of two or more modes in a Fabry-Perot laser (three or more are required for this effect in a ring laser). The nonlinearity intrinsic to the field-atom interaction in a laser generates combination frequencies from the sum and differences of odd numbers of optical field components. The resulting combination tones may be nearly resonant with modes that are already oscillating. If the combination tones can lock to the nearby modes, then the modes become phase locked with a fixed frequency spacing leading to the generation of ultrashort pulses (which become shorter as more modes are involved) [3.19,34]. This is the phenomenon commonly referred to as mode-locking.

Alternatively, the combination tones may become nearly fixed at some frequency offset from the adjacent modes, leading effectively to low frequency modulation of the amplitudes of the modes. In another view, these offset combination tones lead to a low-frequency beating with the adjacent modes, resulting in an overall low-frequency modulation of the short pulses generated by inter-mode beating (a phenomenon sometimes called "breathing") [3.5–7,35,36]. We have shown that this modulation occurs in the laser itself, affecting the dynamical evolution of the modes [3.5–7] and thus that pulsations of this type are not merely low-frequency beats generated by a square-law intensity detector between free-running optical fields.

With all of the above phenomena possible in laser systems, we are nevertheless able to distinguish clearly a separate case which is that of dynamical pulsations in a single-mode system.

3.2 The Single-Mode, Inhomogeneously-Broadened, Unidirectional Ring Laser

3.2.1 Background

The unidirectional, single-mode, ring laser is that simple object favored for theoretical modeling because the field is described as a traveling wave in the laser cavity [3.1–3,13,14,37–50], eliminating the need for details that account for the standing-wave patterns formed in a Fabry-Perot cavity. Of all the theoretical work in this area there are only a few papers which deal directly with the stability questions in a standing-wave system [3.1–3,14,37,39,46,49,50,52] which is by far the easier to study experimentally [3.1,3,4,8–11]. Regardless, the greatest theoretical and numerical progress has been made for this special case of the ring laser and we will describe our work in the experimental studies of such a system.

Initially the inhomogeneous broadening (caused by Doppler shifts) of the $3.51\,\mu$m xenon transition used in our experiments had also seemed to be a formidable barrier to theoretical analyses. However, in the last three to five years, following the earlier pioneering work of Casperson, the wealth of publications indicates that this hurdle has also been surmounted. Two very helpful reviews of this progress have been given by *Casperson* [3.3,14]. (See also the January 1985 issue of J. Opt. Soc. Am. B.)

The principal parameters of the system are the degree of excitation of the medium and the set of linewidths (or equivalently decay rates) that characterize the field and material variables. The inhomogeneously broadened medium is made up of classes of atoms with different resonant frequencies, though each atom within each packet is usually well-characterized by a decay rate for the population difference and a decay rate for the dipole moment (or off-diagonal elements of the density matrix) between the levels involved in the lasing action. Although Casperson has shown that some critical features of the pulsations of a xenon laser are obtained only with a sufficiently complex model involving upper and lower level decay rates and cross-spectral relaxation rates (velocity-changing collisions) [3.13,14], for simplicity, we will speak of the decay rate of the population inversion of a given packet, $\gamma_{\parallel}$, and the decay rate of the polarization, $\gamma_{\perp}$, using the approximation suited to a medium made up of two-level systems. Experimentally, $\gamma_{\perp}$ can be made much larger than $\gamma_{\parallel}$ by increasing the gas pressure with a consequent increase in phase-interrupting collisions. The inhomogeneous broadening leads to a spread in resonant frequencies for the atoms that can be characterized by an effective half linewidth σ. Though there remains some dispute, we quote the following linewidths as full widths at half maximum for the xenon transition: Doppler linewidth: 110 MHz; natural linewidth: 4.6 MHz; pressure broadening: 18.6 MHz/Torr-He and 10.6 MHz/Torr-Xe [3.53].

3.2.2 Experimental Set-Up

In our studies of single-mode instabilities we have used the high-gain infrared transitions in noble gas lasers (3.51 μm in xenon and 3.39 μm in neon). These gas laser lines operate under a variety of excitation and pressure conditions, permitting adjustment of parameters in the exploration of different types of laser behavior. The most notable benefit of the high gain is that it can readily offset the relatively high mirror losses which turn out to be crucial to the observation of most of the interesting phenomena.

Our unidirectional ring laser is shown in Fig. 3.1. The optical path follows the perimeter formed by four mirrors, three 99 % reflecting and the fourth 90 % reflecting. The ring is designed with two long arms and two short arms to permit the maximum filling of the cavity path with active medium which is ac-

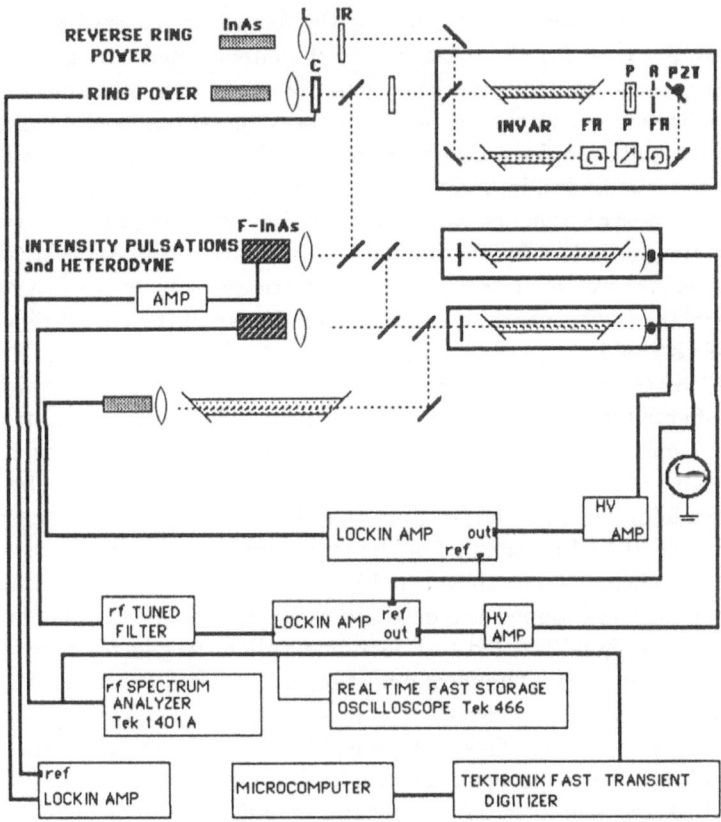

Fig. 3.1. Set up of optics and electronics for unidirectional ring laser (P: wire-grid polarizer; FR: Faraday rotator; A: aperture; F-InAs: high-speed photodiode; InAs: low-speed photodiode; C: chopper; L: quartz lens; IR: 3.51 μm dielectric filter; PZT: piezoelectric mirror translator)

complished with two gas discharge tubes having 19 cm and 11 cm, respectively, of "active" length of discharge confined to two millimeter diameter capillary tubes for efficient excitation. The discharge tubes are excited in a cold cathode configuration and we use a large gas reservoir surrounding each cathode to off-set the notorious "clean-up" phenomenon in low-pressure xenon discharges. A non-excited return path is provided from anode to cathode to maintain pressure equilibrium and to offset the effects of cataphoresis (gas mixture separation) which becomes noticeable at high current densities. The discharge was maintained by stabilized high-voltage power supplies with large series ballast resistors ($100-250\,k\Omega$) and stray capacitance was minimized to suppress unwanted plasma oscillations [3.54]. Within the optical path we inserted two Faraday rotators and two linear polarizers which ensure the unidirectionality of the laser operation with an extinction ratio of better than 40:1 against counter-propagating signals or horizontally polarized signals. The Faraday rotators were made of specially ordered and anti-reflection coated Yttrium-Iron-Garnet crystals [3.55]. The saturable Faraday effect in these crystals meant that using sufficiently strong annular ceramic magnets we could fix the rotation of linearly polarized light by precisely determining the length of the crystal. The two rotators were designed for 45° rotation and the interspersed wire-grid linear polarizers were set to transmit vertically polarized light and light polarized at a 45° angle with respect to the vertical. The only remaining complication with respect to achieving single-mode operation was the possibility of exciting higher-order transverse modes, in addition to the desired lowest-order longitudinal modes. The transverse modes were suppressed by inserting apertures in the cavity, providing more loss for the modes with the larger transverse structure [3.56]. One mirror was mounted on a piezoelectric transducer permitting us to slightly vary the cavity length, which also slightly varied the cavity alignment, but as far as could be determined misalignment effects were negligible, perhaps because of the strong transverse profile of the gain medium which led to focusing effects that stabilized the mode pattern. The cavity round-trip distance of 80 cm meant that the free-space cavity modes were separated by 375 MHz. Even with a factor of three or more in mode-pulling effects the longitudinal mode-spacing remained larger than the 110 MHz Doppler-broadened linewidth (full width at half maximum) of the transition.

The detection scheme is also shown in Fig. 3.1. The laser output was divided and one portion was chopped and detected by a 1 mm diameter InAs photodiode. The resulting signal was fed to a lock-in amplifier to give readings of the average laser output power. The second portion of the signal was focused by a quartz lens onto a $0.01\,mm^2$, reverse-biased InAs photodiode (the small area and reverse-biasing provide high-speed response) mounted on an impedance-matched preamplifier. The resulting signal with its fluctuations within a 1 kHz–100 MHz bandwidth was amplified by a wideband amplifier and then split and displayed on a fast storage oscilloscope and on a power spectrum analyzer to observe snapshots of the intensity pulsations versus time and the time-averaged intensity power spectrum, respectively.

Additional information about the behavior of the laser was obtained through measurements of the optical (electric field amplitude) spectrum of the laser. This is because the power spectrum of the intensity often obscured the physical origin of pulsations which, most fundamentally, have a direct effect on the electric field amplitude. By heterodyning a stabilized reference laser with the output of the unstable laser and observing the pattern of resulting beat notes on a spectrum analyzer, details of the optical electric field amplitude spectrum could be recorded. To provide the needed frequency-stabilized reference laser we adopted a procedure reported previously [3.10,11] using two lasers, one of which was stabilized to the peak of the gain of a reference optical amplifier and a second which was stabilized at a fixed frequency offset from the first laser. Only the first laser was modulated. This generated the error signal for its stabilization when it was detuned from the reference amplifier and generated a modulated beat note between the two lasers which was detected through an rf-tuned filter by a lock-in amplifier to provide the control signal to stabilize the second laser. This procedure provided a frequency-stabilized reference laser which was not modulated and thus the reference laser retained its narrow line width, giving greater clarity to the spectrum when it was heterodyned with the ring laser.

3.2.3 Thresholds for Transitions from Stable to Pulsed Behavior

It has been well established theoretically that for suitable parameter values an inhomogeneously broadened laser will not operate stably (with a constant intensity output) above a critical threshold level of excitation. As with the homogeneously broadened laser [3.3] a necessary condition is the "bad cavity requirement", namely that the decay rate for the electric field inside the cavity exceed the polarization decay rate (or, in other words, that the cavity linewidth exceed the homogeneously broadened linewidth of the medium). Note that for the inhomogeneously broadened case, this does not require that the cavity linewidth be greater than the gain bandwidth of the laser (as is the case for the homogeneously-broadened system). When the system is relatively inhomogeneously broadened then the critical level of excitation for transitions to pulsed output is reduced to very near the threshold for laser action. For very large cavity linewidths or for very large inhomogeneous broadenings this second laser threshold can be given analytically. For arbitrary line shapes (mixed broadening cases) made up of a Gaussian distribution of Lorentzian, homogeneously-broadened components, the thresholds are given by implicit relations which can be evaluated numerically [3.47].

In preliminary studies, we observed the real time displays of snapshots of the intensity and the rf intensity power spectra. The instability threshold was found by increasing the excitation and noting the emergence of intensity pulsations and corresponding peaks in the rf spectrum. To ensure that the laser was operating in single mode, we first plotted the laser power output versus cavity

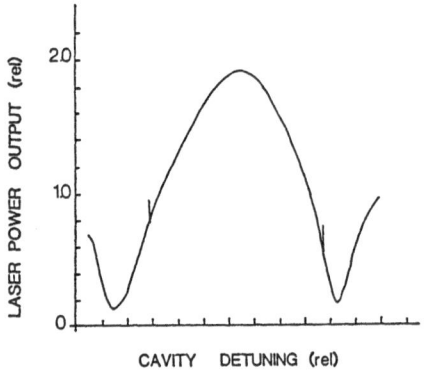

Fig. 3.2. Laser power output vs. detuning. *Vertical lines* denote the boundaries of the instability region (Pressures: 70 mTorr Xe-136 and 380 mTorr He; discharge current: 5 mA)

detuning, as shown for a typical case in Fig. 3.2. The power output dips sharply to zero as the mirror is moved one wavelength and the absence of intermediate peaks or variations from a smooth, bell-shaped output strongly suggests single mode operation. Only in the extremes of the sharp dip in output (when it did not go fully to zero) did we observe a weak pulsation near 100 MHz as might be expected from the mode-pulled beat notes between two competing modes. By removing the aperture we could observe other pulsations at a frequency of about 40 MHz which we presumed resulted from beats between the longitudinal and transverse modes as was confirmed by the presence of a second peak in the power output for detuning through one free spectral range. By selection of a sufficiently small aperture (1.0 mm) we were able to fully suppress this unwanted mode.

Tuning to the peak in the laser output we could be reasonably certain that the laser was operating nearly resonantly on a single longitudinal mode. Differences from exact resonance are the result of gain and dispersion focusing that arise in a medium that has a transverse profile as is common in axial discharges of the type used in our experiments [3.57–60].

The instability threshold noted on Fig. 3.3 was determined as the discharge current value at which peaks occurred in the power spectrum of the photocur-

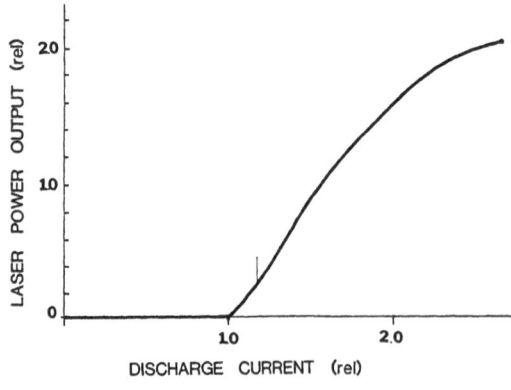

Fig. 3.3. Laser power output versus discharge current. *Vertical line* denotes the instability threshold. Pressures as in Fig. 3.2

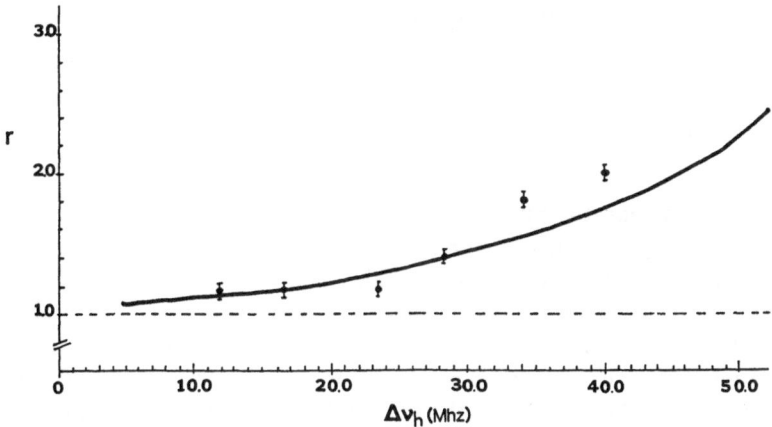

Fig. 3.4. Laser instability threshold (*solid line*-theory after [3.44], *data points* indicate experimental values). Pressures as in Fig. 3.2

rent from the fast detector measuring the intensity of the laser output. There is some uncertainty in this determination, as many stable systems frequently show such peaks resulting from weak and transient sidebands to the main optical carrier frequency. In the presence of noise (spontaneous emission, discharge fluctuations, or mechanical vibrations) peaks of this type appear in advance of the actual transitions to sustained oscillations [3.61–63]. This is because the system undergoes a loss of dynamical stability via a Hopf bifurcation at the pulsation threshold. This means that below the pulsation threshold the system responds to perturbations with a damped oscillation that returns to the constant intensity condition with the damping constant decreasing to zero at the pulsation threshold. Disturbances of the system when it is near (yet below) the pulsation threshold generate numerous damped oscillations which show up in the power spectrum of the intensity. Despite our difficulty in precisely determining the second threshold experimentally, we find that there is good agreement between the experimental determinations of the instability threshold and the theoretical predictions, as indicated in Fig. 3.4.

Data of the type acquired in Fig. 3.3 are compiled in Fig. 3.4 for different discharge pressures. The values of the excitation current appear to be linearly related to the laser output power (at least for low currents) so we have assumed that the discharge current is linearly proportional to the "excitation parameter" of the various models. Thus the specific values of the current are renormalized to the value of the current at which lasing action is first observed giving the parameter "r" plotted on the vertical scale of Fig. 3.4. As the pressure is raised, the homogeneous broadening of the line increases. With fixed cavity parameters (and thus a fixed cavity loss rate), this reduces the degree to which the "bad cavity" condition is satisfied and reduces the degree of inhomogeneous broadening both of which contribute to an increase in the second threshold. For the highest pressure used we were unable to excite the laser sufficiently far above the lasing threshold to reach the regime of pulsations and that result is also consistent with the prediction of a relatively high second threshold.

3.2.4 Changes in the Pulsation Pattern Above the Second Threshold

With increasing excitation the pulsation patterns change. The details are quite sensitive to the cavity tuning and the gas pressures used. We present several examples of the kinds of phenomena which can be observed. Figure 3.5 shows several results taken for the laser with the cavity resonantly tuned. The left-hand panel shows a typical snapshot of the intensity pulsations while the right-hand panel shows the spectrum observed on the spectrum analyzer. By a convenient choice of the frequency of the stabilized reference laser the heterodyne spectrum is separated from the homodyne spectrum. Figures 3.6 and 7 show plots of the peaks of these spectra for increasing excitation compiled from the data of Fig. 3.5. Below the pulsing threshold there is a single beat note corresponding to single optical carrier frequency of constant intensity. There is no corresponding fluctuation in the intensity nor is there a low-frequency component to the spectrum. A weak instability appears at first, revealed by the pulsing intensity and the peak at low frequencies in the intensity power spectrum, but as a weak sideband frequency is not clearly resolved in the optical spectrum, the amplitude modulation must be less than 10 %. The stronger pulsing state that follows for higher discharge current is clearly distinguished in the heterodyne spectrum as the result of bands of frequencies in two parts symmetrically placed with respect to the original line center and other bands at odd harmonics of the spacing. With increasing current these bands become stronger until the bands are larger than the signal at the original main optical carrier frequency. The existence of broad bands of frequencies instead of narrow lines suggests an irregular time dependent behavior which, as we will demonstrate later, has its origin in deterministically chaotic behavior. The apparent subharmonic seen in the intensity power spectrum corresponds to the weak lingering of the original carrier frequency of the CW operation. This peak appears intermittently and with variable amplitude suggesting an intermittent form of chaos involving switching from symmetric to antisymmetric modes of oscillation. At higher values of the excitation, the peaks become narrower suggesting a transition to regular, periodic behavior.

Operation with detuning breaks the symmetry of the spectra presented for resonant conditions, as shown in Fig. 3.8. In this case the sidebands do not emerge symmetrically with the result that the beats between the remnants of the original carrier frequency and the sidebands are different, perhaps serving as the origin of apparently quasi-periodic spectra for the intensity pulsations. At strong excitations the original carrier is apparently completely suppressed in favor of two equally strong optical carrier frequencies. A subharmonic is observed but there is no longer clear evidence in the heterodyne spectrum of its origin.

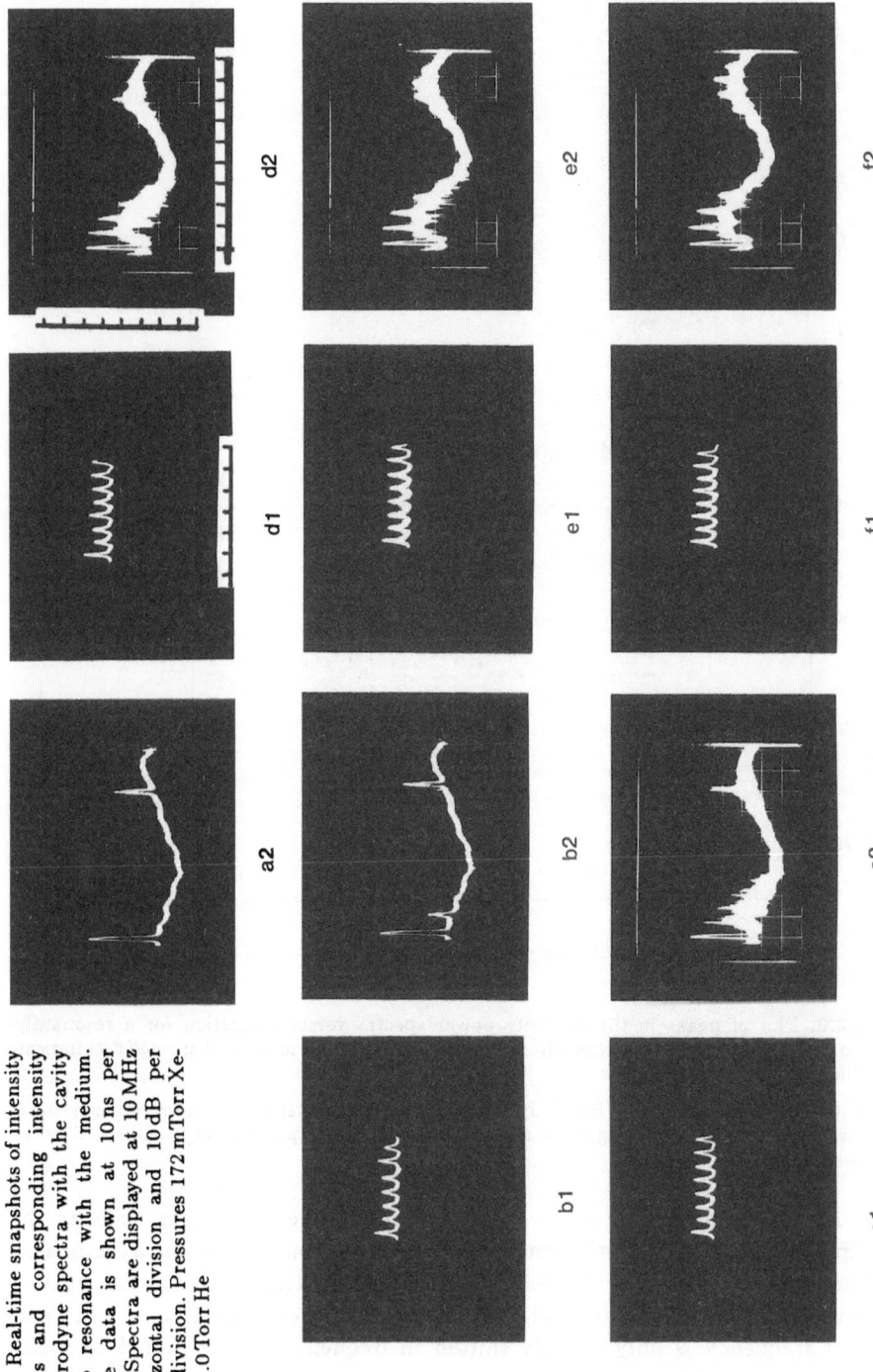

Fig. 3.5. Real-time snapshots of intensity pulsations and corresponding intensity and heterodyne spectra with the cavity tuned to resonance with the medium. Real-time data is shown at 10 ns per division. Spectra are displayed at 10 MHz per horizontal division and 10 dB per vertical division. Pressures 172 mTorr Xe-136 and 1.0 Torr He

59

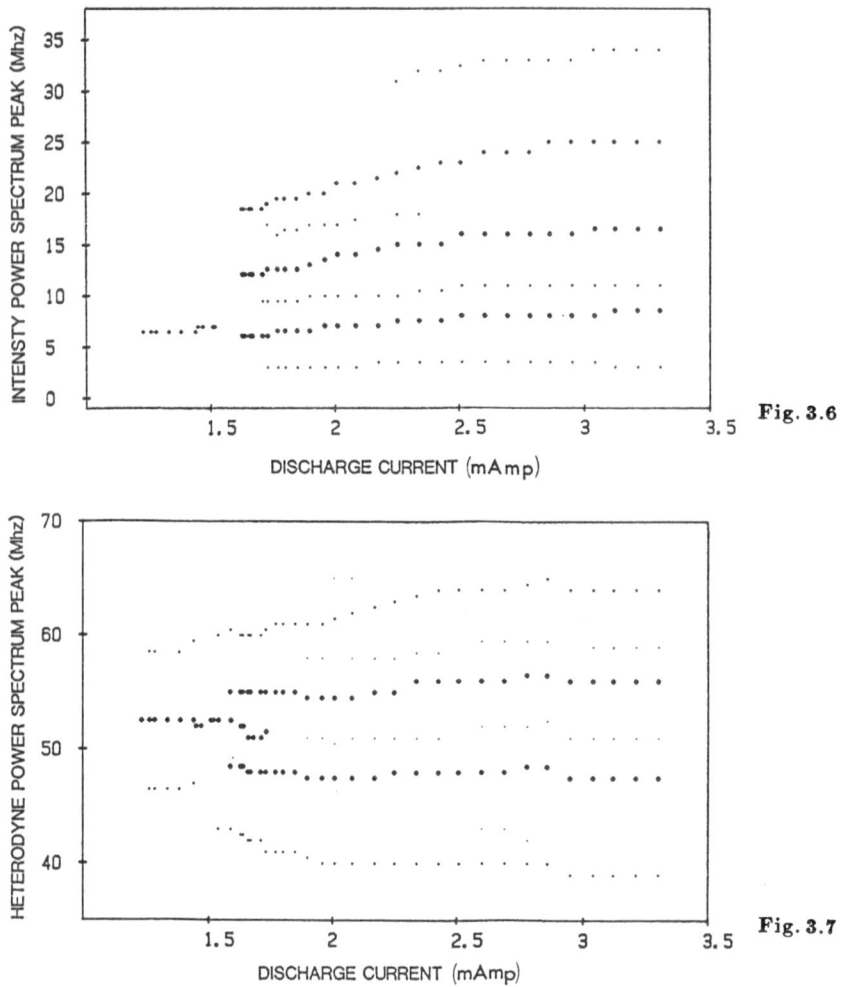

Fig. 3.6. Plot of peaks in the intensity power spectra versus excitation for a resonantly-tuned cavity, compiling data from Fig. 3.5. *Large dots* represent large peaks; *small dots* indicate distinct yet lower peaks

Fig. 3.7. Plot of peaks in the heterodyne spectra versus excitation for the resonantly-tuned cavity, compiling data from Fig. 3.5. *Large* and *small dots* correspond to large and small peaks, respectively

For larger detunings there are other qualitative differences such as the disappearance of the chaotic behavior at the pulsation thresholds, as indicated in the sequence in Fig. 3.9. Instead the laser makes a transition to weak pulsing and this is confirmed by the heterodyne spectra which show that the original carrier frequency is only slightly shifted in frequency and is joined by weak

Fig. 3.8. Pulsations and spectra as in Fig. 3.5 for a cavity detuning of 10 MHz from the material resonance causing a detuning of 3 MHz in the near threshold operating frequency of the laser

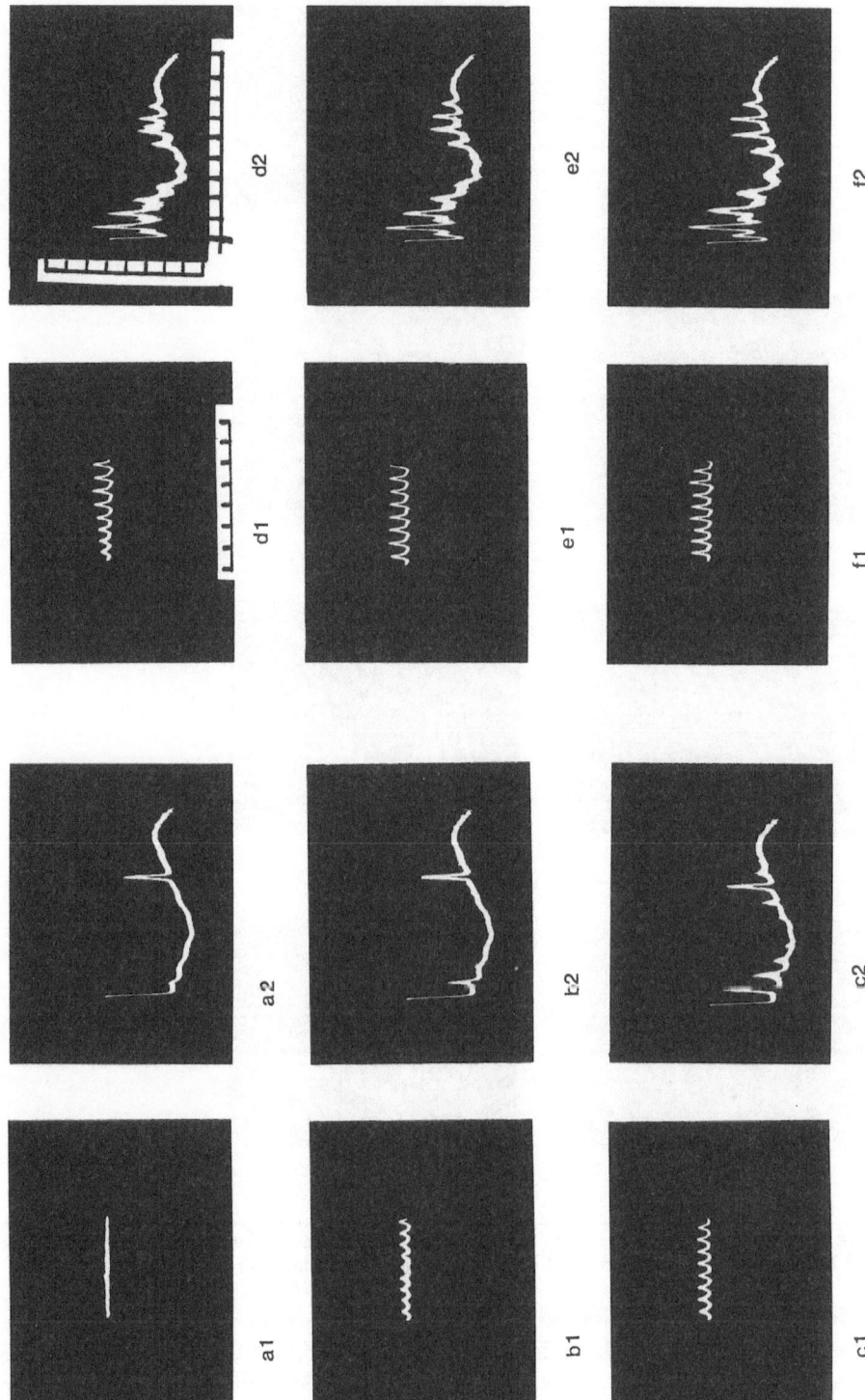

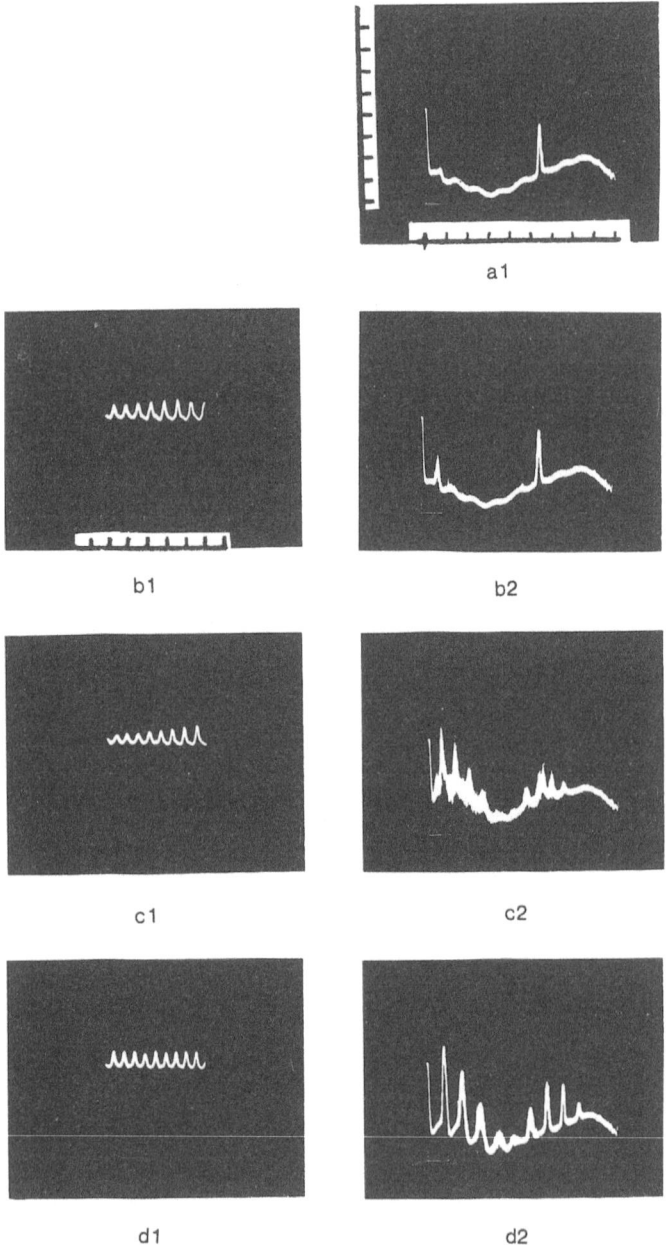

Fig. 3.9. Pulsations and spectra as in Figs. 3.5 and 6 for a cavity detuning of 15 MHz causing a 5 MHz detuning in the threshold operating frequency of the laser

sidebands. With increasing excitation the sidebands grow differently, with those closer to the peak in the gain profile growing more and moving toward the peak in the gain profile, indicating increasing mode pulling as is expected for inhomogeneously broadened lasers.

62

A remarkable feature of these results is that they can be modelled by numerical integration of the equations for the inhomogeneously broadened laser [3.3,14,48,50,52]. Although one hundred to three hundred equations are required to achieve good accuracy [3.47], the results of time-dependent signals, of intensity power spectra, and of heterodyne spectra are in excellent agreement with our experimental results [3.64].

Extension of the experimental work reported here and comparisons with the predictions of the Lugiato-Narducci model for the single-mode inhomogeneously broadened laser have recently been reported elsewhere [3.64].

An interesting way to summarize many of the present experimental results and to show some of the effects of nonresonant tuning of the cavity is presented in Fig. 3.10. Here the peaks in the intensity power spectra and in the hetero-

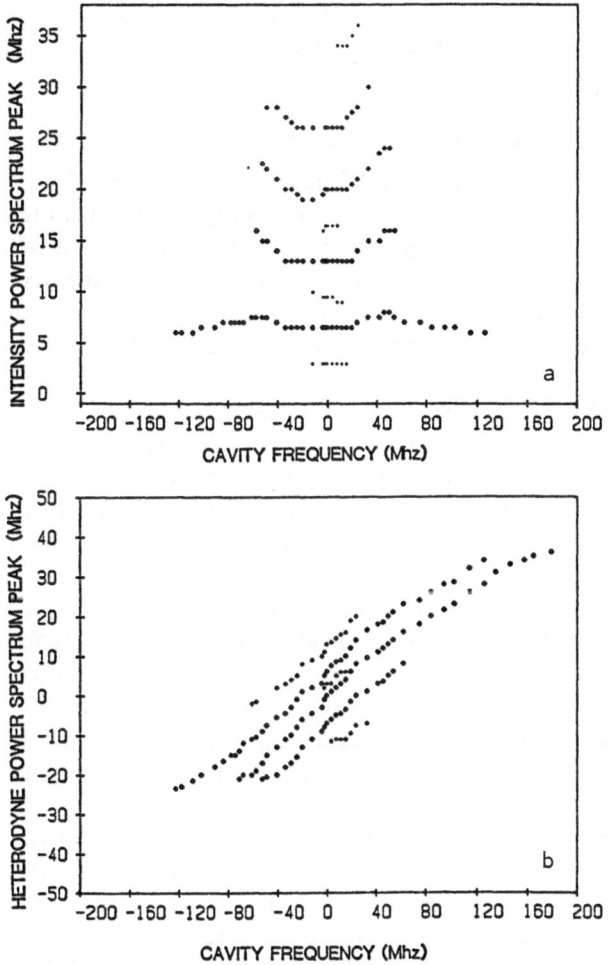

Fig. 3.10. (a) Peaks in the intensity power spectrum versus cavity detuning; **(b)** peaks in the heterodyne spectrum versus cavity detuning. Both data sets for pressure conditions in Figs. 3.5–9 and at a discharge current of 1.73 mA

dyne spectra are shown versus cavity detuning. In the regions of no pulsing there is but a single beat frequency in the heterodyne spectrum. In the wings of the region of pulsations, the pattern is dominated by a central peak and weak sidebands. Closer to the center of the pulsing range the spectrum is dominated by two nearly equal peaks with some weak sidebands equally spaced. Again regions that appear to be period doublings from the intensity power spectra are shown to occur because of the reemergence of the central frequency of the CW state. This leads us to identify what had seemed to be subharmonic operation as a symmetry breaking instead, although it also leads to alternating pulse heights for the intensity pulsations. Bifurcations of this type have recently been discussed theoretically for codimension-two systems [3.65,66], and we believe this may be the first experimental evidence of this type of alternative to the "conventional, Feigenbaum-type" [3.67] subharmonic bifurcation sequences which are found rather universally in one-parameter systems and also appear in many more complex systems such as lasers [3.5,8,68].

3.2.5 Confirming and Characterizing the Chaos

Chaos is now quite a popular subject for research in many fields [3.69–72]. It seems to be an almost inevitable form of operation for highly nonlinear multi-dimensional systems. Theoreticians working with computer models have a relatively easy task in identifying chaos by its lack of periodicity. Quantitative measures can confirm the chaotic behavior through calculation of Lyapunov exponents, dimensionalities of the attracting subset in the phase space, or entropies.

Experimentalists must be equally disciplined and quantitative, particularly for systems such as lasers which have intrinsic noise. One desires indications that the irregular pulsations and broadband spectra result from dynamical effects and not from some amplification of the intrinsic stochastic noise. Techniques for such analyses have recently been proposed [3.73,74] and applied with some success [3.16,75–78].

We have recently established that brief digitized records of intensity pulsations from the ring laser under study here can be analyzed quantitatively for indications that the broadband (or broadened) spectra result from evolution on a low-dimensional strange attractor. Our digitizing was done by a Tektronix fast transient digitizer with an effective 400 ps sampling window and a 4.0 ns interval between samples. Each record had a maximum length of 512 points with up to 10 bit accuracy.

Two selected time series and corresponding intensity power spectra for the ring laser at different cavity detunings are shown in Fig. 3.11. To simulate the attractor, we used the embedding technique of analyzing the data in a multidimensional space formed by taking each sequence of m data values to be an m-dimensional vector [3.79]. The value of this embedding technique is that it has been shown that the topology of the attractor in the fundamental variable space is recovered fully by the reconstructed attractor when the embedding

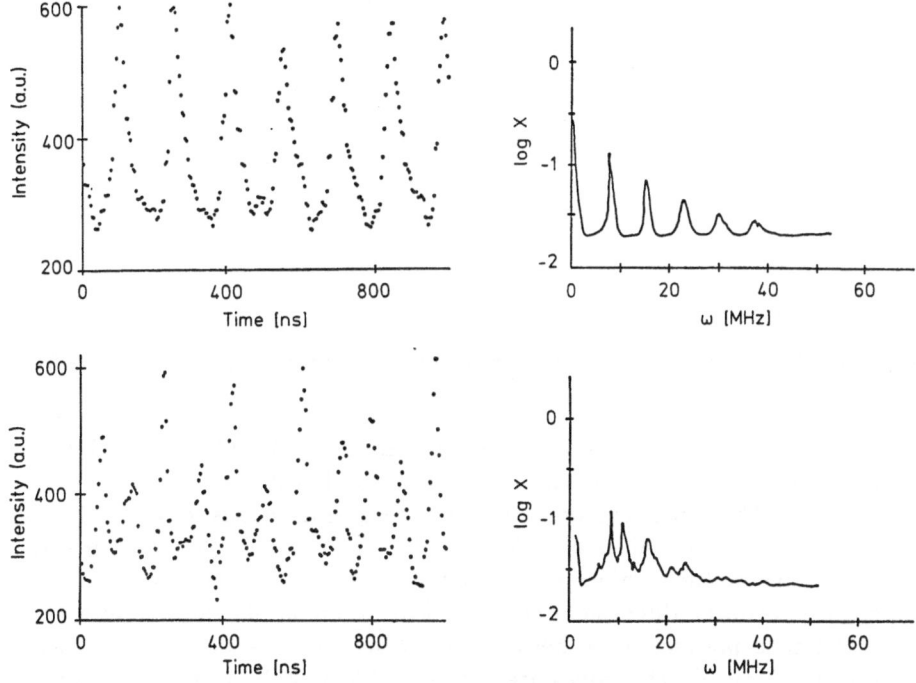

Fig. 3.11a,b. Examples of digitized time series (and corresponding intensity power spectra) analyzed for evidence of strange attractors. Data was taken for discharge pressures of 70 mTorr Xe-136 and 679 mTorr He at 1.73 mA and detunings as indicated

dimension (m) is sufficiently large. Thus we calculate the dimensionality of the attractor for increasing m-values and look for asymptotic convergence.

Specifically we calculate the order-2 information dimension D_2 by first determining the correlation sum defined by *Grassberger* and *Procaccia* [3.73]. This means defining a set of m-dimensional vectors from the data (with components being data values equally spaced in time). Then we determine the number, $N(\varepsilon)$, of intervectorial distances which are less than selected values of the distance ε.

For high ε every inter-vectorial distance between points on the attractor is less than ε. For low ε values there are too few occurrences of such distances and, in addition, spacings between points are dominated by noise which blurs the specific data values. The fractal dimensionality is defined and the self-similarity of the attractor appears for the asymptotic limits of $\varepsilon \rightarrow 0$ and $N \rightarrow \infty$, but the constraints mentioned above limit us to a possible scaling region to be found for intermediate values of ε. In such a scaling region we expect to find that $N(\varepsilon)$ scales as ε^{D_2}. The limited numbers of data points severely limit the scaling region as it is nearly impossible to resolve features on a length scale less than $\langle \varepsilon \rangle / N$ where $\langle \varepsilon \rangle$ is the average distance between pairs of points in the multi-dimensional space. Despite these limitations, we have determined that reliable results can be obtained with as few as 500 points [3.80], a welcome fact as our

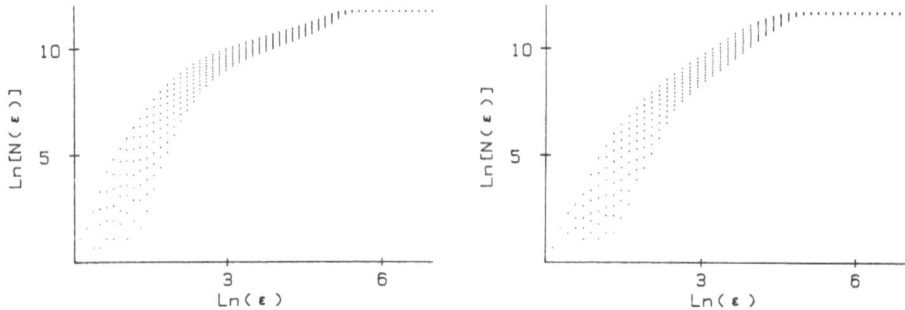

Fig. 3.12. Sample plots of ln $N(\varepsilon)$ vs. ln ε for different dimensions 10–20 using data from Fig. 3.11

digitizer is limited to records of 512 points. Sample log-log plots of $N(\varepsilon)$ versus ε for different embedding dimensions are given in Fig. 3.12.

In Fig. 3.13 we plot the slopes of the log-log plots versus ln $N(\varepsilon)$. Plotting the slopes versus the values of ln $N(\varepsilon)$ shows us the scaling for the same percentage range of the number of interpoint distances. We see that curves for different embedding dimensions converge to a common plateau and in various data sets such plateaus have been found to be near one, near two, or larger than two. (Typically the values are high by 10 % or 20 % as a consequence of the limited number of points.) These are consistent with the expected results that give dimensionality of one for periodic behavior, two for quasi-periodic behavior (two incommensurate pulsing frequencies) and larger than two for chaotic behavior.

The experimental results are summarized for different detunings in Fig. 3.14. As a useful benefit of the calculational technique followed here, we are also able to determine an estimator of the Kolmogorov entropy, K_2, from the successive vertical displacements of the curves in plots such as Fig. 3.12 [3.73]. Values for this estimator are also shown in Fig. 3.14 as are the values of the principal peaks in the intensity power spectrum for reference. The expected results are

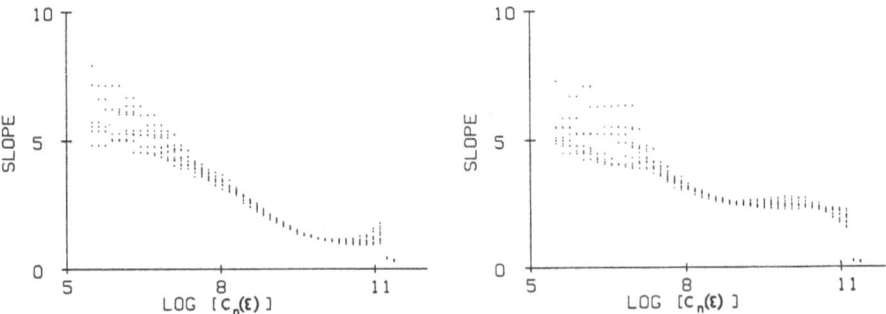

Fig. 3.13. Sample plots of slopes of graphs in Fig. 3.12 vs. ln $N(\varepsilon)$. Plateaus occur at values of 1.1±0.1 for the periodic case and at 2.6±0.3 for the case with broadened, incommensurate peaks in the power spectra

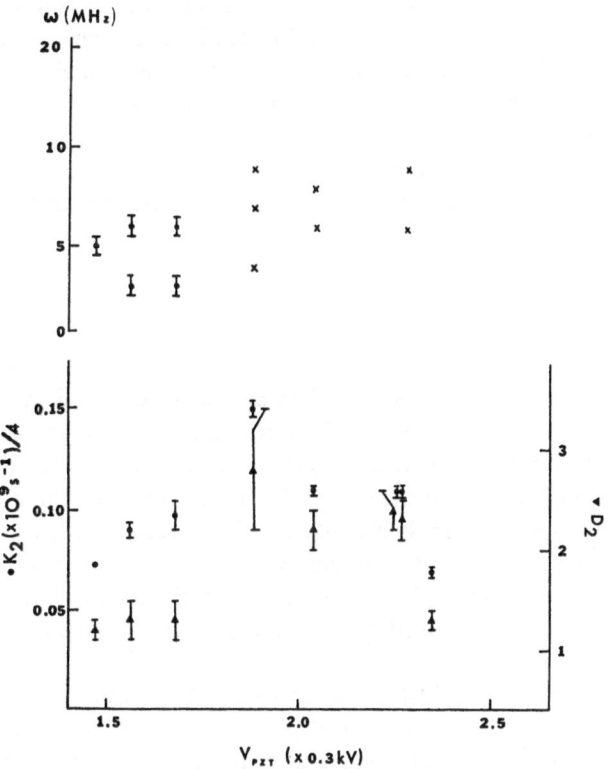

Fig. 3.14. Principal pulsing frequencies ω, order-2 information dimension D_2, and order-2 Kolmogorov entropy K_2, versus voltage across the piezoelectric translator V_{PZT}. On the top graph, signals characterized by a single frequency and its harmonics are represented by a single point at the fundamental frequency. A signal characterized by two frequencies and their harmonics and linear combinations is represented by points at the principal frequencies. The x's denote locations of broad peaks, or of peaks superimposed on a broad background as in the lower part of Fig. 3.11b

values near zero for periodic or quasi-periodic behavior, finite and nonzero for chaotic behavior, and infinite for stochastic behavior. The residual noise and finite number of data points will give a small non-zero contribution, but we see low values in the periodic and quasi-periodic regions (dimensions one and two) and higher, but definitely finite, values in the regions of broader power spectra that we have presumptively called chaos. No asymptotic limit emerged for K_2 in the analysis of random signals. The results of both measures confirm that the transition from discrete line power spectra to broadened lines or broadband power spectra has its origin in deterministically chaotic evolution of our laser system.

We should note that in our studies of Fabry-Perot laser systems we found similar evidence of chaos from dimensionality studies that reinforce our appellation of chaos to spectra with broadened peaks or broadband spectra, in addition to those occurring at the end of period-doubling or quasi-periodic sequences identified from their power spectra [3.16]. In addition, in recent studies of amplified spontaneous emission, we have been unable to establish either an attractor dimensionality (the slopes keep increasing with embedding dimension up to dimension 20) or a finite value of the entropy. Both of these results are consistent with the belief that fluctuations in the ASE signal arise from the stochastic effects of 10^8 (or more) independent spontaneous emissions from the excited atoms in the source [3.81].

3.3 Discussion

Considerable progress has now been made in a field almost abandoned by experimentalists for ten years. The resurgence of interest has inspired searches for more nearly homogeneously broadened laser systems which may match relatively simple models. Recent progress in the study of FIR instabilities (Chap. 5) suggests that this goal may soon be reached [3.82,83]. There too, details of the optical spectrum can be compared to the intensity spectrum to see the full subtleties of the "Lorenz-like" chaos. These examples may finally provide experimental evidence to support the traditional pedagogical discussions of this model.

Nevertheless, the inhomogeneously broadened case retains many special features which will warrant study for a number of years to come. In particular, a full understanding of the strange phenomena seen near the Lamb dips in our work with Fabry-Perot lasers [3.10,11] will require careful study. The unusual pulsation patterns and their dependence on detuning most likely arise from the interaction between atoms and two fields. A complete understanding of changes in pulsation features and shifts to higher or lower instability thresholds may contribute to a number of practical applications.

More current reports of work on optical chaos and laser instabilities can be found in [3.84].

Acknowledgements. This work was supported in part by grants from the Alfred P. Sloan Foundation and the National Science Foundation ECS82-10263. We are particularly grateful for the assistance of our collaborators in the theoretical and numerical analyses, L.M. Narducci, L.A. Lugiato, D.K. Bandy, and T. Isaacs and for the work of our colleagues B. Das and G.C. deGuzman on the study of the existence of dimensionalities (or underlying attractors) for various signals including digitized records from CW amplified spontaneous emission. We also wish to acknowledge helpful discussions on matters of laser stability and laser dynamics with L.W. Casperson, P. Mandel, F.T. Arecchi and J.R. Tredicce. Consultations with I. Procaccia, J. Gollub, H. Swinney, J.D. Farmer, R. Kapral, A. Wolf, S. Ciliberto and M.A. Rubio on the analysis of chaotic signals have greatly aided our own calculations. We would also be remiss without expressing our sincere gratitude for the cooperative spirit of our colleagues in the Department of Physics and Atmospheric Science at Drexel University who have given us nearly unlimited use of their fast transient digitizer.

References

3.1   L.W. Casperson: IEEE J. QE-**14**, 756–761 (1978)
3.2   L.W. Casperson, M.L. Minden: IEEE J. QE-**18**, 1952–1957 (1982)
3.3   L.W. Casperson: In *Laser Physics*, ed. by J.D. Harvey and D.F. Walls, Lecture Notes Phys., Vol. 182 (Springer, Berlin, Heidelberg 1983) pp. 88–106
3.4   N.B. Abraham, T. Chyba, M. Coleman, R.S. Gioggia, N.J. Halas, L.M. Hoffer, S.-N. Liu, M. Maeda, J.C. Wesson: In *Laser Physics*, ed. by J.D. Harvey and D.F. Walls, Lecture Notes Phys., Vol. 182 (Springer, Berlin, Heidelberg 1983) pp. 107–131
3.5   C.O. Weiss, H. King: Opt. Commun. **44**, 59–61 (1982)
3.6   C.O. Weiss, A. Godone, A. Olafsson: Phys. Rev. A**28**, 892–895 (1983)
3.7   N.J. Halas, S.-N. Liu, N.B. Abraham: Phys. Rev. A**28**, 2915–2920 (1983)
3.8   R.S. Gioggia, N.B. Abraham: Phys. Rev. Lett. **51**, 650–653 (1983)
3.9   R.S. Gioggia, N.B. Abraham: Opt. Commun. **47**, 278–282 (1983)
3.10 R.S. Gioggia, N.B. Abraham: Phys. Rev. A**29**, 1304–1309 (1984)
3.11 R.S. Gioggia, N.B. Abraham: In *Coherence and Quantum Optics V*, ed. by L. Mandel and E. Wolf (Plenum, New York 1984) pp. 563–570
3.12 L.E. Urbach, S.-N. Liu, N.B. Abraham: In *Coherence and Quantum Optics V*, ed. by L. Mandel and E. Wolf (Plenum, New York 1984) pp. 593–600
3.13 L.W. Casperson: J. Opt. Soc. Am. B**2**, 62–72 (1985)
3.14 L.W. Casperson: J. Opt. Soc. Am. B**2**, 73–80 (1985)
3.15 L.M. Hoffer, T.H. Chyba, N.B. Abraham: J. Opt. Soc. Am. B**2**, 102–107 (1985)
3.16 A.M. Albano, J. Abounadi, T.H. Chyba, C.E. Searle, S. Yong, R.S. Gioggia, N.B. Abraham: J. Opt. Soc. Am. B**2**, 47–55 (1985)
3.17 N.B. Abraham: *Laser Focus* p. 73–81 (May 1983)
3.18 R.J. Collins, D.F. Nelson, A.L. Schawlow, W. Bond, C.G.B. Garrett, W. Kaiser: Phys. Rev. Lett. **5**, 303–305 (1960);
      R. Hauck, F. Hollinger, H. Weber: Opt. Commun. **47**, 141–145 (1983);
      C.L. Tang, H. Statz, G. deMars: J. Appl. Phys. **34**, 2289–2295 (1963)
3.19 F.J. McClung, R.W. Hellwarth: J. Appl. Phys. **33**, 828 (1962); Proc. IEEE **51**, 604 (1963);
      N.G. Basov, V.S. Zuev, D.G. Krjukov (1962) reprinted in *Lasers*, ed. by J. Weber (Gordon and Breach, New York 1968) p. 257;
      G.H.C. New: Rept. Prog. Phys. **46**, 877–971 (1983)
3.20 H. Statz, G.A. deMars, C.L. Tang: J. Appl. Phys. **38**, 2212–2222 (1967)
3.21 C.L. Tang: J. Appl. Phys. **34**, 2935–2940 (1963)
3.22 F.T. Arecchi, R. Meucci, G.P. Puccioni, J. Tredicce: Phys. Rev. Lett. **49**, 1217 (1982)
3.23 T. Kimura, K. Otsuka: IEEE J. QE-**6**, 764–769 (1970);
      K. Kubodera, K. Otsuka: IEEE J. QE-**17**, 1139–1144 (1981)
3.24 W. Klische, H.R. Telle, C.O. Weiss: Opt. Lett. **9**, 561–563 (1984)
3.25 W.E. Lamb, Jr.: Phys. Rev. **134**, A1429–A1450 (1964)
3.26 S.G. Zeiger, E.E. Fradkin: Opt. Spektrosk. **21**, 386–390 (1966) [Opt. Spectrosc. (USSR) **21**, 217 (1966)];
      F. Aronowitz: In *Laser Applications*, Vol. 1, ed. by Monte Ross (Academic, New York 1971)
3.27 W. Brunner, R. Fischer, H. Paul: J. Opt. Soc. Am. B**2**, 202–210 (1985) and references therein
3.28 A.L. Schawlow, C.H. Townes: Phys. Rev. **112**, 1940–1949 (1958)
3.29 A.G. Fox, T. Li: Bell Sys. Tech. J. **40**, 453–488 (1961);
      A.G. Fox, T. Li: Proc. IRE **51**, 80–89 (1963);
      H. Kogelnik, T. Li: Proc IEEE **54**, 1312–1329 (1966)
3.30 W.R. Bennett: Phys. Rev. **126**, 580–593 (1962);
      W.R. Bennett: Appl. Opt. Supp. **1**, 573–610 (1963); *The Physics of Gas Lasers* (Gordon and Breach, New York 1977)
3.31 L.W. Casperson, A. Yariv: Appl. Phys. Lett. **17**, 259 (1970)
3.32 P.A. Khandokin, Ya.I. Khanin: J. Opt. Soc. Am. B**2**, 225–231 (1985)
3.33 G.L. Lippi, J.R. Tredicce, N.B. Abraham, F.T. Arecchi: Opt. Commun. **53**, 129–132 (1985)
3.34 P.W. Smith: Proc. IEEE **58**, 1342–1357 (1970);

L. Allen, D.G.C. Jones: Prog. Opt. **9**, 179 (1971);
P.W. Smith, M.U. Duguay, E. Ippen: Prog. Quantum Electron. **3**, 105 (1974)

3.35 T. Uchida, A. Ueki: IEEE J. QE-**3**, 17–30 (1967)

3.36 P. Mandel: Physica **82C**, 353–367 (1976)

3.37 L.W. Casperson: Phys. Rev. A**21**, 911–923 (1980) and Phys. Rev. A**23**, 248 (1981)

3.38 P. Mandel: Opt. Commun. **44**, 400–404 (1983); **45**, 269–272 (1983)

3.39 S. Hendow, M. Sargent III: Opt. Commun. **40**, 385–390 (1982); **43**, 59–63 (1982)

3.40 L.A. Lugiato, L.M. Narducci, D.K. Bandy, N.B. Abraham: Opt. Commun. **46**, 115–120 (1984)

3.41 P. Mandel: In *Coherence and Quantum Optics V*, ed. by L. Mandel and E. Wolf (Plenum, New York 1984) pp. 579–584

3.42 L.A. Lugiato, L.M. Narducci, D.K. Bandy, N.B. Abraham: In *Coherence and Quantum Optics V*, ed. by L. Mandel and E. Wolf (Plenum, New York 1984) pp. 217–224

3.43 N.B. Abraham, L.A. Lugiato, P. Mandel, L.M. Narducci, D.K. Bandy: J. Opt. Soc. Am. B**2**, 35–46 (1985)

3.44 D.K. Bandy, L.M. Narducci, L.A. Lugiato, N.B. Abraham: J. Opt. Soc. Am. B**2**, 56–61 (1985)

3.45 R. Graham, Y. Cho: In *Optical Bistability II*, ed. by C.M. Bowden, H.M. Gibbs, S.L. McCall (Plenum, New York 1984) pp. 103–110; Opt. Commun. **47**, 52–56 (1983)

3.46 S. Hendow, M. Sargent: J. Opt. Soc. Am. B**2**, 84–101 (1985)

3.47 M.-L. Shih, P.W. Milonni, J.R. Ackerhalt: J. Opt. Soc. Am. B**2**, 130–136 (1986)

3.48 J.-Y. Zhang, H. Haken, H. Ohno: J. Opt. Soc. Am. B**2**, 141–147 (1985)

3.49 J.C. Englund, W.C. Schieve: J. Opt. Am. B**2**, 81–83 (1985)

3.50 M.L. Minden, L.W. Casperson: J. Opt. Soc. Am. B**2**, 120–129 (1985)

3.51 V.S. Idiatulin, A.V. Uspenskiy: Phys. Lett. **58A**, 161 (1976)

3.52 M.L. Minden, L.W. Casperson: Private Communication

3.53 X. Husson, M. Margerie: Opt. Commun. **5**, 139 (1972);
R. Vetter, E. Marie: J. Phys. B**11**, 2845 (1978)

3.54 S. Chuang, H. Gamo: Appl. Phys. Lett. **19**, 150–152 (1971)

3.55 S.S. Chuang: Ph. D. Thesis, UC Irvine (1971) unpublished

3.56 J. Dembowski, H. Weber: Opt. Commun. **42**, 133–137 (1982)

3.57 L.W. Casperson, A. Yariv: Appl. Opt. **11**, 462 (1972);
P.W. Wolff, N.B. Abraham, S.R. Smith: IEEE J. QE-**13**, 400–403 (1977)

3.58 G. Stephan, M. Trumper: Phys. Rev. A**30**, 1925–1939 (1984)

3.59 S. Asami, H. Gamo, T. Tako: Jap. J. Appl. Phys. **22**, 88–100 (1983)

3.60 A. LeFloch, J.M. LeNormand, R. Le Naour, P. Brun: IEEE J. QE-**19**, 1474–1476 (1983)

3.61 D.E. McCumber: Phys. Rev. **141**, 306 (1966);
H. Haken: *Laser Theory*, Corr. Printing (Springer, Berlin, Heidelberg 1984)

3.62 K.R. Manes, A.E. Siegmann: Phys. Rev. A**4**, 373–385 (1971);
B. Daino, P. Spano, M. Tamburrini, S. Piazzolla: IEEE J. QE-**19**, 266–270 (1983)

3.63 K. Weisenfeld: In *Fluctuations and Sensitivity in NonEquilibrium Systems*, ed. by W. Horsthemke and D.K. Kondepudi, Springer Proc. Phys. 1 (Springer, Berlin, Heidelberg 1984) p. 268; J. Stat. Phys. **38**, 1071–1097 (1985);
See also J.D. Farmer: Phys. Rev. Lett. **47**, 179 (1981)

3.64 M.F.H. Tarroja, N.B. Abraham, D.K. Bandy, L.M. Narducci: Physical Review A**34**, 3148–3158 (1986)

3.65 P. Coullet, C. Vanneste: Hel. Phys. Acta **56**, 813–823 (1983); and in the Synergetics Workshop at Schloss Elmau, 1984

3.66 Y. Kuramoto, S. Koga: Phys. Lett. **92A**, 1–4 (1982)

3.67 M.J. Feigenbaum: J. Stat. Phys. **19**, 25 (1978); **21**, 669 (1979); Phys. Lett. **74A**, 375 (1979)

3.68 M. Giglio, S. Musazzi, U. Perini: Phys. Rev. Lett. **47**, 243 (1981);
A. Arneodo, P. Coullet, C. Tresser, A. Libchaber, J. Maurer, D. D'Humieres: Physica **6D**, 385–392 (1983);
J.P. Gollub, S.V. Benson, J. Steinman: Ann. NY Acad. Sci. **357**, 22 (1980);
M. Gorman, L.A. Ruth, H.L. Swinney: Ann. NY Acad. Sci. **357**, 10 (1980);
J.P. Gollub, S.V. Benson: J. Fluid Mech. **100**, 499 (1980)

3.69 N.B. Abraham, J.P. Gollub, H.L. Swinney: Physica **11D**, 252–264 (1984)

3.70 J.P. Eckmann: Rev. Mod. Phys. **53**, 643 (1981);

M.I. Rabinovich: Sov. Phys. Usp. **21**, 443 (1978);
A.S. Monin: Sov. Phys. Usp. **21**, 429 (1978)

3.71   P. Cvitanovic (ed.): *Universality in Chaos*, (Hilger, London 1984); Acta Physica Polonica **A65**, 203–238 (1984)

3.72   P. Berge, Y. Pomeau, C. Vidal: *L'Ordre dans le chaos: vers une approche deterministe de la turbulence* (Hermann, Paris 1985)

3.73   P. Grassberger, I. Procaccia: Phys. Rev. **A28**, 2591 (1983); Phys. Rev. Lett. **50**, 346 (1983); Physica **9D**, 189 (1983); Physica **13D**, 34 (1984);
A. Ben Mizrachi, I. Procaccia, P. Grassberger: Phys. Rev. **A29**, 975 (1985);
A. Cohen, I. Procaccia: Phys. Rev. **A31**, 1872 (1985)

3.74   J.D. Farmer, E. Ott, J. Yorke: Physica **7D**, 153 (1983);
J.D. Farmer: In *Fluctuations and Sensitivity in NonEquilibrium Systems*, ed. by W. Horsthemke and D.K. Kondepudi, Springer Proc. Phys. 1, (Springer, Berlin, Heidelberg 1984) p. 172

3.75   A. Brandstater, J. Swift, H. Swinney, A. Wolf, D. Farmer, E. Jen, J. Crutchfield: Phys. Rev. Lett. **51**, 1442 (1983);
A. Brandstater, H.L. Swinney: In *Fluctuations and Sensitivity in NonEquilibrium Systems*, ed. by W. Horsthemke and D.K. Kondepudi, Springer Proc. Phys. 1 (Springer, Berlin, Heidelberg 1984) p. 166

3.76   S. Ciliberto, J.P. Gollub: J. Fluid Mechanics **158**, 381–398 (1985)

3.77   B. Malraison, P. Atten, P. Berge, M. DuBois: J. Physique Lett. **44**, L897–L902 (1983)

3.78   M. Giglio, S. Musazzi, U. Perini: Phys. Rev. Lett. **53**, 2402 (1984)

3.79   H. Whitney: Ann. Math. **37**, 645 (1936);
F. Takens: In *Proc. of the Warwick Symposium*, ed. by D. Rand and L.S. Young (Springer, Berlin, Heidelberg 1981)

3.80   N.B. Abraham, A.M. Albano, B. Das, G. deGuzman, S. Yong, R.S. Gioggia, G.P. Puccioni, J.R. Tredicce: Phys. Lett. **114A**, 217–221 (1986);
N.B. Abraham, A.M. Albano, G.C. deGuzman, M.F.H. Tarroja, S. Yong, S.P. Adams, R.S. Gioggia: In *Perspectives in Nonlinear Dynamics*, ed. by M.F. Shlesinger, R.W. Cawley, A.W. Saenz, and W. Zachary (World Scientific, Singapore 1986) pp. 214–229

3.81   N.B. Abraham, A.M. Albano, B. Das, T. Mello, M.F.H. Tarroja, N. Tufillaro, R.S. Gioggia: In *Optical Chaos*, ed. by J. Chrostowski and N.B. Abraham, SPIE Proceedings Volume 667 (SPIE, Bellingham 1986) pp. 2–9

3.82   M. Lefebvre, D. Dangoisse, P. Glorieux: Phys. Rev. **A29**, 758 (1984);
N.B. Abraham, D. Dangoisse, P. Glorieux, P. Mandel: J. Opt. Soc. Am. **B2**, 23–34 (1985); and private communication from P. Glorieux

3.83   C.O. Weiss: J. Opt. Soc. Am. **B2**, 137–140 (1985);
C.O. Weiss, W. Klische, P.S. Ering, M. Cooper: **52**, 405 (1985)

3.84   J. Opt. Soc. Am. B, January 1985, Special Issue on Instabilities in Active Optical Media, ed. by N.B. Abraham, L.A. Lugiato and L.M. Narducci;
J.R. Ackerhalt, P.W. Milonni, M.-L. Shih: "Chaos in Quantum Optics", Physics Reports, **128**, 205–300 (1985);
R.G. Harrison, D.J. Biswas: "Pulsating instabilities and chaos in lasers", Prog. Quantum Electron. **10**, 147–228 (1985);
Optical Instabilities, ed. by R.W. Boyd, M.G. Raymer and L.M. Narducci, Proceedings of the International Conference on Instabilities and Dynamics in Nonlinear Optical Media, Rochester, June 1985 (Cambridge U. Press, Cambridge, 1986);
Quantum Optics IV, Proceedings of the Fourth International Symposium, Hamilton, NZ, February 1986, ed. by J.D. Harvey and D.F. Walls (Springer, Berlin, Heidelberg 1986);
Optical Chaos, Proceedings of the SPIE Symposium 667 in Quebec, June 1986, ed. by J. Chrostowski and N.B. Abraham (SPIE, Bellingham 1986)

4. Single- and Multi-Mode Operation of a Laser with an Injected Signal

D.K. Bandy, L.A. Lugiato, and L.M. Narducci

With 18 Figures

This chapter contains several new results on the steady-state and linear stability properties of a laser with an injected signal, and a brief review of selected time-dependent studies. The steady-state analysis assumes a plane-wave profile for the laser field and for the incident signal, but is more general than previous models because it applies to a ring-cavity resonator whose mirrors have an arbitrary reflectivity, and to active media with an arbitrary small-signal gain per pass. The results include a survey of the dependence of the output on the input field for a number of typical operating conditions, and a study of the longitudinal variations of the field modulus inside the cavity. The linear stability analysis generalizes earlier investigations to include the case of a multi-mode ring cavity in the mean-field limit, a restriction that appears unavoidable at the present time. Off-resonant modes display wide domains of instability and suggest that earlier single-mode dynamical studies should be revisited. Our survey of the time-dependent output oscillations is limited to currently published single-mode operation results because work is currently in progress on multi-mode extensions. However, this survey does include discussion of several different techniques to quantify the temporal response of the system. These techniques are generally applicable to arbitrary cavity configurations.

4.1 Background Information

The design of a highly stable laser oscillator is a significant tour de force because even small changes in environmental factors are likely to cause considerable variations in the operating parameters. Thus, it is virtually impossible to assemble two or more nominally identical lasers in the expectation that their output charcteristics will match one another for any reasonable length of time.

Apparently *Oraevskii* was one of the first researchers to foresee the stabilizing effects of an injected signal on the output of a maser oscillator [4.1]. His proposal was followed by a number of other investigations that focused on the requirements for stable, locked response as a function of the injected signal strength and detuning parameters [4.2,3]. The successful application of the so-called injection-locking technique in producing stable single-mode output from a Doppler-broadened He-Ne laser [4.4] and in reducing the spectral width of a dye laser [4.5] stimulated additional efforts to extend this approach to many other systems from high-power CO_2 [4.6] to semiconductor lasers [4.7].

When stripped down to its essential components, a laser with an injected signal (LIS) consists of an ordinary laser oscillator with two partially transmitting windows; one of these mirrors functions as the entrance port for the external beam while the second is the output coupler. As a way of reducing the analytic complexity of the problem, the laser oscillator is usually modeled as a unidirectional ring-cavity device, often operating in a single transverse and longitudinal mode. The active medium may have a homogeneously or inhomogeneously broadened gain profile, although the former has been the usual choice in most theoretical studies.

From a practical viewpoint, the injection of a strong driving signal appears as an attractive way to force frequency and phase synchronization in separate lasers by taking advantage of the well-known affinity of nonlinear oscillators to lock to a periodic forcing element. From a theoretical viewpoint, the LIS is the active counterpart of an optically bistable system that is known to display a variety of pulsing modes of operation, such as a single-mode (SM) or a multimode (MM) device [4.8]. The potential for interesting dynamical effects is even greater for a LIS because of the natural competition between the free-running laser (FRL) and the external driving field.

When the laser oscillator operates above threshold in a free-running stable mode, one can easily visualize two simple limiting configurations: if the injected signal strength is weak, the oscillator is essentially unaffected and the resulting output displays the characteristic beat pattern of two independent coherent sources with slightly different carrier frequencies; at higher input power levels, the oscillator locks its frequency and phase to the injected signal and becomes indistinguishable from it. At intermediate operating points between these two extremes, one expects a large variety of pulsed behaviors for different settings of the control parameters.

A first-principles theoretical description of the principles of operation of a LIS was advanced by *Spencer* and *Lamb* [4.9]. On the basis of their analysis and other subsequent advances [4.10], the existence of a locked mode of operation, even in the MM laser action, was discovered. In this setting, the injected field is visualized as an additional pump for one of the free-running modes of the laser whose intensity is made to grow. If this increase is sufficiently large, the gain of the neighboring modes may be depressed to the point that they are driven below threshold. Thus, the spectral output of an injection-locked laser may be not only stable relative to the driving field, but also considerably narrower than that of the free-running configuration. Experiments, in fact, have confirmed this expectation [4.11].

Studies of dynamical instabilities in LIS were advanced by *Lugiato* [4.12], *Yamada* and *Graham* [4.13], *Scholz* and collaborators [4.14], and extended more recently by several other investigators [4.15–18]. The papers cited in [4.13,14] deal with configurations where either the input intensity [4.13] or the pump parameter [4.14], are modulated sinusoidally in time. The work desribed in [4.15] focuses on the bahvior of a one-mode ring cavity with constant parameters and reveals a wide range of pulsation phenomena. The subsequent investigations by *Arecchi* and collaborators [4.16,18c] are based on realistic selections of parameters for a typical CO_2 laser system (Chap. 2). Under these conditions, the various rates of decay are such that the atomic polarization can be eliminated adiabatically; still, even with only three dynamical degrees of freedom, these studies have shown the persistence of interesting temporal oscillations.

The experimental research on instabilities in the LIS is not yet developed to a level comparable to that of the theory because of the stringent stability requirements that must be imposed on the relative frequency and phase differences between the driver and the driven laser. Results have been reported with a driven and modulated NMR laser in the radio frequency range [4.19a], and with a CO_2 laser [4.19b]. The NMR laser has also been used in a LIS configuration with constant parameters [4.19a]. Recent reviews of various aspects of the laser with an injected signal have been provided by *Abraham* et al. [4.20,21], and *Haken* [4.22]. Numerous technical contributions and useful up to date informations can be found in [4.23].

In assembling this contribution, we have decided to maintain our discussion as general as possible within the confines of the homogeneously broadened, two-level atomic model. We have managed to do so for the steady-state analysis where the only significant approximation is the plane-wave assumption for both the cavity and the injected field. We have confined the linear stability analysis to the mean-field limit (MFL) because, to our knowledge, no MM extension for arbitrary mirror reflectivity and atomic gain has been developed [4.24]. The dynamical studies reflect work that was done by members of our group in recent years in connection with the SM model [4.15,18a]. A MM extension of the time-dependent calculations using the full Maxwell-Bloch equations was developed by *Narducci* et al. [4.25].

This chapter is organized as follows. In Sect. 4.2 we set up the appropriate equations of motion for the unidirectional ring-cavity model, derive the steady-state relation between input and output fields and construct the longitudinal field profile inside the cavity. In Sect. 4.3 we develop a treatment of the linear response of the system for an arbitrary MM configuration in the MFL and derive and discuss the stability criteria for several operating parameters. The treatment of Sects. 4.2 and 3 is parallel to that adopted in [4.8a] for the case of optical bistability. When dealing with a MM system, some care must be exercised in carrying out an adiabatic elimination of fast-relaxing variables. We discuss this topic in detail in Sect. 4.4 and point out some pitfalls that are to be avoided. Section 4.5 is devoted to the study of the dynamical evolution of a SM system. The emphasis is on the identification of the major features that one may expect from a slow scan of the driving field. Finer details can be extracted from power spectral studies and from the construction of Lyapunov exponents; these are reviewed in Sect. 4.6. Several appendices contain useful technical details.

4.2 Equations of Motion
and the Steady-State Configuration

The LIS is modeled as a unidirectional ring cavity containing a collection of two-level atoms inside a cylindrical volume of length L. The active medium is kept in a state of inversion between two homogeneously broadened lasing levels spaced by a frequency ω_A; the atomic polarization and population difference are characterized by relaxation rates $\gamma_\perp$ and $\gamma_\parallel$, respectively. The cavity is shaped in the form of a ring of length $\mathcal{L}$ with input and exit mirrors having an arbitrary reflectivity coefficient R. The additional mirrors that are needed to close the ring are assumed to be ideal reflectors. An external CW signal with a carrier frequency ω_0 is injected into the laser cavity in a copropagating direction relative to the circulating laser field. The unsaturated gain of the active medium is arbitrary, as long as it is sufficiently high to maintain the system above threshold for laser action. On the other hand, we ignore transverse effects by operating in the framework of the plane-wave approximation for both the cavity and the injected fields. An analysis of the one-mode LIS with a Gaussian transverse profile has been given in [4.26].

The equations of motion are the well-known Maxwell-Bloch equations:

$$\frac{\partial \mathcal{F}}{\partial z} + \frac{1}{c}\frac{\partial \mathcal{F}}{\partial t} = -\alpha P \tag{4.1a}$$

$$\frac{\partial P}{\partial t} = \gamma_\perp[\mathcal{F}D - (1 + i\tilde{\Delta})P] \tag{4.1b}$$

$$\frac{\partial D}{\partial t} = -\gamma_\parallel\left[\frac{1}{2}(\mathcal{F}P^* + \mathcal{F}^*P) + D + 1\right] \tag{4.1c}$$

for the complex field envelope $\mathcal{F}(z,t)$, the atomic polarization envelope $P(z,t)$ and the population difference $D(z,t)$; α represents the small-signal gain constant per unit length, $\tilde{\Delta} = (\omega_A - \omega_0)/\gamma_\perp$ is the scaled detuning of the atomic transition frequency with the carrier frequency of the injected field taken as a reference. With some minor modifications, (4.1) is appropriate for the description of a LIS, an ordinary FRL or a bistable system [4.27].

With the inclusion of the boundary conditions

$$\mathcal{F}(0,t) = TY + R\mathcal{F}(L, t - \Delta t)e^{-i\delta_0} \tag{4.2}$$

and the requirement $\alpha > 0$, the system is forced to operate as a LIS; Y is proportional to the amplitude of the incident field and is chosen, for definiteness, to be real and positive; $\Delta t = (\mathcal{L} - L)/c$ is the transit time of light beween the exit and entrance ports of the active medium, and δ_0 is the accumulated phase difference per round trip $[\delta_0 = \mathcal{L}(\omega_C - \omega_0)/c]$ due to a possible mismatch between the reference cavity resonance ω_C and the injected carrier frequency; T is the mirror transmittivity, $T = 1 - R$.

In the steady state, the atomic variables are given by

$$P_{st}(z) = -\mathcal{F}_{st}\frac{1 - i\tilde{\Delta}}{1 + \tilde{\Delta}^2 + |F_{st}(z)|^2} \quad , \tag{4.3a}$$

$$D_{st}(z) = -\frac{1 + \tilde{\Delta}^2}{1 + \tilde{\Delta}^2 + |\mathcal{F}_{st}(z)|^2} \quad . \tag{4.3b}$$

In general, both P_{st} and D_{st} depend on the longitudinal coordinate z; this dependence becomes negligible only in the MFL, as we show later in this section. The longitudinal field profile can be calculated from the field equation (4.1a) in the steady state:

$$\frac{d}{dz}\mathcal{F}_{st}(z) = \alpha\frac{1 - i\tilde{\Delta}}{1 + \tilde{\Delta}^2 + |\mathcal{F}_{st}(z)|^2}\mathcal{F}_{st}(z) \quad . \tag{4.4}$$

In solving (4.4), it is convenient to represent the complex field amplitude in the form

$$\mathcal{F}_{st}(z) = \varrho(z)e^{i\theta(z)} \quad , \tag{4.5}$$

where the modulus and phase satisfy the coupled equations

$$\frac{d}{dz}\varrho(z) = \frac{\alpha\varrho(z)}{1 + \tilde{\Delta}^2 + \varrho^2(z)} \quad , \tag{4.6a}$$

$$\frac{d}{dz}\theta(z) = -\frac{\alpha\tilde{\Delta}}{1 + \tilde{\Delta}^2 + \varrho^2(z)} \quad . \tag{4.6b}$$

Form the identity

$$\frac{1}{\varrho(z)}\frac{d}{dz}\varrho(z) = -\frac{1}{\tilde{\Delta}}\frac{d\theta(z)}{dz} \quad , \tag{4.7}$$

we obtain a relation between ϱ and θ

$$\ln\left(\frac{\varrho(z)}{\varrho(0)}\right) = -\frac{1}{\tilde{\Delta}}[\theta(z) - \theta(0)] \quad , \tag{4.8}$$

while from (4.6a) we derive the transcendental equation for $\varrho(z)$ in terms of $\varrho(0)$

$$(1 + \tilde{\Delta}^2)\ln\left(\frac{\varrho(z)}{\varrho(0)}\right) + \frac{1}{2}[\varrho^2(z) - \varrho^2(0)] = \alpha z \quad . \tag{4.9}$$

The required connection between the input and output fields in the steady state is provided by

$$\varrho(0)e^{i\theta(0)} = TY + R\varrho(L)e^{i\theta(L)}e^{-i\delta_0} \qquad (4.10)$$

after suitable elimination of the phase variables $\theta(0)$ and $\theta(L)$ and the field modulus $\varrho(0)$ at the input of the active medium. This is best accomplished as follows: first we write (4.10) in the form

$$T^2Y^2 = \varrho^2(0)\left\{1 + R^2\frac{\varrho^2(L)}{\varrho^2(0)} - 2R\frac{\varrho(L)}{\varrho(0)}\cos\left[\theta(L) - \theta(0) - \delta_0\right]\right\} ; \qquad (4.11)$$

next we eliminate the phase change $\theta(L) - \theta(0)$ with the help of (4.8), and finally we eliminate $\varrho(0)$ between (4.11 and 9), after setting $z = L$. The elimination of $\varrho(0)$ cannot be carried out by analytic means because of the transcendental nature of (4.9). The fastest course of action is probably to select the required cavity and atomic parameters, to solve numerically for $\varrho(L)$ using (4.9) in correspondence with a given positive value of $\varrho(0)$, and to substitute both $\varrho(0)$ and $\varrho(L)$ into (4.11) to calculate the amplitude of the incident field Y. This procedure can be repeated for arbitrary positive values of $\varrho(0)$ in order to trace out the state equation linking the output and the input fields. The longitudinal profile of the cavity field $\varrho(z)$ can be calculated directly from (4.9) after fixing $\varrho(0)$ (and the corresponding value of Y). Note that the correspondence between Y and $\varrho(0)$ is not always of the one-to-one type, so that care must be exercised in deciding whether the calculated longitudinal profile corresponds to the upper or lower operating branch. Typical state equations are shown in Figs. 4.1 and 2. In each figure, we plot three steady-state curves corresponding to different values of the intermode spacing (or, equivalently, of the cavity length $\mathcal{L}$), which controls the shape of the state equation rather critically. For a sufficiently high reflectivity, the general trend is for the S-shape of the steady-state curve to disappear, as the cavity modes are brought closer to one another.

The cavity field, under the conditions of Figs. 4.1,2, displays significant variations along the axis of the laser. Understandably, the most-obvious lack of longitudinal uniformity occurs under weak driving field conditions; in the high intensity part of the state equation, saturation effects produce more-uniform field profiles. An illustration of this behavior is shown in Fig. 4.3 where we have overlapped the state equation in Fig. 4.2a with two curves displaying the percentage variation of the cavity field as a function of the injected signal strength. A large increase in field uniformity (small percentage variation) is apparent for operating conditions in the upper branch of the state equation.

The field uniformity is a sensitive function of the mirror reflectivity, the gain of the active medium, and the accumulated phase lag δ_0 per round trip. These parameters can be adjusted to improve the longitudinal field uniformity. In fact, ideally, steady-state configurations with a perfectly uniform field can be obtained in the so-called MFL. This is defined by the simultaneous conditions

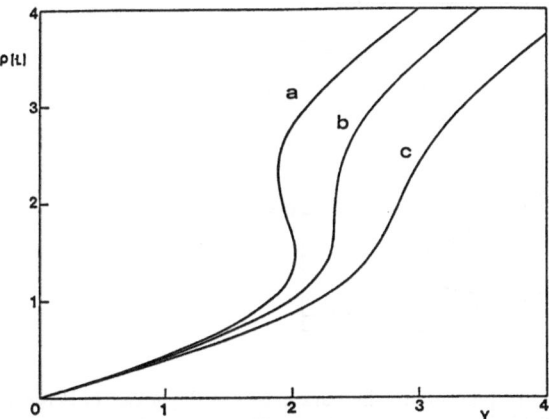

Fig. 4.1. Steady-state equation linking the modulus of the output field to the amplitude of the driving signal. The parameters are as follows: $R = 0.8$, $\alpha L = 1.2$, $\tilde{\Delta} = 1.0$; the frequency separation between the atomic and cavity frequency in units of $\gamma_\perp$ is zero; the intermode spacing, also in units of $\gamma_\perp$ is (a) 500, (b) 100, (c) 50

$$\alpha L \to 0 \ , \quad R \to 1 \ , \quad \delta_0 \to 0 \tag{4.12}$$

with the constraints

$$\frac{\alpha L}{|\ln R|} \to \frac{\alpha L}{T} \equiv 2C = \text{const} \ , \tag{4.13a}$$

$$\delta_0 \equiv \theta T \ , \quad \theta = \text{const} \ . \tag{4.13b}$$

Under these conditions, (4.8,10) can be combined to yield the state equation

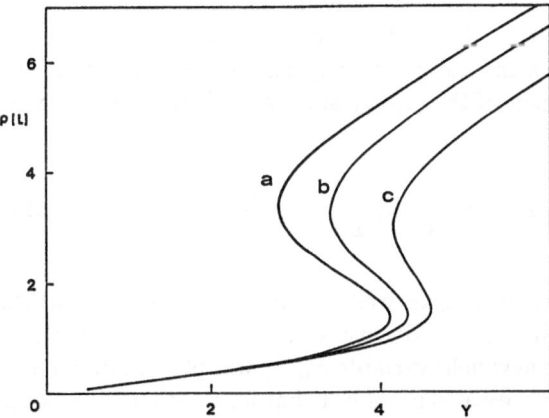

Fig. 4.2. Steady-state equation linking the modulus of the output field to the amplitude of the driving signal. The parameters are as follows: $R = 0.95$, $\alpha L = 0.5$, $\tilde{\Delta} = 1.0$; the frequency separation between the atomic and cavity frequency in units of $\gamma_\perp$ is 0.8; the intermode spacing, also in units of $\gamma_\perp$ is (a) 500, (b) 100, (c) 50

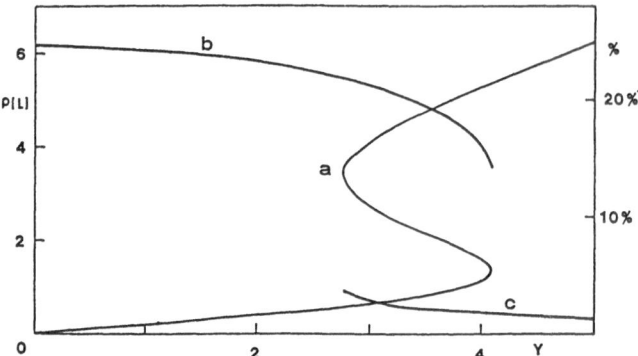

Fig. 4.3. *(a)* Steady-state equation corresponding to the parameters of Fig. 4.2a. *(b)* Percentage variation of the internal cavity field $\mathcal{F}_{st}$ over the lower branch of the state equation. *(c)* Percentage variation of the internal cavity field over the upper branch of the state equation.

$$Y = \mathcal{F}\left[1 - \frac{2C}{1 + \tilde{\Delta}^2 + |\mathcal{F}|^2} + i\left(\theta + \frac{2C\tilde{\Delta}}{1 + \tilde{\Delta}^2 + |\mathcal{F}|^2}\right)\right] \tag{4.14}$$

well known from earlier work on optical bistability and the LIS. A sketch of this procedure leading to (4.14) is given in Appendix 4.A.

Field uniformity, of course, is a considerable advantage for theoretical studies of this problem; in fact, all recent contributions to this subject have adopted this approximation, more or less explicitly. From a practical viewpoint, it is especially useful to know how far one can venture away from the MFL before the assumption of longitudinal field uniformity becomes untenable. A study of the exact steady-state equations shows that variations of the modulus of the cavity field $\mathcal{F}$ of the order of a few percent or better require impractical constraints on the design of the cavity elements and on the selection of the active medium. The field variable $\mathcal{F}$ turns out to be inappropriate for this purpose. Anticipating some of the results contained in the next section, we propose that a much more convenient framework for a theoretical description of some aspects of the LIS is provided by the replacement of the steady-state field $\mathcal{F}_{st}(z)$ with the new dependent variable

$$F_{st}(z) = \mathcal{F}_{st}(z) \exp\left(-\frac{z}{L}|\ln R|\right) \exp\left(-i\delta_0 \frac{z}{L}\right) + TY \frac{z}{L}\ . \tag{4.15}$$

The significance of this transformation goes well beyond the purpose for which we introduce it at this point, and it will be discussed further in the next section.

Here we anticipate that the new field variable $F_{st}(z)$ will play an important role in our subsequent discussions; we also point out that for a rather wide range of values of αL, R, and δ_0, the modulus of F_{st} is highly uniform, regardless of Y. An illustration of this statement is shown in Fig. 4.4.

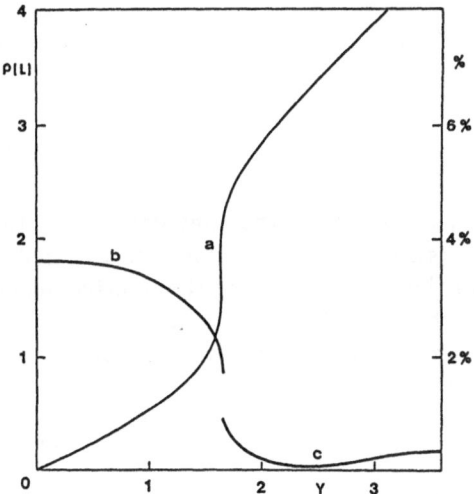

Fig. 4.4. (a) Steady-state equation corresponding to the parameters $R = 0.8$, $\alpha L = 1.0$, $\hat{\Delta} = 1.0$, atomic-cavity spacing $= 0$, intermode spacing (in units of $\gamma_\perp$) $= 500$. (b) Percentage variation of the internal cavity field F_{st} over the lower branch of the state equation. (c) Percentage variation of the internal cavity field over the upper branch of the state equation

4.3 Linear Stability Analysis in the Mean-Field Limit

The main objective of this section is to investigate the linear stability of the steady-state configurations of a LIS. Even within the framework of the plane-wave approximation, the linear response of a homogeneously broadened MM LIS is a problem of considerable complexity for arbitrary values of the system parameters. The main source of mathematical difficulties is the appearance of space-dependent coefficients in the linear equations for the space-time dependent deviations from the steady state.

We consider a simpler version of the linearization analysis by selecting laser parameters such that the modulus of the cavity field is sufficiently uniform over the entire range $0 \le z \le L$. At the end of the previous section, we have already mentioned that this restriction is far less severe than it may seem at first, if one adopts a suitable new set of field and atomic variables.

Actually, there is another even more compelling reason why the original set of variables $\mathcal{F}$, P, and D is not ideal for the study of a MM stability study. Except in the very near neighborhood of the MFL, the notion of a cavity mode as a stable resonating structure is not well defined because of the losses associated with a finite reflectivity. A convenient way out of this difficulty is provided by a procedure that was originally suggested by *Benza* and *Lugiato* [4.28], and which is adopted here with a few modifications.

We first introduce the new independent variables

$$z' = z , \quad t' = t + \frac{\mathcal{L} - L}{c}\frac{z}{L} \tag{4.16}$$

and the new set of field and atomic functions:

$$F(z',t') = \mathcal{F}(z',t') \exp\left[\frac{z'}{L}\ln(Re^{-i\delta_0}) + \frac{TY}{L}z'\right] , \tag{4.17a}$$

81

$$P(z',t') = \tilde{P}(z',t') \exp\left[\frac{z'}{L}\ln\left(Re^{-i\delta_0}\right)\right] , \tag{4.17b}$$

$$D(z',t') = \tilde{D}(z',t') \exp\left[\frac{2z'}{L}\ln R\right] . \tag{4.17c}$$

The task of the transformation (4.16) is to remove the time delay from the boundary conditions (4.2). Equation (4.17a), instead, eliminates the multiplicative factors R and $\exp\left(-i\delta_0\right)$ and the term containing the injected signal amplitude from the right-hand side of the boundary condition.

The field variable $F(z',t')$ satisfies the familiar isochronous periodicity condition

$$F(0,t') = F(L,t') , \tag{4.18}$$

which becomes a pivotal point in the MM analysis that follows. It is immediately clear that the new field amplitude $F(z',t')$ can be expanded in a normal modal expansion of the Fourier type, as in the standard one-dimensional vibrating-string problem. The time-dependent coefficients of this expansion represent the natural mode amplitudes of the lossy cavity. In addition, the steady-state profile $F_{st}(z')$ is far more uniform than $\mathcal{F}_{st}(z)$, even if the physical parameters of the cavity and the active medium are rather removed from the mean-field values. The original Maxwell-Bloch equations can easily be transformed in terms of the new dynamical variables (4.27) for arbitrary values of αL, T, and δ_0. In the following, however, we consider only the case of the MFL (4.12) where the equations for F, P, and D take the form

$$\frac{\partial F}{\partial t'} + \frac{cL}{\mathcal{L}}\frac{\partial F}{\partial z'} = -\kappa[(1+i\theta)F - Y + 2CP] , \tag{4.19a}$$

$$\frac{\partial P}{\partial t'} = \gamma_\perp[FD - (1+i\tilde{\Delta})P] , \tag{4.19b}$$

$$\frac{\partial D}{\partial t'} = -\gamma_\parallel\left[\frac{1}{2}(F^*P + FP^*) + D + 1\right] . \tag{4.19c}$$

The field damping rate κ is defined as $\kappa = cT/\mathcal{L}$ and the appropriate boundary conditions are given by (4.18). We note that F and P are complex functions and that the system of equations (4.19) must be completed by the complex conjugate equations of (4.19a) and (4.19b); $D(z',t')$, of course, is a real variable. Next we introduce the modal decomposition

$$\begin{pmatrix} F(z',t') \\ P(z',t') \\ D(z',t') \end{pmatrix} = \sum_{n=-\infty}^{\infty} e^{ik_n z'} e^{-i\alpha_n t'} \begin{pmatrix} f_n(t') \\ p_n(t') \\ d_n(t') \end{pmatrix} \tag{4.20a}$$

with $k_n = 2\pi n/L$ and $\alpha_n = 2\pi nc/\mathcal{L}$ and the corresponding expansions for F^* and P^*

$$\begin{pmatrix} F^*(z',t') \\ P^*(z',t') \end{pmatrix} = \sum_{n=-\infty}^{\infty} e^{-ik_n z'} e^{+i\alpha_n t'} \begin{pmatrix} f_n^*(t') \\ p_n^*(t') \end{pmatrix} . \tag{4.20b}$$

The selection of the wave numbers $k_n = 2\pi n/L$ ensures that $F(z',t')$ is indeed consistent with the boundary conditions (4.18); the frequencies α_n can be recognized as multiples of the cavity intermode spacing α_1; they are, of course, just the eigenfrequencies of the empty ring resonator. The modal amplitudes d_n^* and d_{-n} are equal to one another because $D(z',t')$ is a real function. The modal function $\exp(ik_n z')$ form a complete, orthonormal set over $(0,L)$ in the sense that

$$\frac{1}{L} \int_0^L dz' e^{ik_n z'} e^{-ik_m z'} = \delta_{n,m} \tag{4.21}$$

It is now a simple matter to substitute (4.20) into (4.19), and to derive the infinite set of coupled equations for the modal amplitudes. Their explicit form is given by

$$\dot{f}_n = -\kappa[(1+i\theta)f_n - Y\delta_{n,0} + 2Cp_n] , \tag{4.22a}$$

$$\dot{f}_n^* = -\kappa[(1-i\theta)f_n^* - Y\delta_{n,0} + 2Cp_n^*] , \tag{4.22b}$$

$$\dot{p}_n = i\alpha_n p_n + \gamma_\perp [\sum_{n'} f_{n'} d_{n-n'} - (1+i\tilde{\Delta})p_n] , \tag{4.22c}$$

$$\dot{p}_n^* = -i\alpha_n p_n^* + \gamma_\perp [\sum_{n'} f_{n'}^* d_{n-n'}^* - (1-i\tilde{\Delta})p_n^*] , \tag{4.22d}$$

$$\dot{d}_n = i\alpha_n d_n - \gamma_\parallel [\tfrac{1}{2} \sum_{n'} (f_{n'}^* p_{n+n'} + f_{n'} p_{n'-n}^*) + d_n + \delta_{n,0}] , \tag{4.22e}$$

where the dots indicate differentiation with respect to t'. A steady-state configuration that is consistent with a spatially uniform field amplitude F_{st} must be such that

$$f_n^{st} = 0 , \quad p_n^{st} = 0 , \quad d_n^{st} = 0 , \quad n \neq 0 . \tag{4.23}$$

The steady-state amplitudes of the $n = 0$ mode can be calculated at once with the result

$$p_0 = -f_0 \frac{1 - i\tilde{\Delta}}{1 + \tilde{\Delta}^2 + |f_0|^2} , \tag{4.24a}$$

83

$$d_0 = -\frac{1 + \tilde{\Delta}^2}{1 + \tilde{\Delta}^2 + |f_0|^2} , \tag{4.24b}$$

where f_0 is the solution of the state equation

$$(1 + i\theta)f_0 - Y - 2C\frac{1 - i\tilde{\Delta}}{1 + \tilde{\Delta}^2 + |f_0|^2}f_0 = 0 . \tag{4.25}$$

This can be written in the more familiar form

$$Y = f_0\left[\left(1 - \frac{2C}{1 + \tilde{\Delta}^2 + |f_0|^2}\right) + i\left(\theta + \frac{2C\tilde{\Delta}}{1 + \tilde{\Delta}^2 + |f_0|^2}\right)\right] , \tag{4.26}$$

which coincides with (4.14) because in the MFL, $\mathcal{F}_{st}$ becomes spatially uniform and identical to f_0. The nonlinear coupled equations (4.22) would offer a formidable challenge even in their linearized form. The spatial uniformity of the steady-state, however, introduces a remarkable simplification because with the help of (4.23) the infinite-dimensional matrix associated with the linearized form of (4.22) breaks up into an infinite set of disconnected five-dimensional blocks, one for each of the mode indices. In fact, it is easy to show that the linearized equations associated with (4.22) take the form

$$\dot{\delta f_n} = -\kappa[(1 + i\theta)\delta f_n + 2C\delta p_n] , \tag{4.27a}$$

$$\dot{\delta f^*_{-n}} = -\kappa[(1 - i\theta)\delta f^*_{-n} + 2C\delta p^*_{-n}] , \tag{4.27b}$$

$$\dot{\delta p_n} = i\alpha_n\delta p_n + \gamma_\perp[\delta f_n d_0^{st} + f_0^{st}\delta d_n - (1 + i\tilde{\Delta})\delta p_n] , \tag{4.27c}$$

$$\dot{\delta p^*_{-n}} = i\alpha_n\delta p^*_{-n} + \gamma_\perp[\delta f^*_{-n}d_0^{st} + f_0^{*st}\delta d_n - (1 - i\tilde{\Delta})\delta p^*_{-n}] , \tag{4.27d}$$

$$\dot{\delta d_n} = i\alpha_n\delta d_n - \gamma_\parallel[\tfrac{1}{2}(\delta f^*_{-n}p_0^{st} + f_0^{*st}\delta p_n \\ + \delta f_n p_0^{*st} + f_0^{st}\delta p^*_{-n}) + \delta d_n] . \tag{4.27e}$$

for $n = 0, \pm 1, \pm 2$.

The ansatz

$$\begin{pmatrix} \delta f_n \\ \delta f^*_{-n} \\ \delta p_n \\ \delta p^*_{-n} \\ \delta d_n \end{pmatrix} = e^{\lambda t}\begin{pmatrix} \delta f_n^{(0)} \\ \delta f_{-n}^{(0)*} \\ \delta p_n^{(0)} \\ \delta p_{-n}^{(0)*} \\ \delta d_n^{(0)} \end{pmatrix} \tag{4.28}$$

substituted into (4.27) leads to a system of five linear homogeneous equations
and to the fifth-order characteristic equation in λ. As usual

$$\sum_{i=0}^{5} A_i \lambda^i = 0 \ . \tag{4.29}$$

As usual, the roots of (4.29) provide direct information on the stability of the
system; a selected steady-state configuration is stable if and only if all roots of
(4.29) have a negative real part, for all values of n.

Past investigations of the stability properties of a LIS have focused only on
the behavior of the $n = 0$ mode in the context of SM theory. The characteristic
equation (4.29) provides an extension of previous work and allows the investigation of the stability properties of any sideband. From a technical point of view,
an unpleasant feature of the new characteristic equation is that the coefficients
A_i are no longer real, thus precluding the use of the Hurwitz criterion to assess
the sign of the real parts of the eigenvalues [a listing of the coefficients of (4.29)
is given in Appendix 4.B, for completeness]. On the other hand, one can always
calculate the roots λ_i by standard numerical techniques, as we have done in this
case. A typical display of the real parts of the eigenvalues is shown in Fig. 4.5
for two selected values of the driving field Y. Figure 4.5b corresponds to an
unstable state in the lower branch, while Fig. 4.5c corresponds to a stable state
of the upper branch. A more detailed scan of this set of parameters shows that
the entire lower branch is unstable, not only for the resonant mode $(n = 0)$,
but also for a band of side modes roughly contained between $\alpha_{min} \simeq -1.5$ and
$\alpha_{max} \simeq 1.5$. The width of the instability domain changes little as Y varies along
the lower branch, however, it does depend more significantly on $\tilde{\kappa}$, displaying
a monotonic growth as $\tilde{\kappa}$ increases.

An example of the behavior of the eigenvalues in the case of a single-valued
state equation is shown in Fig. 4.6. For the selected parameters, the resonant
mode is unstable over the range $0 \lesssim Y \lesssim 1.55$ of the incident field amplitude. On
this basis, and for a SM system, one would expect injection locking to occur
at position (A) of Fig. 4.6a. The linear stability analysis, shows, instead, that
properly placed sidebands can be unstable even beyond the stability boundary of the resonant mode. This is shown explicitly in Fig. 4.6b where we have
plotted the maximum unstable frequency of the sidebands as a function of the
driving field strength, and where it is clear that over the range $1.55 \lesssim Y \lesssim 1.58$
the resonant mode is stable, but certain sidebands are not. Thus, apart from
hard mode instabilities that cannot be predicted on the basis of the linearized
approach, the stable operating range of the driven laser begins at somewhat
higher values of the injected-field strength that would be predicted on the basis
of the SM theory. An example of the real parts of eigenvalues for a configuration yielding a stable resonant mode and an unstable range for the sidebands
is shown in Fig. 4.6c.

It is important to stress that the existence of unstable off-resonant sidebands requires a good quality cavity in the sense that $\tilde{\kappa}$ must be sufficiently

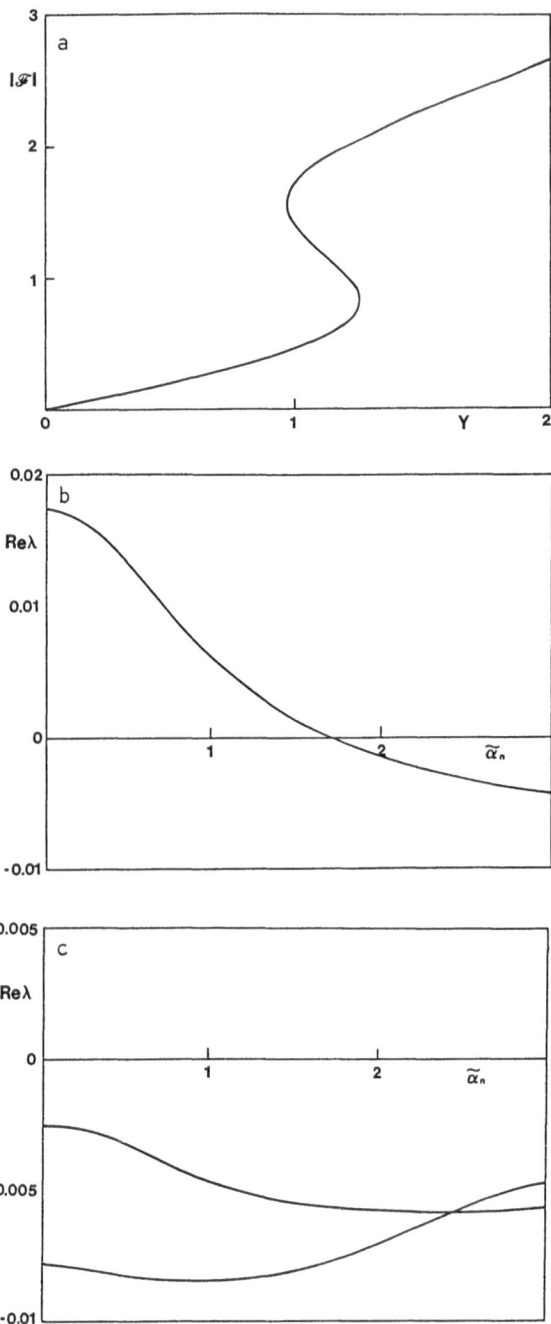

Fig. 4.5. (a) State equation corresponding to the parameters $C = 2$, $\tilde{\Delta} = 0.0004$, $\tilde{\theta} = 0.60$ and $\omega_A = \omega_C$ (resonance). **(b)** The largest real part of the linearized eigenvalues is plotted as a function of $\tilde{\alpha}_n$ regarded as a continuous variable for $\tilde{\kappa} = 6.63 \times 10^{-3}$, $\tilde{\gamma} = 2$ and $|\mathcal{F}| = 0.2$ (lower branch of the state equation). The intermode spacing is $\tilde{\alpha}_1 = 0.5$. **(c)** Same as **(b)** for $|\mathcal{F}| = 1.8$ (upper branch). The system is completely stable for these operating conditions

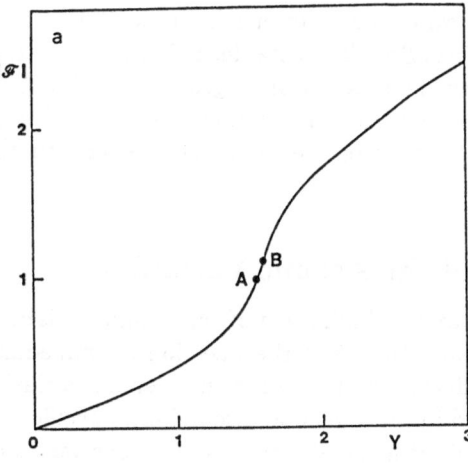

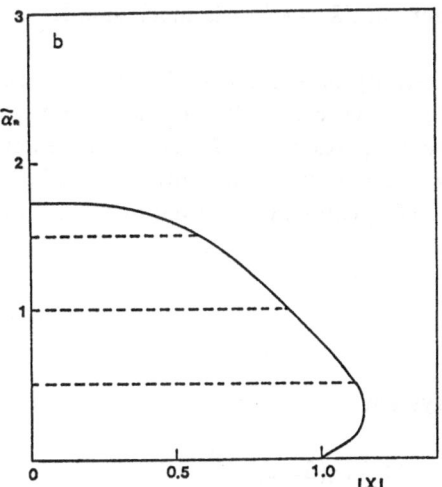

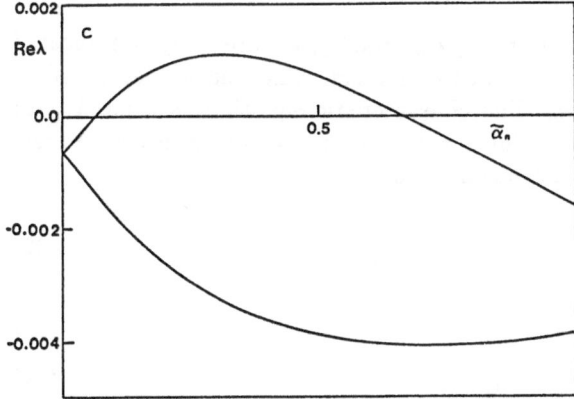

Fig. 4.6. (a) State equation corresponding to the parameters $R = 0.92$, $\alpha L = 0.333$, $\tilde{\alpha}_1 = 0.5$, $\tilde{\Delta} = 0.008$. The points (A) and (B) mark the end of the instability domains for the resonant mode and for the first sideband, respectively. Of course, point B is the actual injection-locking threshold. **(b)** Instability range for the sidemode frequency $\tilde{\alpha}_n$ viewed as a continuous variable [interior region bounded by the (——)]. (- - -) mark the position of the cavity modes. The parameters correspond to the state equation shown in **(a)**. **(c)** the two largest real parts of the linearized eigenvalues are plotted as functions of the sidemode frequency $\tilde{\alpha}_n$ viewed as a continuous variable for $|\mathcal{F}| = 1.05$. In this case, the system is unstable at a frequency corresponding to the first side band $\tilde{\alpha}_1$

smaller than unity. This is a consequence of the fact that typical instability ranges extend to maximum values of $\tilde{\alpha}_n$ of the order of a few units. Thus, if we require that the sideband at $\tilde{\alpha}_1 = 2\pi c/\mathcal{L}\gamma_\perp$ be unstable, it is necessary that the ratio $c/\mathcal{L}\gamma_\perp$ be smaller than unity. This can be arranged by selecting large enough values of $\mathcal{L}$ or $\gamma_\perp$. On the other hand, this implies that $\tilde{\kappa} = cT/\mathcal{L}\gamma_\perp$ is considerably smaller than unity because of the MFL requirement on the transmittivity.

4.4 Adiabatic Elimination of the Atomic Variables

We now review the matter of the adiabatic elimination of the atomic variables. This is a problem of much wider scope than just the question of immediate interest; however, we shall limit our discussion to the LIS in an effort to clarify some issues that have been overlooked in more general discussions. Specifically, we seek the correct procedure for eliminating adiabatically the atomic variables in situations where $\gamma_\perp$ and $\gamma_\parallel$ are much larger than κ. (This is often called the good cavity limit in reference to the fact that $\tilde{\kappa} = \kappa/\gamma_\perp$ is much smaller than unity.)

In this limit, and subject to the validity of a number of conditions that have been discussed in some detail in [4.29], we can replace the left-hand sides of (4.27c–e) by zero, solve the algebraic equations (4.27c–e) for the atomic fluctuation variables and substitute the appropriate results into (4.27a,b). The linear differential equations for δf_n and δf^*_{-n} can now be solved by introducing the usual ansatz

$$\begin{pmatrix} \delta f_n \\ \delta f^*_{-n} \end{pmatrix} = e^{\lambda t} \begin{pmatrix} \delta f_n^{(0)} \\ \delta f_{-n}^{*(0)} \end{pmatrix} \; ; \tag{4.30}$$

The result is the linear homogeneous system

$$[\tilde{\lambda} + \tilde{\kappa}(1 + i\theta) + 2C\tilde{\kappa}T_1(\tilde{\alpha}_n)]\delta f_n^{(0)} + 2C\tilde{\kappa}T_2(\tilde{\alpha}_n)\delta f_{-n}^{*(0)} = 0 \; ,$$

$$2C\tilde{\kappa}T_2^*(-\tilde{\alpha}_n)\delta f_n^{(0)} + [\tilde{\lambda} + \tilde{\kappa}(1 - i\theta) + 2C\tilde{\kappa}T_1^*(-\tilde{\alpha}_n)]\delta f_{-n}^{*(0)} = 0 \; , \tag{4.31}$$

where $\tilde{\alpha}_n = \alpha_n/\gamma_\perp$ and the functions $T_1(\tilde{\alpha}_n)$ and $T_2(\tilde{\alpha}_n)$ are fairly complicated but explicit expressions of the system parameters, which have been listed in Appendix 4.C for completeness. The characteristic equation associated with (4.31) is a simple quadratic equation for $\tilde{\lambda}$ that can be solved in closed form without difficulty. The required eigenvalues are

$$\tilde{\lambda}_\pm = \tfrac{1}{2}[-(U_1 + U_2) \pm \sqrt{(U_1 - U_2)^2 - 4V}] \; , \tag{4.32}$$

where

$$U_1(\tilde{\alpha}_n) = \tilde{\kappa}(1 + i\theta) + 2C\tilde{\kappa}T_1(\tilde{\alpha}_n) \; , \tag{4.33a}$$

$$U_2(\tilde{\alpha}_n) = \tilde{\kappa}(1 - i\theta) + 2C\tilde{\kappa}T_1^*(-\tilde{\alpha}_n) \ , \tag{4.33b}$$

$$V(\tilde{\alpha}_n) = -(2C\tilde{\kappa})^2 T_2(\tilde{\alpha}_n)T_2^*(-\tilde{\alpha}_n) \ . \tag{4.33c}$$

the explicit solution (4.32) is too complicated to allow a straightforward examination of the behavior of the eigenvalues in the adiabatic limit. However, a graphical analysis of the roots (4.32) shows that in the limits $\kappa/\gamma_\perp, \kappa/\gamma_\parallel \to 0$, two of the eigenvalues of the exact characteristic equation (4.29) converge smoothly to the two eigenvalues of the adiabatic equation (Fig. 4.7) [4.30].

In some instances, the magnitudes of $\gamma_\parallel$ and $\gamma_\perp$ may differ significantly from one another; we consider a situation where the inequality $\kappa, \gamma_\parallel \ll \gamma_\perp$ is well satisfied. One can then simplify the exact stability analysis by focusing

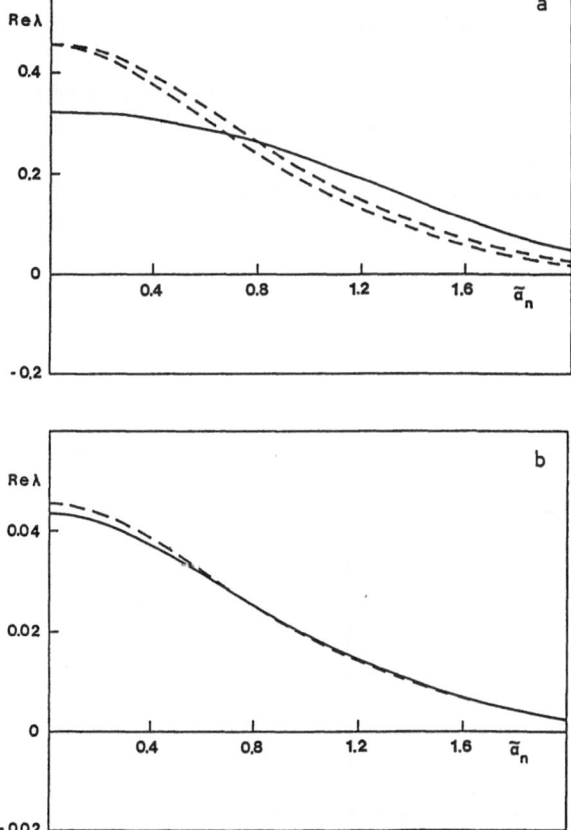

Fig. 4.7. (a) Comparison between the real parts of the roots of the exact secular equation (——) and of (4.32) (- - -). The parameters are as follows: $C = 3$, $\tilde{\Delta} = 0.04$, $\theta = 0.4$, $\tilde{\gamma} = 2$, $|f_0| = 0.2$ and $\tilde{\kappa} = 0.1$. The agreement is satisfactory only in a qualitative sense. (b) By decreasing $\tilde{\kappa}$ by one order of magnitude ($\tilde{\kappa} = 0.01$), the agreement between the exact and approximate eigenvalues becomes considerably more satisfactory. In this simulation we have chosen a smaller value of $\tilde{\kappa}$ by increasing $\gamma_\perp$ and holding θ fixed in both scans (a) and (b); $\tilde{\Delta}$ is equal to 0.004 in the latter case

on the reduced set of three coupled equations that follow from (4.27) after adiabatic elimination of the polarization. In this case also, the calculation is straightforward: first we replace the left-hand side of (4.27c,d) by zero, solve for δp_n and δp^*_{-n}, and then seek solutions of the remaining equations according to the ansatz

$$
\begin{pmatrix} \delta f_n \\ \delta f^*_{-n} \\ \delta d_n \end{pmatrix} = e^{\lambda t} \begin{pmatrix} \delta f_n^{(0)} \\ \delta f_{-n}^{(0)*} \\ \delta d_n^{(0)} \end{pmatrix} .
\tag{4.34}
$$

The resulting cubic equation for the eigenvalue λ is easily soluble by numerical methods. A typical comparison between the real parts of the exact eigenvalues and those of the cubic equation is shown in Fig. 4.8. Again, as with the example displayed in Fig. 4.7, if the inequality between the decay rates is well satisfied, the agreement between exact and adiabatic solutions is quite satisfactory, at least in the linear regime.

We now raise the question: what if the adiabatic elimination is carried out directly from the Maxwell-Bloch equations (4.19)? We show that for a MM system this is not possible.

In fact, if we perform the adiabatic elimination directly on (4.19b,c), we obtain

$$
P = -F \frac{1 - i\tilde{\Delta}^2}{1 + \tilde{\Delta}^2 + |F|^2} ,
\tag{4.35a}
$$

$$
D = -\frac{1 + \tilde{\Delta}^2}{1 + \tilde{\Delta}^2 + |F|^2} .
\tag{4.35b}
$$

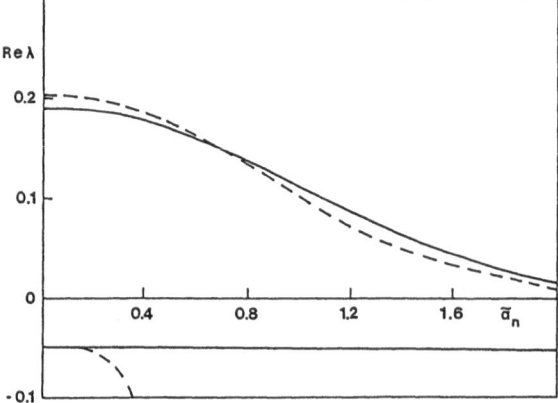

Fig. 4.8. Comparison between the real parts of the roots of the exact secular equation (——) with the cubic equation that results from the adiabatic elimination of the polarization. The parameters are: $C = 3$, $\tilde{\gamma} = 0.02$, $\theta = 0.04$, $\tilde{\kappa} = 0.05$, $|f_0| = 0.2$, and $\tilde{\kappa} = 0.05$

If we substitute (4.35a) into (4.29a), the field equation takes the form

$$\frac{\partial F}{\partial t'} + \frac{cL}{\mathcal{L}} \frac{\partial F}{\partial z'} = -\tilde{\kappa}\left[(1 + i\theta)F - Y - 2CF\frac{1 - i\tilde{\Delta}}{1 + \tilde{\Delta}^2 + |F|^2}\right] . \tag{4.36}$$

It appears that at this point one can proceed in two different ways:

a)   One can linearize (4.36) and its complex conjugate, and then carry out the modal expansion (4.20);

b)   One can expand the field amplitude according to (4.20), and then linearize the resulting modal equations.

As one can verify, the resulting eigenvalue equations are identical in both cases (a and b), so that we choose to illustrate the point by the algebraically simplest route (method a).

The linearized field equations are

$$\frac{\partial}{\partial t'}\delta F + \frac{cL}{\mathcal{L}}\frac{\partial}{\partial z'}\delta F = -\kappa\left[\left(1 + i\theta - 2C\frac{(1 - i\tilde{\Delta})(1 + \tilde{\Delta}^2)}{(1 + \tilde{\Delta}^2 + |F_{st}|^2)^2}\right)\delta F\right.$$
$$\left. + 2C\frac{(1 - i\tilde{\Delta})F_{st}^2}{(1 + \tilde{\Delta}^2 + |F_{st}|^2)^2}\delta F^*\right] \tag{4.37a}$$

$$\frac{\partial}{\partial t'}\delta F^* + \frac{cL}{\mathcal{L}}\frac{\partial}{\partial z'}\delta F^* = -\kappa\left[2C\frac{(1 + i\tilde{\Delta})F_{st}^{*2}}{(1 + \tilde{\Delta}^2 + |F_{st}|^2)^2}\delta F\right.$$
$$\left. + \left(1 - i\theta - 2C\frac{(1 + i\tilde{\Delta})(1 + \tilde{\Delta}^2)}{(1 + \tilde{\Delta}^2 + |F_{st}|^2)^2}\right)\delta F^*\right] . \tag{4.37b}$$

We expand the field fluctuations in the form

$$\delta F(z', t') = \sum_n e^{ik_n z'}e^{-i\tilde{\alpha}_n t'}\delta f_n(t') , \tag{4.38a}$$

$$\delta F^*(z, t') = \sum_n e^{-ik_n z'}e^{i\tilde{\alpha}_n t'}\delta f_n^*(t') \tag{4.38b}$$

and seek elementary solutions of the type

$$\begin{pmatrix} \delta f_n \\ \delta f_n^* \end{pmatrix} = e^{\lambda t}\begin{pmatrix} \delta f_n^{(0)} \\ \delta f_n^{*(0)} \end{pmatrix} . \tag{4.39}$$

The resulting characteristic equation is identical to the one associated with (4.31) after replacing $T_1(\tilde{\alpha}_n)$ and $T_2(\tilde{\alpha}_n)$ by $T_1(0)$ and $T_2(0)$. Hence, this procedure leads to the correct linearized eigenvalues for the resonant mode $(n = 0)$, but to the wrong answer for all modes with $n \neq 0$ (in fact, in this case, the eigenvalues are altogether independent of α_n).

4.5 Dynamical Behavior of the Single-Mode Model

The complexity of the MM problem, even in the MFL, is still a serious obstacle to a systematic investigation of the temporal behavior of a LIS. The SM model, instead, has been the subject of fairly extensive studies, beginning with the first-principles theoretical description by *Spencer* and *Lamb* [4.9] and continuing with the more recent analyses contained in [4.12–18].

The investigations discussed in [4.15] are based on the coupled equations of motion

$$\dot{f}_0 = -\kappa[(1 + i\theta)f_0 - Y + 2Cp_0] \ , \tag{4.40a}$$

$$\dot{f}_0^* = -\kappa[(1 - i\theta)f_0^* - Y + 2Cp_0] \ , \tag{4.40b}$$

$$\dot{p}_0 = \gamma_\perp[f_0^* d_0 - (1 + i\Delta)p_0] \ , \tag{4.40c}$$

$$\dot{p}_0^* = \gamma_\perp[f_0^* d_0 - (1 - i\Delta)p_0^*] \ , \tag{4.40d}$$

$$\dot{d}_0 = -\gamma_\parallel[\tfrac{1}{2}(f_0^* p_0 + f_0 p_0^*) + d_0 + 1] \ , \tag{4.40e}$$

which follow immediately from (4.22) if one neglects every modal amplitude corresponding to indices $n \neq 0$.

Under very high gain conditions, self-pulsing, breathing, frequency locking, period doubling, and chaos have been identified over various ranges of the driving field strength [4.15]. The richness of behavior associated with the simplified model was further emphasized by *Arecchi* and collaborators [4.16] using parameters that are typical of a CO_2 laser (Chap. 2). Even under conditions that suggest the validity of the adiabatic elimination of the polarization, very interesting dynamical effects were shown to persist.

In this section, we review some of the predicted behaviors for a SM high gain system. While the accessible gain range of a laboratory device is likely to be substantially lower than that we used in our numerical simulations, it is probably useful to identify some of the most interesting phenomena as a guide for future experimental searches.

As already noted, in the steady state, the input and output fields are related by the state equation

$$Y = X\left[\left(1 - \frac{2C}{1 + \tilde{\Delta}^2 + X}\right)^2 + \left(\theta + \frac{2C\tilde{\Delta}}{1 + \tilde{\Delta}^2 + X}\right)^2\right]^{1/2} \ , \tag{4.41}$$

where $X \equiv |f_0^{st}|$. The appearance of triple-valuedness for certain values of the parameters raises the possiblity of bistable behavior between the low and high transmission branches. Naturally, the occurrence of bistability is predicated on the existence of coexisting stable states. For moderate values of the gain parameter, when the state equation is triple-valued, the lower branch tends to be unstable, with the injection locking threshold lying at the upper turning point.

A careful scan of detuning and mistuning parameters shows, however, the existence of reasonable ranges of $\tilde{\Delta}$ and θ values where the domain of instability and self-pulsing is confined to a segment of the lower branch, which lies outside the triple-valued domain of the driving field. Everywhere else the system is stable, with the obvious exception of the negative slope region of the state equation. Under these conditions, a slow scan of the driving field amplitude gives clear evidence of hysteretic and bistable behavior (Fig. 4.9a). We must keep in mind, on the other hand, that the linearized stability analysis (Fig. 4.9b), while confirming the stability of the resonant mode, shows the existence of a wide range of instability for the off-resonant modes; thus, only cavities with a sufficiently large intermode spacing will be able to support the kind of behavior displayed in Fig. 4.9a.

It is interesting to observe that for the selected detuning and mistuning parameters, the width of the self-pulsing region depends rather critically on the cavity damping rate $\tilde{\kappa}$, and, in particular, it can be reduced by selecting smaller values of $\tilde{\kappa}$. In the case shown in Fig. 4.9, a value $\tilde{\kappa} \simeq 0.10$ eliminates the domain of instability. This is not surprising because for $\tilde{\kappa} \simeq 0.6$ the entire branch is unstable, in line with the fact that $C = 6$ becomes progressively higher than the laser threshold value, as $\tilde{\kappa}$ grows. In fact, the threshold gain for the FRL $(Y = 0)$ is given by

$$2C_{thr} = 1 + \left(\frac{\Delta - \theta\tilde{\kappa}}{1 + \tilde{\kappa}}\right)^2 . \tag{4.42}$$

Quite frequently, however, the lower branch is entirely unstable, and for sufficiently high gain values, it can yield complicated sequences of pulsing states. As shown by Fig. 4.9a, a bird's-eye view of the main features of the unstable behavior can be gained by solving the SM equations with a time-dependent driving field amplitude. Naturally, at this point, the new set of equations is no longer autonomous, and very few precise statements can be made in analytic terms. Strictly speaking, even the notion of a state equation is no longer meaningful in this context; still, empirical evidence from our numerical solutions shows that when the time-dependent driving field sweeps through a stable domain, the output field actually traces a curve that is essentially identical to the state equation corresponding to the selected operating parameters. This is true, of course, only if the sweep rate is sufficiently small, otherwise, lag effects become significant. Thus, if some care is applied, the pulsation patterns that one would observe under unstable conditions with a fixed value of Y are reproduced with a reasonable accuracy under swept conditions. The global picture

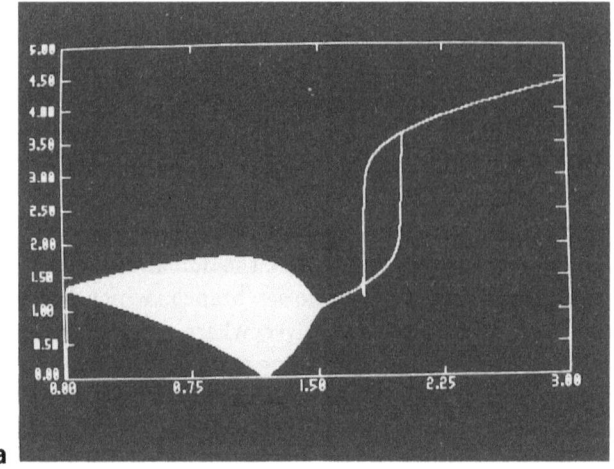

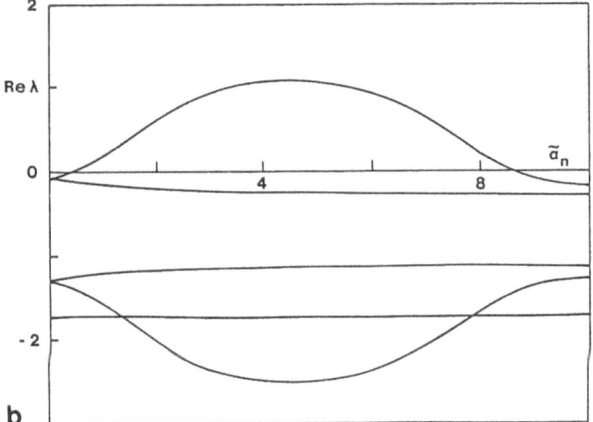

Fig. 4.9. (a) Temporal evolution of the output field $|f_0(t)|$ according to (4.40), but with a slowly varying input field amplitude $Y(\tau) = Y_0 + V\tau$. The forward and backward scans are overlapped in this figure. The parameters are: $C = 6$, $\tilde{\Delta} = 3.5$, $\theta = -1.5$, $\tilde{\kappa} = 0.3$, $\tilde{\gamma} = 2$, $V = 10^{-3}$. The resonant mode is unstable over the approximate range $0 < Y(\tau) \lesssim 1.5$. For $Y(\tau) \gtrsim 1.5$, the output field amplitude traces essentially the state equation because of the very small scan rate. **(b)** Real parts of the linearized eigenvalues as functions of $\tilde{\alpha}_n$ for $|f_0| = 1.5$

that emerges in this way is a useful one because it corresponds closely to the procedure followed in experimental observations.

An example of rich dynamical behavior is shown in Figs. 4.10a,b. In both forward and backward scans one can observe regions of chaotic oscillations and either continuous or discontinuous changes in the amplitude of self-pulsing. The transition from one pattern of pulsations to another is a consequence of the existence of different domains of attraction which often overlap over certain ranges of values of the driving field. As long as the pulsations are produced

94

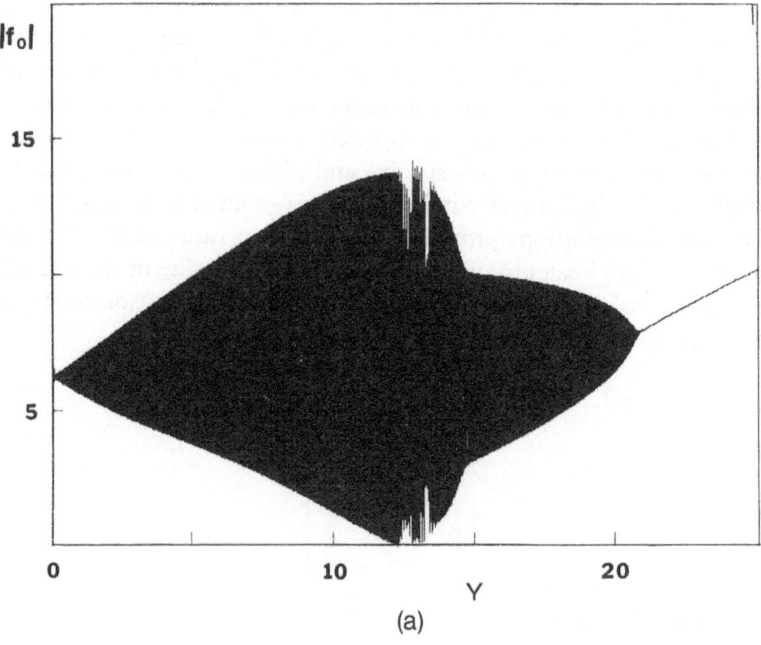

(a)

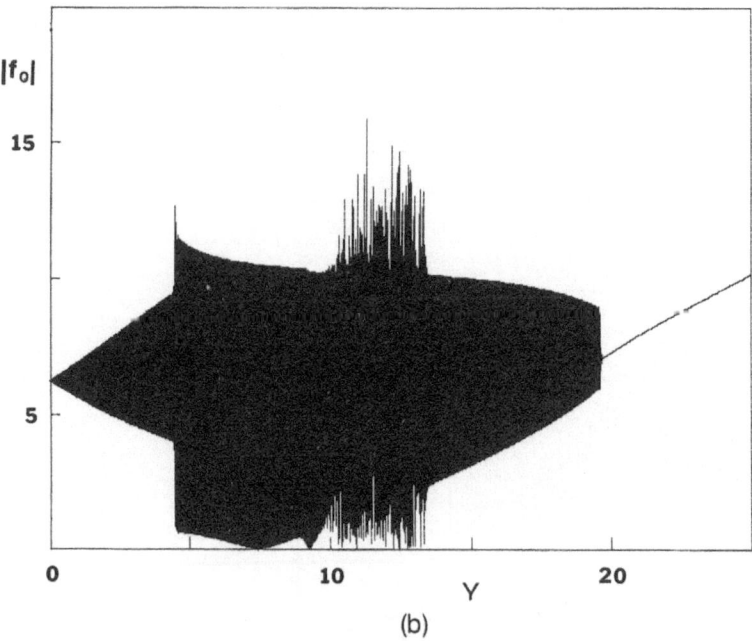

(b)

Fig. 4.10. (a) Output oscillations of $|f_0|$ corresponding to a slowly varying driving field of the type $Y(\tau) = Y_0 + V\tau$ and to the parameters $C = 20$, $\tilde{\Delta} = 1$, $\theta = 2$, $\tilde{\kappa} = 0.5$, and $\tilde{\gamma} = 0.05$. This figure shows the forward scan with $V = 2.5 \times 10^{-3}$. (b) Backward scan with the same sweep rate as in (a)

by the same attractor, the observed patterns are the same for both forward and backward scans (see Figs. 4.11a,b, for example). It is easy, however, to find control parameters where this is not the case; switching between different domains and hysteretic behavior can then be observed. A priori, the presence of hysteresis could be ascribed to lag effects. Actually, this is not so for the scans shown here, as the analysis of the power spectra and of the Lyapunov exponents show conclusively. Additional direct confirmation is provided in Figs. 4.12a,b, where we show output oscillations produced for the same value of the driving field. These solutions have been obtained by gradually increasing or decreasing the value of Y, and by using as initial conditions for each new solution the final configuration of the previous solution (adiabatic scan).

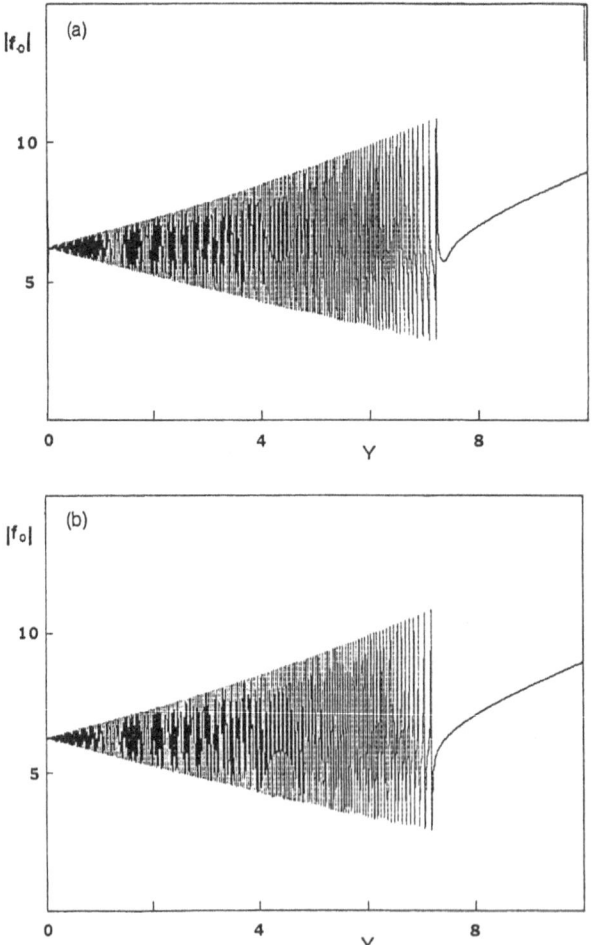

Fig. 4.11. (a) Output oscillations of $|f_o|$ corresponding to a slowly varying driving field with a sweep rate $V = 2 \times 10^{-3}$ and parameters $C = 20$, $\bar{\Delta} = 0.4$, $\theta = 0.8$, $\bar{\kappa} = 0.5$, $\bar{\gamma} = 0.05$. **(b)** Backward scan with the same sweep rate as in **(a)**

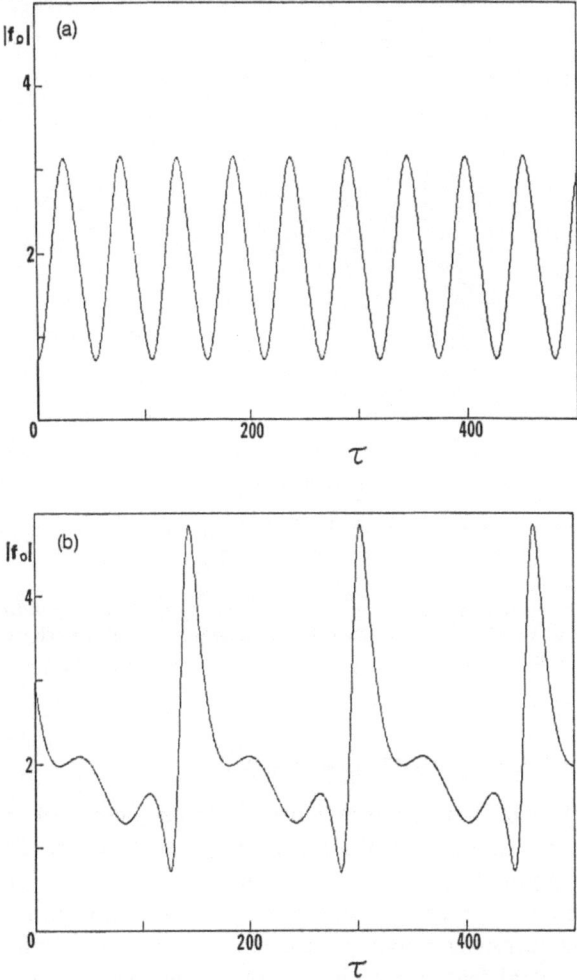

Fig. 4.12. (a) Output field as a function of time for $U = 3$, $\tilde{\Delta} = 0.5$, $\theta = 0.5$, $\tilde{\kappa} = 0.1$, and $\tilde{\gamma} = 0.01$. The selected value of Y is 1.7. (b) Output field as a function of time for the same parameters as in (a). The solution evolves under the influence of a different basin of attraction

When the state equation is single-valued, the driven laser is normally kept in a self-pulsing state over a range of the external field amplitude that extends from zero to a certain locking threshold Y_{thr}. There are instances, however, where a scan of the driving field shows an additional small region of instability for higher values of the injected field. An example is shown in Fig. 4.13.

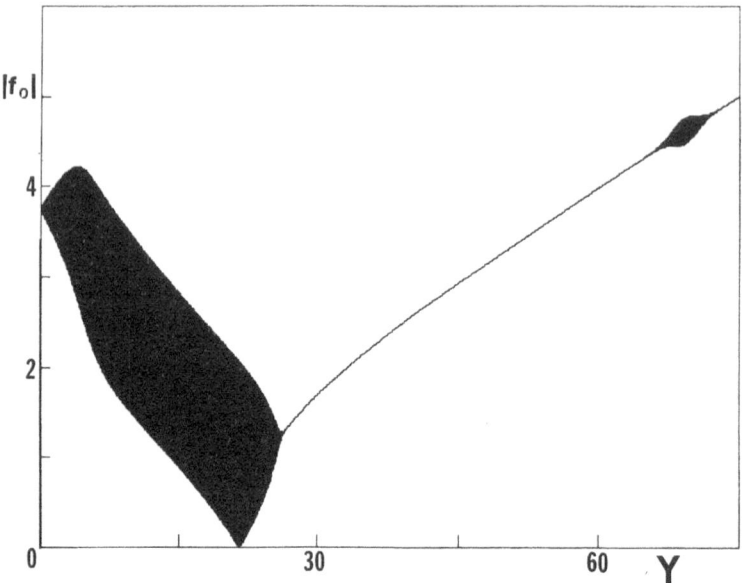

Fig. 4.13. Output oscillations corresponding to a slowly varying driving field corresponding to the parameters $C = 20$, $\tilde{\Delta} = 0$, $\theta = 15$, $\tilde{\kappa} = 0.5$, and $\tilde{\gamma} = 2$. The forward and backward scans are identical; the sweep rate is $V = 0.005$

4.6 Power Spectra and Lyapunov Exponents

While a slow scan of the injected field provides a useful global picture of the pulsations, its resolution is very limited. A more convenient way to quantify finer details is to monitor the power spectrum of the output oscillations. The fundamental pulsation frequency is a rather sensitive function of the type of attractor in which a given solution is moving; additional quantitative information is brought out by spectral features such as subharmonic components, incommensurate frequencies, broadband structures and noisy background.

An example of the type of information that can be extracted from the spectral analysis is shown in Fig. 4.14b. Here, the frequency of the fundamental component shows a rather complicated dependence on the strength of the incident field. By adopting a procedure that simulates an adiabatic scan of the injected field [4.31], we note at first a monotonic dependence of the fundamental frequency on Y over the range $0 < Y \gtrsim 11.12$, followed by a discontinuous jump to a higher value. This is a clear indication that the basin of attraction corrsponding to the lower frequency has lost its stability, and that the solution has been "caught" by a different attractor. By reversing the direction of the scan, we can follow the behavior of the system down to $Y \gtrsim 5.88$, where, now, the second attractor loses stability and forces a return to the earlier configuration. A downward scan that begins at the injection-locking threshold and proceeds to lower values of the driving field yields a fundamental frequency dependence of the type shown by curve 3 in Fig. 4.14b.

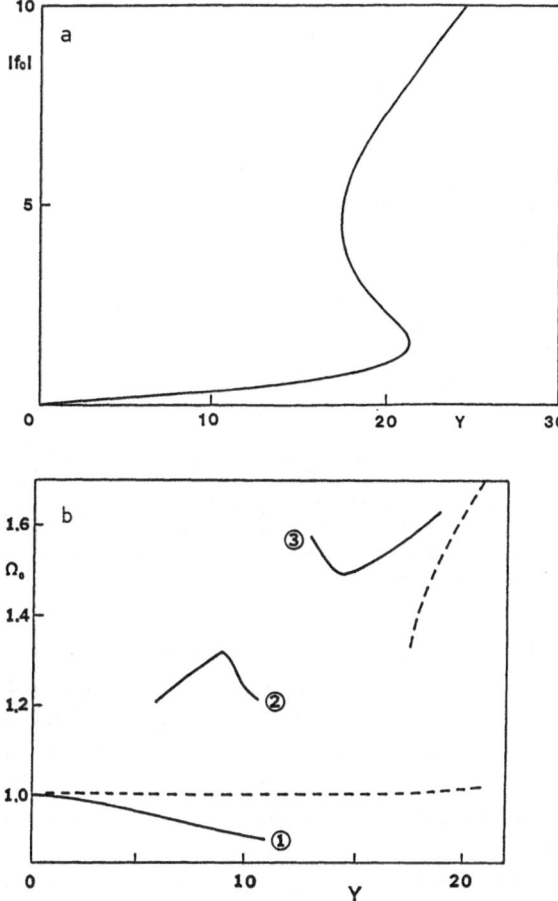

Fig. 4.14. (a) Steady-state equation correspondong to $C = 20$, $\tilde{\Delta} = 1$, and $\theta = 2$. (b) The behavior of the fundamental frequency component Ω_0 in the power spectrum of the output oscillations as a function of the driving field strength for $C = 20$, $\tilde{\Delta} = 1$, $\theta = 2$, $\tilde{\kappa} = 0.5$ and $\tilde{\gamma} - 0.05$. The fundamental frequencies 1 and 2 were recorded by adiabatic scans along two different attractors. The fundamental frequency 3 was calculated by an adiabatic scan that begins at the injection-locking threshold and proceeds to lower values of Y. (---) correspond to the imaginary parts of the linearized eigenvalues along the lower and upper intensity branches of the state equation which has an S-shape for the chosen parameters

The structure of the attractor that is responsible for the fundamental frequency marked 2 in Fig. 4.14b is very complicated. With the help of a fine scan of the injected field, one can monitor the emergence of new frequency components at $1/2$, $1/3$, $1/6$, $1/12$, ... of the fundamental frequency. The indication is for the existence of an infinite sequence of bifurcations that begin with period doubling and continue with a period of the type 3×2^n. An example of a spectrum of this type is shown in Fig. 4.15. Beyond the onset of chaotic behavior, we have observed periodicity windows that appear to correspond to frequency-locked oscillations.

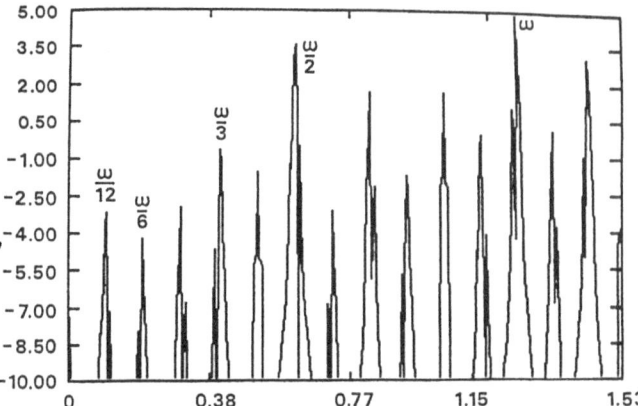

Fig. 4.15. Power spectrum of a solution displaying numerous subharmonic components. The solution evolves under the influence of the attractor *2* in Fig. 4.14 and corresponds to the parameters $C = 20$, $\tilde{\Delta} = 1$, $\theta = 2$, $\tilde{\kappa} = 0.5$, $\tilde{\gamma} = 0.05$, and a driving field $Y = 10.43$. The main spectral components are identified in the figure. The others are recognizable as combination frequencies. Only a portion of the frequency axis is displayed for clarity. The vertical scale is labelled by the natural logarithm of the power spectrum

A well-known signature of the existence of chaotic behavior (motion on a strange attractor) is the appearance of exponential divergence of irregular trajectories whose origins in phase space are arbitrarily close to one another. As a test of this behavior, we have monitored as a function of time the cartesian distance of two nearby trajectories. On a semilogarithmic plot, the average linear growth of the logarithm of the distance is a good indication of the presence of deterministic chaos (Fig. 4.16).

In spite of the wealth of information provided by the power spectra of the output signal, important finer details still remain inaccessible. A more powerful procedure, although considerably more time consuming, is based on the calculation of the Lyapunov exponents and on the accompanying analysis of the phase-space portraits of the solutions. A brief summary of the information that can be obtained from the calculation of the Lyapunov exponents and some technical details on a possible way to construct them as reported in [4.32]. Here we remind the reader of some relevant definitions [4.33] and summarize the results of one detailed scan [4.32].

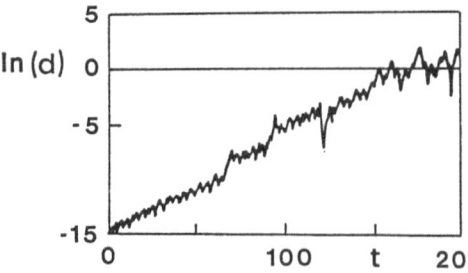

Fig. 4.16. The logarithm of the cartesian distance between two nearby trajectories evolving in the presence of a strange attractor diverges linearly in time. The plot gives evidence of the so-called exponential divergence

Given a set of autonomous (usually nonlinear) differential equations of the type

$$\dot{X}_i = F_i(X_1, X_2, \ldots, X_N) \,, \quad i = 1, 2, \ldots, N \tag{4.43}$$

one can obtain the trajectory of a representative point $P \equiv (X_1, X_2, \ldots, X_N)$ in phase space by standard integration techniques. To test the stability of the trajectory, one may select a point in its vicinity and follow its evolution according to the linearized equations of motion. If δX_i represents the initial deviation of the chosen point from a value X_i on the actual trajectory, and if W denotes a vector whose components are proportional to $(\delta X_1, \delta X_2, \ldots, \delta X_N)$, respectively, the linearized evolution of W is described by the equation

$$\dot{W} = \hat{J} W \,, \tag{4.44}$$

where $\hat{J}$ is the Jacobian matrix whose elements are given by

$$J_{ij} \equiv \frac{\partial F_i(X)}{\partial X_j} \,. \tag{4.45}$$

The stability in the neighborhood of the trajectory can be measured by the rate of change of the length of W. If $\hat{J}$ is a time-independent matrix, the Lyapunov exponents are the real parts of the eigenvalues of $\hat{J}$. [This happens only if the selected displaced point lies in the neighborhood of a steady state of (4.43).] In general, of course, $\hat{J}$ will be time dependent because the point P itself is moving in phase space. Let

$$W(t) = T \left[\exp \int_{t_0}^{t} dt' \hat{J}(t') \right] W(t_0) \equiv \hat{U}(t, t_0) W(t_0) \tag{4.46}$$

be the solution of (4.44), where T is the usual time-ordering operator. The matrix $\hat{U}(t, t_0)$ can be written as the product $\hat{O}(t, t_0) \hat{S}(t, t_0)$ of an orthogonal $(\hat{O})$ and a nonnegative $(\hat{S})$ matrix, whose action is to induce rotations and dilatations (or contractions) of the N-dimensional vector space. Denote with λ_i the ith eigenvalue of $S(t, t_0)$; then, by definition, the Lyapunov exponent is given by

$$E_i^{\mathrm{L}} \equiv \lim_{t \to \infty} \frac{1}{t - t_0} \ln \lambda_i(t, t_0) \,. \tag{4.47}$$

In the following discussion, we adopt the usual ordering rule $E_1^{\mathrm{L}} \geq E_2^{\mathrm{L}} \geq \ldots \geq E_N^{\mathrm{L}}$.

The main reason why the Lyapunov exponents are a useful tool for stability considerations can be appreciated at once from the following synopsis of some of their properties.

1) Different attractors exhibit different combinations of Lyapunov exponents. For autonomous dynamical systems, for example, we have

Type of attractor	Lyapunov Exponent
Steady state	$E_1^L < 0$
Limit cycle	$E_1^L = 0, \; E_2^L < 0$
m-dimensional torus	$E_1^L = E_2^L = \ldots = E_m^L = 0, \; E_{m+1}^L < 0$
Chaos	$E_1^L > 0$

2) The nature of certain bifurcations of a limit cycle can be understood by tracing the dependence of the Lyapunov exponents upon the control parameters. Thus:

a)   If the limit cycle undergoes a period-doubling or symmetry-breaking bifurcation (the original basin of attraction splits into two domains of attraction), then E_2^L will show a maximum of zero at the bifurcation point.

b)   If the limit cycle undergoes a tangent bifurcation, E_2^L will first increase from a negative value to zero; then, above the bifurcation point, either E_1^L becomes positive, while E_2^L remains zero, or E_2^L undergoes a sudden jump. The latter case is indicative that the trajectory has jumped into another coexisting attractor.

c)   If a limit cycle undergoes a Hopf bifurcation, the second and third Lyapunov exponents will be equal to one another over a finite range of parameters before the bifurcation point. The nature (subcritical or supercritical) of the Hopf bifurcation can be recognized by the behavior of E_2^L and E_3^L above the bifurcation point. Thus, if E_1^L and E_2^L remain zero, while E_3^L becomes negative again, the bifurcation is supercritical; if E_1^L becomes positive or E_2^L and E_3^L undergo a sudden change, the bifurcation may be expected to be subcritical.

d)   If the second Lyapunov exponent of a limit cycle shows a sudden change, without the first becoming equal to zero, this may be the symptom of a collision of the limit cycle with the boundary of its basin, followed by the sudden disappearance of the limit cycle.

By means of the adiabatic scan technique, we have calculated the five Lyapunov exponents for a configuration characterized by $C = 20$, $\tilde{\Delta} = 1$, $\theta = 2$, $\tilde{\kappa} = 0.5$ and $\tilde{\gamma} = 0.05$ in the range $Y = 15$ to $Y = 5$. At the appearance of changes in dynamical behavior, as indicated by the Lyapunov exponents and the phase-space portraits, we have always reversed the direction of variation of Y and explored the existence of hysteretic behavior. With this procedure, we have produced a global view of the dynamical features of our system as shown synthetically in Fig. 4.17. Many of the large-scale features of this plot provide

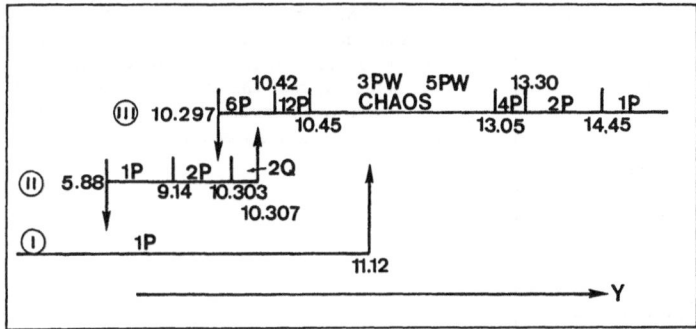

Fig. 4.17. Global behavior of the system in the range $0 < Y \lesssim 15$. $1P$ denotes a simple periodic solution, $2P$ is a doubly periodic solution (i.e., the spectrum contains a fundamental frequency ω and its subharmonic $\omega/2$), etc; $2Q$ denotes quasi-periodic motion with two incommensurate frequencies (two-dimensional torus), W denotes a window in chaos (thus $3PW$ labels a window with a triply periodic solution); the values of Y, e.g., $Y = 11.12$ at the end of domain I, denote the approximate thresholds

direct confirmation of the results obtained by the spectral analysis. It is clear, on the other hand, that finer details had been missed by the spectra. Each of the three branches in Fig. 4.17 protrays the bifurcation scheme of separate attractors (note that for the chosen parameters, up to three attractors can coexist over a range of values of the driving field).

Branch I begins at $Y = 0$ and ends at $Y \cong 11.12$. The Lyapunov exponents support the relation

$$0 = E_1^L > E_2^L = E_3^L > E_4^L > E_5^L$$

over the entire range. In the vicinity of $Y = 11.12$, E_2^L and E_3^L both approach zero, indicating the occurrence of a Hopf bifurcation. No stable torus appears beyond this point. However, by selecting different initial points in the vicinity of the $1P$ limit cycle for $Y \lesssim 11.12$, and by monitoring the transient evolution of the system in a Poincaré surface of section, we have identified the existence of an unstable torus which contracts into the limit cycle as Y approaches the Hopf bifurcation from below. On the basis of this evidence, we conclude that the Hopf bifurcation at $Y = 11.12$ is of a subcritical type. Similar evidence supports the conclusion that a subcritical Hopf bifurcation also arises when Y approaches the value 5.88 from above along branch II.

Another Hopf bifurcation occurs in branch II for $Y \simeq 10.303$. Beyond this bifurcation point, a stable torus $(2Q)$ develops; evidence for its appearance had been provided earlier by the observation of a second frequency in the power spectra. This bifurcation is supercritical, as indicated schematically in Fig. 4.18. A systematic search for the basin of attraction of the $2Q$ torus reveals the existence of a second (unstable) torus surrounding the first. The stable and unstable tori approach each other as Y increases and eventually collide at $Y \simeq 10.307$, suggesting the existence of a saddle-node type of bifurcation.

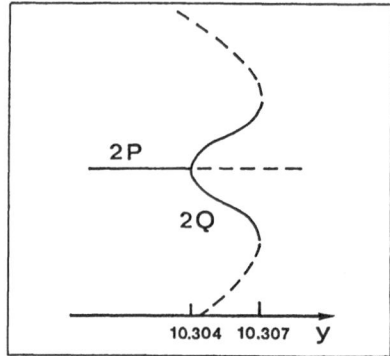

Fig. 4.18. Schematic representation of the supercritical bifurcation occurring at $Y = 10.307$ on branch II. (———) denote the stable $2P$ solutions and the $2Q$ torus, respectively. (---) denote the unstable limit cycle and the unstable torus, respectively

Branch III begins with a $6P$ limit cycle at $Y = 10.297$. The numerical evaluation of the Lyapunov exponents show that in the neighborhood of and above $Y = 10.297$ we have

$$|E_1^L| < 10^{-3} , \quad E_2^L \simeq E_3^L = -0.003 , \quad E_4^L \simeq E_5^L \simeq -1.552 .$$

Because E_2^L remains negative when Y approaches the value 10.297 from above, the sudden disappearance of the limit cycle can be explained only by the possible loss of stability on some points in the cycle itself (in contrast to the loss of stability of the entire trajectory in an ordinary bifurcation). Immediately after the disappearance of the $6P$ cycle, the basin of the original $6P$ attractor becomes the basin of the $2P$ attractor of branch II. This suggests the reasonable assumption that the loss of stability of the $6P$ attractor is caused by its collision with the unstable torus of branch II whose presence was revealed by the bifurcation of the $2Q$ torus at $Y = 10.307$.

The unfortunate drawback of the type of search that can be advanced with the help of the Lyapunov exponents is the amount of computational time and labor that is required for these investigations. As revealing and useful as this procedure undoubtedly is in extracting very fine details of the bifurcation structures, the entire process is impractical without a large mainframe computer [4.34].

4.A Appendix: Derivation of (4.14)

The starting point is the set of equations (4.8 and 9), specialized to $z = L$, plus the boundary conditions (4.10). In the limit (4.12) we can set

$$\varrho(L) \simeq \varrho(0) + \delta\varrho , \tag{4.A.1a}$$

$$\theta(L) \simeq \theta(0) + \delta\theta \tag{4.A.1b}$$

to obtain

$$\frac{\delta\varrho}{\varrho(0)} = -\frac{1}{\tilde{\Delta}}\delta\theta \tag{4.A.2a}$$

from (4.8), and

$$\frac{\delta\varrho}{\varrho(0)}[1 + \tilde{\Delta}^2 + \varrho^2(0)] = \alpha L \tag{4.A.2b}$$

from (4.9). If we now substitute (4.A.1 and 2) into (4.10), and retain only first-order terms in αL and T, we obtain

$$Y = \varrho(0)e^{i\theta(0)}\left\{\frac{1 - e^{-i\delta_0}}{T} + e^{-i\delta_0}\right.$$

$$\left. \times\left(1 - \frac{\alpha L/T}{1 + \tilde{\Delta}^2 + \varrho^2(0)} + i\tilde{\Delta}\frac{\alpha L/T}{1 + \tilde{\Delta}^2 + \varrho^2(0)}\right)\right\} . \tag{4.A.3}$$

This result is not completely consistent yet with the requirements of the MFL (4.12,13) because δ_0 is itself a quantity of order T; we have exhibited this intermediate step only to emphasize the role played by the accumulated phase lag in arriving at the mathematical MFL. If we now set $\delta_0 = \theta T$ ($\theta = $ const), and again retain only the leading terms on each side of (4.A.3), we obtain the required result (4.14). Clearly, at this point, it becomes unnecessary to specify the value of the z-coordinate where $\mathcal{F}$ is to be evaluated because, in this limit, $\mathcal{F}$ is uniform over $0 \leq z \leq L$.

4.B Appendix: List of the Coefficients of (4.29)

$$A_5 = 1 , \tag{4.B.1}$$

$$A_4 = \tilde{\gamma} + 2 + 2\tilde{\kappa} - 3i\tilde{\alpha}_n , \tag{4.B.2}$$

$$A_3 = 2(\tilde{\gamma} - i\tilde{\alpha}_n)(1 + \tilde{\kappa} - i\tilde{\alpha}_n) + (1 - i\tilde{\alpha}_n)^2 + 4\tilde{\kappa}(1 - i\tilde{\alpha}_n)$$
$$+ \tilde{\Delta}^2 + \tilde{\kappa}^2(1 + \theta^2) + \tilde{\gamma}|f_0|^2 - \frac{2C\tilde{\kappa}}{1 + \tilde{\Delta}^2 + |f_0|^2}2(1 + \tilde{\Delta}^2) , \tag{4.B.3}$$

$$A_2 = (\tilde{\gamma} - i\tilde{\alpha}_n)[(1 - i\tilde{\alpha}_n)^2 + \tilde{\Delta}^2 + 4\tilde{\kappa}(1 - i\tilde{\alpha}_n) + \tilde{\kappa}^2(1 + \theta^2)]$$
$$+ 2\tilde{\kappa}[(1 - i\tilde{\alpha}_n)^2 + \tilde{\Delta}^2] + 2\tilde{\kappa}^2(1 + \theta^2)(1 - i\tilde{\alpha}_n) + \tilde{\gamma}|f_0|^2(1 - i\tilde{\alpha}_n)$$
$$+ 2\tilde{\kappa}\tilde{\gamma}|f_0|^2 + \frac{2C\tilde{\kappa}}{1 + \tilde{\Delta}^2 + |f_0|^2}[\tilde{\gamma}|f_0|^2$$
$$- 2(1 + \tilde{\Delta}^2)(1 + \tilde{\gamma} - 2i\tilde{\alpha}_n) - 2\tilde{\kappa}(1 + \tilde{\Delta}^2)] , \tag{4.B.4}$$

$$A_1 = (\tilde\gamma - i\tilde\alpha_n)[2\tilde\kappa((1 - i\tilde\alpha_n)^2 + \tilde\Delta^2) + 2\tilde\kappa^2(1 + \theta^2)(1 - i\tilde\alpha_n)] + \tilde\kappa^2(1 + \theta^2)$$
$$\times [(1 - i\tilde\alpha_n)^2 + \tilde\Delta^2] + 2\tilde\kappa\tilde\gamma|f_0|^2(1 - i\tilde\alpha_n) + \tilde\gamma|f_0|^2\tilde\kappa^2(1 + \theta^2)$$
$$+ \frac{2C\tilde\kappa}{1 + \tilde\Delta^2 + |f_0|^2}\Big[\tilde\kappa\tilde\gamma|f_0|^2 - 2\tilde\kappa(1 + \tilde\Delta^2)(1 + \tilde\gamma - 2i\tilde\alpha_n)$$
$$- 2(1 + \tilde\Delta^2)(1 - i\tilde\alpha_n)(\tilde\gamma - i\tilde\alpha_n) - i\tilde\alpha_n\tilde\gamma|f_0|^2$$
$$+ \tilde\kappa\theta\tilde\delta(2(1 + \tilde\Delta^2) + |f_0|^2)\Big] + \left[\frac{2C\tilde\kappa}{1 + \tilde\Delta^2 + |f_0|^2}(1 + \tilde\Delta^2)\right]^2, \qquad (4.B.5)$$

$$A_0 = (\tilde\gamma - i\tilde\alpha_n)\tilde\kappa^2(1 + \theta^2)[(1 - i\tilde\alpha_n)^2 + \tilde\Delta^2] + \tilde\gamma|f_0|^2(1 - i\tilde\alpha_n)\tilde\kappa^2(1 + \theta^2)$$
$$+ \frac{2C\tilde\kappa}{1 + \tilde\Delta^2 + |f_0|^2}\Big[\tilde\kappa\theta\tilde\Delta(2(1 + \tilde\Delta^2)(\tilde\gamma - i\tilde\alpha_n) - i\tilde\alpha_n\tilde\gamma|f_0|^2)$$
$$- 2\tilde\kappa(1 + \tilde\Delta^2)(1 - i\tilde\alpha_n)(\tilde\gamma - i\tilde\alpha_n) - i\tilde\alpha_n\tilde\kappa\tilde\gamma|f_0|^2\Big] + \left(\frac{2C\tilde\kappa}{1 + \tilde\Delta^2 + |f_0|^2}\right)^2$$
$$\times [(1 + \tilde\Delta^2)^2(\tilde\gamma - i\tilde\alpha_n) - \tilde\gamma(1 + \tilde\Delta^2)|f_0|^2] \qquad (4.B.6)$$

In the above equations, the symbols $\tilde\alpha_n$, $\tilde\gamma$, and $\tilde\kappa$ denote the scaled parameters $\alpha_n/\gamma_\perp$, $\gamma_\|/\gamma_\perp$, and $\kappa/\gamma_\perp$, respectively.

4.C Appendix: List of Coefficients in (4.31)

In order to construct the coefficients $T_1(\tilde\alpha_n)$ and $T_2(\tilde\alpha_n)$ that appear in (4.31), we first define the auxiliary coefficients

$$C_1(\tilde\alpha_n) = 1 + i(\tilde\Delta - \tilde\alpha_n) + \frac{1}{2}\tilde\gamma|f_0^{st}|^2\frac{1}{\tilde\gamma - i\tilde\alpha_n}, \qquad (4.C.1)$$

$$C_2(\tilde\alpha_n) = \frac{1}{2}\tilde\gamma\frac{f_0^{st\,2}}{\tilde\gamma - i\tilde\alpha_n}, \qquad (4.C.2)$$

$$C_3(\tilde\alpha_n) = d_0^{st} - \frac{1}{2}\tilde\gamma\frac{f_0^{st}p_0^{*st}}{\tilde\gamma - i\tilde\alpha_n}, \qquad (4.C.3)$$

$$C_4(\tilde\alpha_n) = -\frac{1}{2}\tilde\gamma\frac{f_0^{st}p_0^{st}}{\tilde\gamma - i\tilde\alpha_n}, \qquad (4.C.4)$$

where p_0^{st} and d_0^{st} are listed in (4.24).

In terms of $C_1(\tilde\alpha_n), \ldots, C_4(\tilde\alpha_n)$ the required functions of $\tilde\alpha_n$ are given by

$$T_1(\tilde\alpha_n) = \frac{C_1^*(-\tilde\alpha_n)C_3(\tilde\alpha_n) - C_2(\tilde\alpha_n)C_4^*(-\tilde\alpha_n)}{C_1(\tilde\alpha_n)C_1^*(-\tilde\alpha_n) - C_2(\tilde\alpha_n)C_2^*(-\tilde\alpha_n)} \qquad (4.C.5)$$

and

$$T_2(\tilde{\alpha}_n) = \frac{C_1^*(-\tilde{\alpha}_n)C_4(\tilde{\alpha}_n) - C_2(\tilde{\alpha}_n)C_3^*(-\tilde{\alpha}_n)}{C_1(\tilde{\alpha}_n)C_1^*(-\tilde{\alpha}_n) - C_2(\tilde{\alpha}_n)C_2^*(-\tilde{\alpha}_n)} \ . \tag{4.C.6}$$

Acknowledgements. We are indebted to many of our colleagues for useful conversations and comments, to Professors N.B. Abraham and J.R. Tredicce for providing much insight in the experimental side of this problem, to Dr. P. Mandel for proposing and demonstrating the advantages of slow scans of the driving field, and to Professor H. Haken for sharing with us, on numerous occasions, the wealth of his knowledge of nonlinear dynamical phenomena.

Special thanks are due to Professors Yan Gu and J.M. Yuan for introducing us to the study of the Lyapunov exponents and for their invaluable collaboration. We are also grateful to Ms. C.A. Pennise, Mr. H. Sadiky and Mr. F. Narducci for their help with innumerable intricate numerical analyses.

This work was partially supported by a contract from the U.S. Army Research Office, by the National Research Council (CNR) of Italy and, during the original phases, by a grant from the Research Laboratories of the Martin-Marietta Corporation. This research was carried out in the framework of an operation launched by the Commission of the European Community Stimulation Action (1983–85). Partial travel support was provided by a NATO grant.

References

4.1   A.N. Oraevskii: Radio Elektroniikkalab. Tek. Korkeakoulu (Kertomus) **4**, 718 (1959)
4.2   A.N. Oraevskii, A.V. Uspenskii: In Proc. Lebedev Inst., Vol. 31, ed. by D.V. Sko-
         bel'tsyn (Consultants Bureau, New York 1968) p. 87;
         N.G. Basov, A.Z. Grazyuk, I.G. Zubarev, L.V. Tevelev: ibid, p. 67
4.3   J.P. Gordon: Proc. IRE **50**, 1898 (1962);
         A.S. Agabekjan, A.Z. Grazyuk, I.G. Zubarev, A.N. Oraevskii, V.I. Svergun: Radio
         Elektroniikkalab. Tek. Korkeakoulu (Kertomus) **9**, 2156 (1964)
4.4   H.L. Stover, W.H. Steier: Appl. Phys. Lett. **8**, 91 (1966);
         R.W. Dunn, S.T. Hendow, W.W. Chow, J. Small: Opt. Lett. **8**, 319 (1983);
         W. Annovazzi, S. Donati: IEEE J. QE-**16**, 859 (1980)
4.5   L.E.Erickson, A. Szabo: Appl. Phys. Lett. **18**, 433 (1981);
         P. Burlamacchi, R. Salimbeni: Opt. Commun. **17**, 6 (1976)
4.6   A. Girard: Opt. Commun. **11**, 346 (1974);
         J.L. Lachambre, P. Lavigne, G. Otis, M. Noel: IEEE J. QE-**12**, 756 (1976);
         C.J. Buczek, R.J. Freiberg: IEEE J. QE-**8**, 643 (1972)
4.7   R. Lang: IEEE J. QE-**18**, 979 (1982) and references therein
4.8   L.A. Lugiato: "Theory of Optical Bistability", in *Progress in Optics,* **21**, 71 ed. by E.
         Wolf (North-Holland, Amsterdam 1984);
         L.A. Orozco, A.T. Rosenberger, H.J. Kimble: Phys. Rev. Lett. **53**, 2547 (1984)
4.9   M.B. Spencer, W.E. Lamb, Jr.: Phys. Rev. A**5**, 884 (1972)
4.10   See, for example, U. Ganiel, A. Hardy, D. Treves: IEEE J. QE-**12**, 704 (1976);
         R. Flamant, G. Megie: IEEE J. QE-**16**, 653 (1980)
4.11   S. Blit, U. Ganiel, D. Treves: Appl. Phys. **12**, 69 (1977);
         Y.K. Park, G. Giuliani, R.L. Byer: Opt. Lett. **5**, 96 (1980)
4.12   L.A. Lugiato: Lett. Nuovo Cimento **23**, 609 (1978) and references therein
4.13   T. Yamada, R. Graham: Phys. Lett. **53A**, 77 (1975)
4.14   M.J. Scholz, T. Yamada, H. Brand, R. Graham: Phys. Lett. **82A**, 321 (1981)
4.15   L.A. Lugiato, L.M. Narducci, D.K. Bandy, C.A. Pennise: Opt. Commun. **46**, 64 (1983);
         D.K. Bandy, L.M. Narducci, C.A. Pennise, L.A. Lugiato: In *Coherence and Quantum
         Optics V,* ed. by L. Mandel, E. Wolf (Plenum, New York 1984)p. 585;
         L.A. Lugiato, L.M. Narducci: ibid. p. 941
4.16   F.T. Arecchi, G. Lippi, G. Puccioni, J.R. Tredicce: In *Coherence and Quantum Optics
         V,* ed. by L. Mandel, E. Wolf (Plenum, New York 1984) p. 1227

4.17 K. Otsuka, H. Iwamura: Phys. Rev. A**28**, 3153 (1983)

4.18 D.K. Bandy, L.M. Narducci, L.A. Lugiato: J. Opt.Soc. Am. B**2**, 148 (1985);
K. Otsuka: ibid. p. 168;
J.R. Tredicce, F.T. Arecchi, G.L. Lippi, G.P. Puccioni: ibid. p. 173

4.19 E. Brun, B. Derighetti, D. Meier, R. Holzner, M. Ravani: J. Opt. Soc. Am. B**2**, 156 (1985) and references therein;
J.L. Boulnois, G.P. Puccioni, F.T. Arecchi, J.R. Tredicce: private communication

4.20 N.B. Abraham, L.A. Lugiato, L.M. Narducci: J. Opt. Soc. Am. B**2** (1985)

4.21 N.B. Abraham, P. Mandel, L.M. Narducci: " Dynamical Instabilities and Pulsations in Lasers", in *Progress in Optics*, ed. by E. Wolf (North Holland, Amsterdam) to be published

4.22 H. Haken: *Light*, Vol. 2 (North Holland, Amsterdam 1985) p. 208

4.23 R.W. Boyd, M.G. Raymer, L.M. Narducci: *Optical Instabilities*, (Cambridge U. Press, London 1986)

4.24 Earlier stability studies for optically bistable systems have been carried out for parameter values that lied outside the MFL. See, for example, R. Bonifacio, L.A.Lugiato: Lett. Nuovo Cimento **21**, 510 (1978);
K. Ikeda: Opt. Commun. **30**, 257 (1979);
L.A. Lugiato, M.L. Asquini, L.M. Narducci: Opt. Commun. **45**, 450 (1982)
The results of these studies can be applied to the case of a LIS by a simple change in sign of the pump parameter

4.25 L.M. Narducci, J.R. Tredicce, L.A. Lugiato, N.B. Abraham, D.K. Bandy: Phys. Rev. A**32**, 1588 (1985)

4.26 L.A. Lugiato, R.J. Horowicz, G. Strini, L.M. Narducci: Phys. Rev. A**30**, 1366 (1984)

4.27 In the case of a FRL, of course, the injected field is zero and the reference frequency ω_0 is replaced by the operating frequency ω_L of the laser, whose value is to be calculated from the state equations

4.28 V. Benza, L.A. Lugiato: Z. Phys. B**35**, 383 (1979)

4.29 L.A. Lugiato, P. Mandel, L.M. Narducci: Phys. Rev. A**29**, 1438 (1984)

4.30 The degree of agreement or disagreement between the exact and the approximate imaginary parts of the linearized eigenvalues is roughly the same as with the real parts. For this reason, and to save space, we have omitted plots of the imaginary parts of the eigenvalues

4.31 For a given value of Y, we use as initial conditions the final values of the variables corresponding to the preceeding run. In this way, a given trajectory is never far removed from its asymptotic configuration except when a given basin of attraction loses stability

4.32 Y. Gu, D.K. Bandy, J.M. Yuan, L.M. Narducci: Phys. Rev. A**31**, 354 (1985)

4.33 An extensive discussion of Lyapunov exponents can be found in G. Benettin, L. Galgani, A. Giorgilli, J. Strelcyn: Phys. Rev. A**14**, 2338 (1976); Meccanica **15**, 9 (1980);
I. Shimada, T. Nagashima: Prog. Theor. Phys. **61**, 1605 (1974)

4.34 Our own codes were developed on a PDP 11/23 minicomputer and executed on a PRIME 850 mainframe at the cost of several weaks of undivided attention by Professor Y. Gu; see [Ref. 4.32]

5. Experimental Observations of Single Mode Laser Instabilities in Optically Pumped Molecular Lasers

D.J. Biswas, R.G. Harrison, C.O. Weiss, W. Klische, D. Dangoisse, P. Glorieux, and N.M. Lawandy

With 8 Figures

Haken [5.1] predicted in 1975 that due to the mathematical equivalence of the equations describing a 2-level homogeneously-broadened single-mode laser and the Lorenz equations [5.2] describing convection and turbulence in heated fluids, chaotic emission should occur from these lasers. These equations describe the simplest of autonomous laser systems with the minimum number of degrees of freedom, namely three, for the generation of chaotic pulsations. However, the onset of instabilities in such systems requires not only a bad cavity condition but also a gain considerably above lasing threshold, thus making the experimental realization of this unfeasible for most lasers of this type.

Consequently attention has been given to alternative, though more complex systems (Chaps. 2–4) for which such restrictions are in part relaxed. Investigations here have yielded a wealth of identifiable dynamic instability phenomena some in reasonable agreement with theoretical predictions.

5.1 Background

Realization of such phenomena in the systems prescribed by Haken nevertheless remains especially appealing in view of its fundamental simplicity and recently, optically pumped far infrared lasers have been identified by *Weiss* and *Klische* [5.3] as perhaps the most promising candidates in this regard. These lasers were first obseved to exhibit persistent and damped self pulsing behaviour by *Lawandy* and *Koepf* in 1980 [5.4] in standing-wave systems in both the homogeneously broadened and inhomogeneously broadened limits of operation. The former case will be briefly discussed in Sect. 5.2. Other earlier observations of pulsing and damped oscillations in optically pumped CW molecular lasers have been summarized in [5.5]. As discussed below, in these systems the bad cavity conditions is automatically satisfied without the usual requirement of high-loss resonators due to their extremely narrow line broadening. Furthermore, such systems normally exhibit the high gain necessary for the generation of instability phenomena in single-mode homogeneously-broadened lasers.

However, as suggested by *Lawandy* [5.5] and shown by *Dupertuis* et al. [5.6], their equivalence to 2-level incoherently excited systems occurs only under limited conditions since optical pumping as distinct from electrical excitation involves coherent interaction between the pump and lasing transitions (see note added in proof [5.31]). The effect of coherent excitation results in several factors which must be considered carefully when drawing conclusions concerning routes to chaos and use of certain atom-field models. Notably pump-field induced Rabi splitting of the emission lineshape of a resonant, coherently pumped, homogeneously broadened three-level system has been shown by *Mehendale* and *Harrison* [5.7] to give rise to a new kind of spontaneous mode splitting leading to oscillatory instabilities in emission. In the absence of this splitting it is surprisingly found that at least for a three-level scheme in which the relaxation rates are equal the emission is stable even for relatively high excitation [5.8]. Furthermore coherent pump depletion and complicated velocity relaxation may result in absorptive contributions to the dispersion relation for mode frequencies.

Notwithstanding, these systems provide easy and sensitive control of operating parameters and are amenable to relatively straightforward analysis. Three modes of operation are readily classified; that of resonant pumping which under suitable operating conditions is identifiable with the Lorentz system; off-resonant pumping which gives two-photon Raman-laser action and thirdly, specific to Fabry-Perot system in which pump absorptions weak, instabilities may arise from excitation of molecules belonging to two distinct velocity groups when the pump laser is slightly detuned from the absorption line centre. These various aspects have been recently considered with regard to instability phenomena in both far-infrared systems and in mid-infrared systems, and are summarized below.

5.2 Optically Pumped Far Infrared Lasers

5.2.1 Resonant Pumping

The pressure at which these lasers operate optimally is very low because rotational relaxation that competes with the FIR laser emission is fast owing to the low transition energies. In the FIR the homogeneous linewidth is determined by pressure broadening alone since spontaneous emission plays no significant role. In these lasers the homogeneous linewidths then lie typically in the range of 100 kHz to a few megahertz. We note that the pump laser does not excite a pure single-velocity group but rather pumps molecules with a spread of velocities that are determined approximately by the power-broadened homogeneous linewidth of the pump transition and by the angular spread of the pump radiation. Thus each velocity group corresponds to a spread of several homogeneous linewidths of the IR frequencies. However, when this velocity spread is translated into the broadening of the FIR transition, the relative broadening is reduced by the ratio of the FIR to the IR frequencies and is small. As the homogeneous linewidths of the FIR and IR transitions are essentially equal, the relative inhomogeneous broadening of the FIR transition is negligible. Thus lasing occurs under essentially homogeneously broadened conditions. In such lasers bad cavity conditions are therefore easily satisfied without the requirement of high loss resonators due to their extremely narrow line broadening. This fact was first noted by *Weiss* and *Klische* [5.3].

Particularly interesting is the case of the optically pumped far-infrared NH_3 lasers, which due to the low partition function of the NH_3 molecule possesses a small signal gain 1–2 orders of mangitude larger than other FIR lasers and is thus ideal for the manifestation of Lorenz instabilities.

Appropriate to ensuring line-centre pumping, *Weiss* and *Klische* considered the $aQ(8,7)$ pump transition of $^{14}NH_3$ resonantly excited by the $10P(13)$ line of the N_2O laser which leads to emission at $81.5\,\mu m$. The major source of inhomogeneous broadening in this laser is the AC Stark effect brought about by the coherent pumping [5.7,13]; dynamic splitting varying for the different M components of the transition dependent on their dipole strength. It has been shown that the broadening due to this effect is less for the backward gain line and is of the order of $2\,MHz$. However, this effect can be overcome by raising the NH_3 pressure to $10\,Pa$, corresponding to $2\,MHz$ homogeneous linewidth. For experimental realisation of this *Weiss* and *Klische* also suggested the use of a FIR ring-laser cavity which could be made to emit in the forward as well as backward direction.

Tuning the pump-laser frequency allows one, via the Doppler effect, to choose the group velocity of molecules providing the laser gain. Pumping substantially off the pump-absorption line centre creates a situation where only the forward- or backward-emitting laser mode interacts with the inverted molecules, and the laser oscillates in a single travelling wave. Pumping close to or at pump line centre allows both modes to interact with the inverted molecules [5.14]. In this case, under CW conditions only one mode will oscillate due to competition. However, in the presence of instabilities both modes may emit. Thus, single-mode operation in the presence of instabilities requires off-centre pumping and a gain linewidth (which in general is AC Stark broadened, in addition to homogeneously broadened), substantially narrower than the FIR Doppler width. This gives an upper limit for the operating pressure and the pump intensity. Recently, they have confirmed these predictions with the observation of what are likely to be Lorenz instabilities from this laser [5.9]. Their scheme, a single-mode travelling-wave (ring cavity) FIR NH_3 laser is illustrated in Fig. 5.1.

At $9\,Pa$ pressure, when tuning the N_2O-lase frequency somewhat off line centre of the NH_3 (8,7) absorption, the forward and backward gain line profiles do not overlap anymore. The FIR laser could therefore be tuned across its whole (backward) oscillation bandwidth (of the TEM_{00} mode) at this pressure. Progression to chaos of the laser output, as the resonator is tuned towards gain line centre is sequentially shown in Fig. 5.2. Trace (a) shows the self-pulsation spectrum consisting of the fundamental frequency and its harmonics. Trace (b) shows a first series of subharmonics appearing when tuning is closer to line centre. Trace (c) shows a second series of subharmonics and trace (d) a third series of subharmonics. Trace (e) shows the appearance of noise, and trace (f) broad-

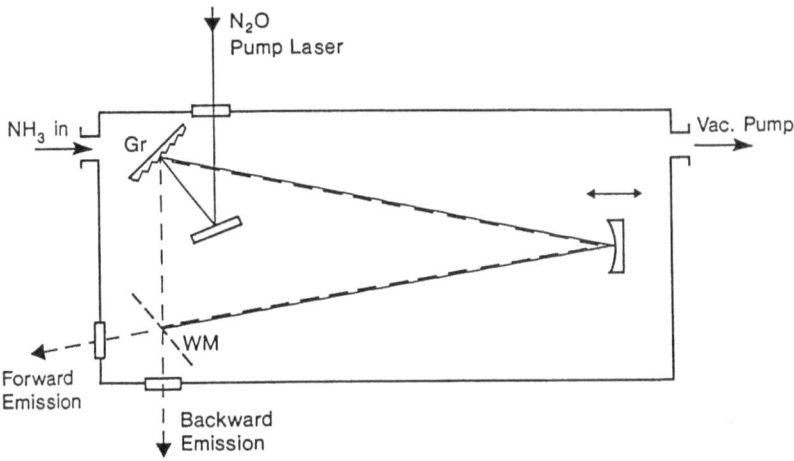

Fig. 5.1. Far infrared NH_3 ring laser. Perimeter of ring is 2 m, radius of curvature of mirror is 2 m. Grating has 80 grooves per mm, wire mesh WM has 30 wires per mm

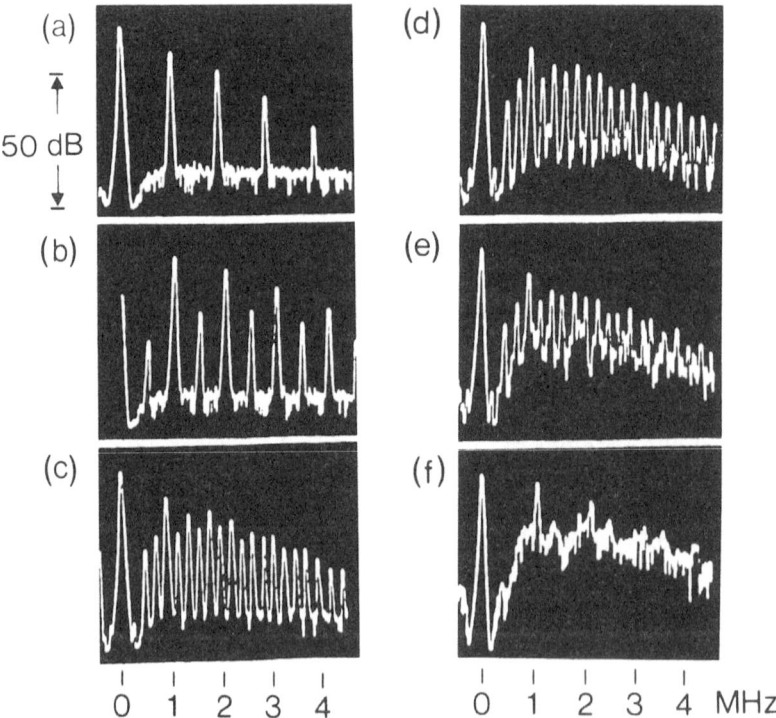

Fig. 5.2a–f. Sequence of subharmonic instabilities leading to chaos observed for NH_3 ring laser at 13 Pa NH_3 pressure and 4 W N_2O-laser pump power. Spectra (**a–f**) are recorded when tuning the FIR laser resonator progressively towards the gain line centre

band noise on which the pulse frequency is superimposed. This corresponds to tuning of the FIR resonator to the gain line centre. Some subharmonic transitions to chaos were found when tuning the cavity mode towards the line centre from either side. At pressures above 13 Pa no instability was observed though the laser output power was at a maximum at 25 Pa. This is attributed to the larger homogeneous broadening at higher pressure which ultimately precludes bad cavity conditions.

These results are intriguing since the observed period doubling sequences to chaos seems to conflict with most known solutions of the Lorenz model [5.1] which predicts abrupt changes from regular to chaotic emission (crisis [5.15]). On the other hand, *Zeghlache* and *Mandel* [5.16] have shown that a model including detuning of the laser yields a transition to chaos via a period doubling sequence. Furthermore, in a recent theoretical re-examination of the Lorenz model, *Narducci* et al. [5.17] predicted that for certain operating conditions periodic and chaotic emission may be observed for values of laser gain lower than those originally predicted. Subsequent observations of such instabilities by *Weiss* and collaborators [5.18a,b] for a system operated under conditions similar to those analysed by *Narducci* et al. are further support that their observed phenomena are of the Lorenz type (see note added in proof [5.31]).

5.2.2 Detuned Pumping

Conventional FIR laser cavities, in general, comprise Fabry-Perot cavity systems within which for typical operating pressures absorption of the pump radiation is weak. For such systems considerations must then also be given to effects arising from both the forward and the backward propagation of the pump. As discussed by *Lefebvre* et al. [5.19] bidirectional pumping brings about a two-peaked gain when the pump frequency is detuned with respect to the absorption line centre ensuring that the active molecules have a significant velocity component along the laser axis. When the FIR laser operates under conditions in which the homogeneous broadening is smaller than or comparable to the Doppler width of the transition on which the laser oscillates and the cavity is tuned to the centre frequency of the lasing transition ω_{FIR}^0, then the FIR gain curve exhibits two maxima when the FIR oscillation frequency is varied. This limit, which occurs when the Rabi frequency is not in excess of the narrowed forward gain spike width has been observed by harmonic mixing experiments by *Lawandy* and *Koepf* [5.4]. The heterodyne spectrum of a cavity scan of a Doppler-broadened CH_3F laser (496 μm) is shown in Fig. 5.3. In a Fabry-Perot cavity because of standing-wave effects, these maxima are both due to the two velocity groups mentioned above and occur approximately at frequencies $\pm\omega$ given by

$$\omega = \omega_{\text{FIR}}^0 \left(1 \pm \frac{\nu}{c}\right) = \omega_{\text{FIR}}^0 \left(1 \pm \frac{\Delta\omega_{\text{IR}}}{\omega_{\text{IR}}}\right) ,$$

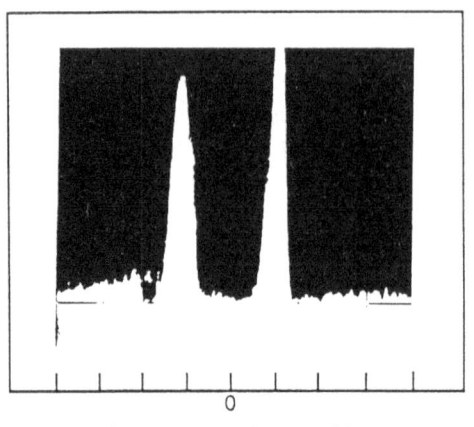

Fig. 5.3. Doppler-gain spike in bidirectionally pumped $^{12}CH_3F$ at 0.009 Torr. CO_2 laser is detuned by 30 MHz from the $Q(12, 2)$ transition

Frequency (0.5 MHz / Div)

where $\Delta\omega_{IR}$ is the detuning between the IR-pump radiation at frequency ω_{IR} and the IR-absorption transition through which population inversion is generated. As discussed earlier, each of the gain peaks is, to an excellent approximation, homogeneously broadened. From the associated unsaturated dispersion of the medium (Fig. 5.4b) it is easy to see that several different steady-state solutions are possible for the same number of wavelengths in the cavity thereby giving rise to the possibility of instabilities in laser emission.

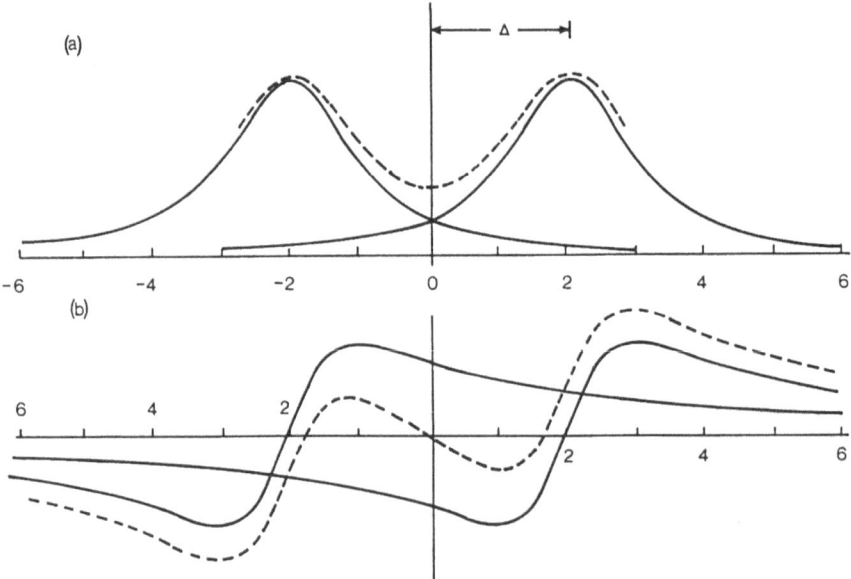

Fig. 5.4. (a) Unsaturated and (b) corresponding dispersion for two resonant groups of atoms with their resonant frequencies separated by a detuning parameter value of $\Delta = 2$. *Solid curves* show contributions of the two groups separately, and the *dashed curves* show the total gain and dispersion

114

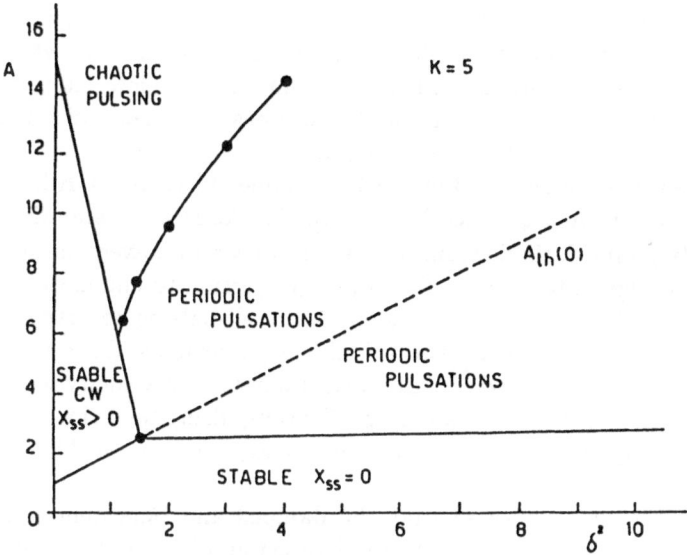

Fig. 5.5. Phase diagram in the A, δ^2 space showing regions of zero intensity or constant non-zero (CW) intensity as well as regions of stable and chaotic pulsing for $K = 5$

In such a system *Abraham* et al. [5.10] have recently reported oscillatory instabilities when the FIR laser resonator is tuned midway between the gain peaks. Their system, a $H^{12}COOH$ laser emitting at $742\,\mu m$ wavelength, was optically pumped by the $9R(40)$ line of CW CO_2 laser. Their results show that the period of oscillation decreases with decreasing CO_2 laser detuning, increasing FIR detuning, increasing CO_2 pump power, or increasing gas pressure. Theoretical modelling of these results places primary emphasis on the two symmetric resonances as an explanation for the observed instabilities. Although no account is taken of the complications of detuning, the standing-wave cavity, or coherent pump effects [5.20], agreement with the experimental results is excellent.

The analysis also predicts abrupt transitions from CW to large-amplitude pulsations which may be tested experimentally, perhaps by fixing the CO_2-laser frequency and varying the pumping intensity with attenuators. For large values of cavity decay rates (k) and $\delta^2 = 1.5$ where δ is far-infrared cavity detuning, the prediction is that chaos should be observed two to three times above the power levels required to see the initial laser action (which is unstable). The details are shown in the phase diagram plotted in Fig. 5.5. No intermediate bifurcations were observed in numerical results from a first scan of the parameter space, and this is as would be expected for chaos of the form of Type III intermittency. In fact, some chaotic behaviour has recently been experimentally observed although not exactly in the predicted region. However, in the Type III intermittency predicted, chaos appears through the growth of the second-harmonic content of the limit cycle while the experimental signals present no such second-harmonic content but indicate rather a transition to chaos through a tangent bifurcations.

In subsequent linear stability analysis of this system by *Wu* and *Mandel* [5.30] they determine up to four bifurcation diagrams. For one case the whole branch of finite intensity is unstable and in a second case the low intensity domain of the finite steady state is unstable. In both cases the laser displays periodic or chaotic output without ever reaching a stable steady output.

Self-pulsing has also been observed in the 500 μm line of the HCOOH laser in the heavily homogeneously broadened limit [5.5]. This laser was a standing-wave, bidirectionally pumped, single transverse mode, dielectric waveguide system. The laser was designed to utilize off-axis pump injection to eliminate instabilities due to pump feedback effects. The 1–5 MHz self-pulsing observed in these lasers cannot be explained via mode splitting arguments as above since they were pumped on line centre and the laser transition was not velocity selective due to the large homogeneous broadening. The self-pulsing behaviour has been observed for homogeneous to inhomogeneous broadening ratios as high as 5.

Recently, *Lawandy* has suggested that the damped and persistent self-pulsing observed in these systems is due to the interaction of a standing-wave mode and a distributed feedback mode (DFB) [5.21]. The high gain in these lasers, along with the low saturation intensities, leads to strongly modulated population gratings. This susceptibility grating is automatically self-adjusting to the additional boundary conditions imposed by the cavity as it is generated

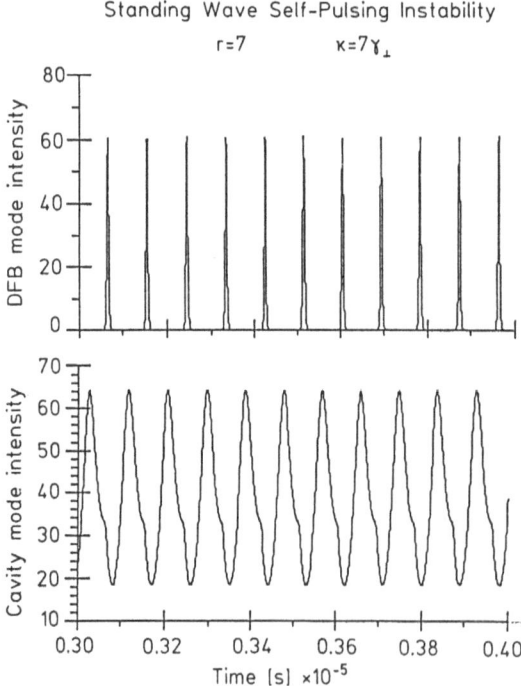

Fig. 5.6. Standing-wave self-pulsing instability. Cavity decay rate $K = 7\gamma_\perp$

by the cavity mode which satisfies these conditions. Calculations based on a modified DFB analysis which includes frequency-dependent susceptibilities indicate that when a DFB mode exists, its frequency is essentially clamped at line centre even when the cavity mode is detuned. A simple model for this interaction in the rate equation limit is given in [5.22] which shows several regimes of behaviour when the cavity decay rate is greater than $\gamma_{\parallel}$. At low pumping and a line centre tuned cavity mode, the system chooses to operate only on the cavity mode. When the cavity mode is detuned, there is a coexistence regime when the two types of modes exists simultaneously indicating that two frequencies will be simultaneously emitted by the laser. This frequency difference increases with detuning of the cavity mode but is always limited to a linewidth. Such a mode of operation is consistent with the persistent oscillations observed in $^{13}CH_3F$. Figure 5.6 shows a numerically generated example of the self-pulsing, which occurs due to the interaction of the cavity mode generated grating and its associated DFB mode. The parameters apply to a $^{13}CH_3F$ laser at 0.025 Torr of pressure and a cavity linewidth of 70 MHz. Nevertheless the model used here is crude and ignores coherence effects of two fields interacting with the atomic ensemble. Inclusion of these effects will generate new polarization components which will result in population pulsation phenomena and possibly a multi-frequency route to chaos.

5.2.3 Single-Mode Laser Instability
in Near Resonantly Pumped Mid-Infrared Systems

It has been noted earlier that a Raman process is responsible for the gain in many of the optically-pumped molecular lasers where chance coincidence of the pump signal frequency with the pump transition seldom ensures resonant conditions. Recently *Harrison* and *Biswas* [5.11,12] have demonstrated instabilities leading to chaos, in a single-mode homogeneously-broadened laser belonging to this class viz. an optically pumped mid-infrared laser. These effects have been obtained on two independent emitting transitions over a wide range of operating conditions, including those for optimum lasing, suggesting that this behaviour is, indeed, general for this class of laser. The operating pressure of the system as for most MIR lasers is much higher than for FIR systems, ensuring good homogeneously broadened conditions, providing the differences in AC Stark shift for the different M components of the transition are small. Also at these pressures due to high absorption of the pump, the possibility of bidirectional pumping considered earlier for Fabry-Perot systems is ruled out. As in the work of *Weiss* and his collaborators NH_3 gas was used as the active medium, selected on the basis of its well-documented spectroscopy and efficient lasing action. Mid-infrared lasing involves vibration-rotation transition and here the $aP(8,0)$ transition at 812 cm^{-1} optically pumped on the $aR(6,0)$ transition 1.3 GHz below line centre by the $9R(16)$ CO$_2$ lasing emission at 1076 cm^{-1} was selected [5.23]. This laser transition has been clearly identified as Raman in origin [5.24] for NH_3 pressures ~ 1–20 Torr and pump intensity of ~ 0.6 MHz/cm^2. The pressure-broadened bandwidth of the pump

200 ns

Fig. 5.7. Cavity tuning scan across the region ($\sim 1/4$ FSR) over which NH_3 lasing is obtained

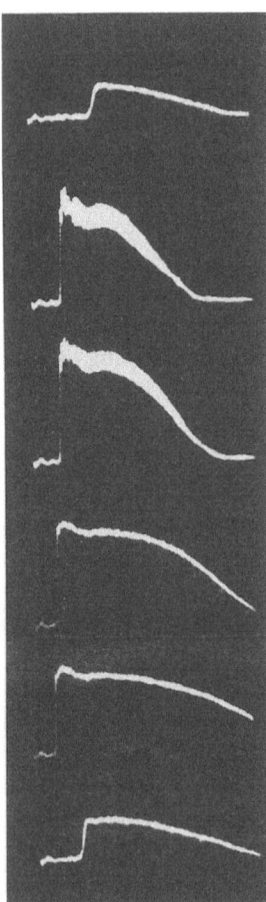

and lasing transition are 17.76 and 17.049 MHz/Torr, respectively [5.25], which for the typical operating pressure of ~ 8 Torr, considerably exceeds the Doppler bandwidth (74 MHz). A transversely excited atmospheric (TEA) CO_2 laser was used as the pump source and operated on a single transverse and axial mode to generate temporally smooth long pulses ($2\,\mu s$ FWHM with 250 kW peak power). The short optically pumped Fabry-Perot cavity (length ~ 25 cm) was provided with a piezoelectric tuning (PZT) facility; details of the cavity optics are provided in the appropriate figure captions.

Transverse scans of the NH_3 emission confirmed an essentially Gaussian spatial profile; indicating operation on the lowest order of longitudinal mode. For a pump intensity of ~ 500 kW/cm^2 lasing was obtained up to a pressure of 11 Torr corresponding to a pressure broadened gain-bandwidth of ~ 187 MHz, substantially smaller than the free spectral range (FSR) of the laser cavity viz. 600 MHz, thus ensuring a single mode condition.

Chaotic and periodic pulsation behaviour in the NH₃ emission was sensitive to cavity length tuning and occurred over NH₃ pressures of 5–9 Torr, smaller than the total range 3–11 Torr for lasing emission, most pronounced effects occurring at a pressure ~ 8 Torr. A typical PZT scan (over FSR) at this pressure is shown in Fig. 5.7. Asymmetrical occurrence of these effects with respect to cavity tuning is similar to the observation of *Weiss* et al. [5.26] for a far-infrared laser.

Within the narrow tuning range over which instability prevailed, two fundamental pulsation periods occurred for different PZT settings; one at ~ 3.8 ns period and the other of relatively long period ~ 18 ns. Straightforward mode-pulling considerations show that the high-frequency pulsation which leads to eventual chaos on fine cavity tuning is consistent with intermode beating of two cavity modes (cavity round-trip time 1.6 ns). More interestingly the slow periodic modulation (Fig. 5.8a) which exhibits distinct period doubling (Fig. 5.8b) with fine cavity length tuning before going into high period chaos (Fig. 5.8c) [5.27] is obtained under single-mode conditions. The cavity linewidth consider-

100ns
| |

50ns
| |

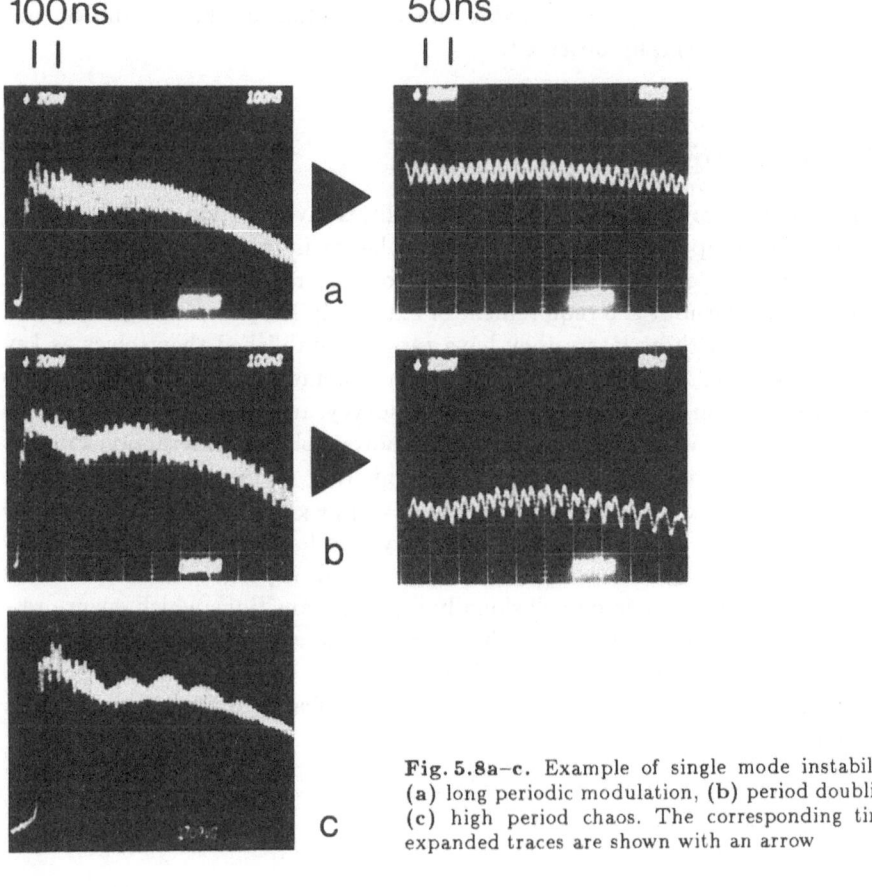

Fig. 5.8a–c. Example of single mode instability: (a) long periodic modulation, (b) period doubling, (c) high period chaos. The corresponding time-expanded traces are shown with an arrow

ations show that the phenomena are obtained only under bad cavity conditions.

The instabilities in this Raman system occur at a considerably lower threshold than that predicted for two-level systems, results indicating a gain of ~ 2 times above threshold for lasing at typical operating pressure of ~ 10 Torr. The mechanism of the instabilities obsrved here may be identified with mode splitting effects. From cavity tuning across the region of the FSR over which lasing persisted instabilities were found to occur only when the mode was detuned by ≥ 30 MHz from the line centre; suggesting that induced mode splitting was not responsible for the observed phenomena. Rather passive mode splitting, which occurs far from the line centre for a sufficiently large value of the mode splitting factor (β) would appear more likely [5.28].

Test experiments, conducted using a dispersive Fabry-Perot system, which could sustain only NH_3 emission, also gave laser instabilities confirming that these effects arise from the active optical cavity alone.

Quantitative understanding of Raman lasers is in a somewhat embryonic state. Aspects of their operation have been recently discussed by *Lawandy* [5.5] and *Mehendale* et al. [5.8,29]. Earlier, various authors have quantified the gain dynamics of these system under steady-state conditions. Extension of such approaches to real-time analysis of these molecular pumped systems contained within optical resonators will enable characterisation of the instability phenomenon experimentally observed.

5.3 Conclusion

In the rapidly growing field of laser-instabilities investigation of these effects in optically pumped mid- and far-infrared lasers is in its infancy. However, with the recognition that bad cavity conditions are readily satisfied in most of these systems without the requirement of extremely lossy resonators and given the high gains of these lasers they have rapidly established themselves as key candidates in this field. As relatively simple systems they also provide highly versatile operation, as indicated in this brief survey, and are readily amenable to quantitative theoretical analysis. The two-photon coherent interaction implicit to the operation of many of these systems identify them as distinct in many respects from more conventional lasers involving, e.g., electrical excitation. As such they provide an exciting new class of system for continued investigation. However, pertaining to the simple Haken system this is perhaps most effectively realised in these lasers using collisionally relaxed transitions which ensure resonant lasing emission and in which coherent interaction between the pump and lasing signal is automatically eliminated. This will undoubtedly prove a profitable avenue for further investigation in far infrared and perhaps also mid infrared systems.

References

5.1 H. Haken: Phys. Lett. **53A**, 77 (1975)
5.2 E.N. Lorenz: J. Atoms Sci. **20**, 130 (1963)
5.3 C.O. Weiss, W. Klische: Opt. Commun. **51**, 47 (1984)
5.4 N.M. Lawandy, G.A. Koepf: IEEE J. QE-**16**, p. 701 (1980)
5.5 N.M. Lawandy: J. Opt. Soc. Am. B**2**, 108 (1985)
5.6 M.A. Dupertuis, R.R.E. Salomaa, M.R. Seigrist: Opt. Commun. **57**, 410 (1986);
   M.A. Dupertuis, M.R. Seigrist, R.R.E. Salomaa: In *Optical Chaos*, ed. by J. Chrostowski, N.B. Abraham, SPIE Proc. **667** (1986)
5.7 S.C. Mehendale, R.G. Harrison: Phys. Rev. A**34** (1986)
5.8 S.C. Mehendale, R.G. Harrison: In *Optical Chaos*, ed. by J. Chrostowski, N.B. Abraham, SPIE Proc. **667** (1986)
5.9 C.O. Weiss, W. Klische, P.S. Ering, M. Cooper: Opt. Commun. **52**, 405 (1985)
5.10 N.B. Abraham, D. Dangoisse, P. Glorieux, P. Mandel: J. Opt. Soc. Am. B**2**, 23 (1985)
5.11 R.G. Harrison, D.J. Biswas: Phys. Rev. Lett. **55**, 63 (1985)
5.12 D.J. Biswas, R.G. Harrison: Appl. Phys. Lett. **47**, 198 (1985) and Opt. Commun. **54**, 112 (1985)
5.13 J. Heppner, C.O. Weiss, U. Hubner, G. Schinn: IEEE J. QE-**16**, 392 (1980)
5.14 J. Heppner, C.O. Weiss: Appl. Phys. Lett. **33**, 590 (1978)
5.15 C. Grebogi, E. Ott, J.A. Yorke: Phys. Rev. Lett. **48**, 1507 (1982)
5.16 H. Zeghlache, P. Mandel: J. Opt. Soc. Am. B**2**, 18 (1985)
5.17 L.M. Narducci, H. Sadiky, L.A. Lugiato, N.B. Abraham: Opt. Commun. **55**, 370 (1985)
5.18 E.H.M. Hogenboom, W. Klische, C.O. Weiss, A. Godone: Phys. Rev. Lett. **55**, 2571 (1985);
   C.O. Weiss: In *Optical Chaos*, ed. by J. Chrostowski, N.B. Abraham, SPIE Proc. **667** (1986)
5.19 M. Lefebvre, D. Dangoisse, P. Glorieux: Phys. Rev. A**29**, 758 (1984)
5.20 N.M. Lawandy, G.A. Koepf: Phys. Rev. A**25**, 433 (1982)
5.21 N.M. Lawandy, W.S. Rabinovich: Conf. on Instabilities and Dynamics of Laser and Nonlinear Optical Systems, Rochester NY (1985)
5.22 N.M. Lawandy, W.S. Rabinovich, C. Adler: Opt. Commun.
5.23 T.Y. Chang, J.D. McGee: Appl. Phys. Lett. **29**, 725 (1976)
5.24 H.D. Morrison, B.K. Garside, J. Reid: IEEE J. QE-**20**, 1060 (1984)
5.25 F.W. Taylor: J. Quant. Spectrosc. Radiat. Transfer **13**, 1181 (1973)
5.26 C.O. Weiss, W. Klische: Opt. Commun. **50**, 413 (1984)
5.27 N.B. Abraham et al.: In *Laser Physics*, ed. by J.D. Harvey, D.E. Walls, Lectur Notes Phys. **182**, 88 (Springer, Berlin, Heidelberg 1983)
5.28 S.T. Hendow, M. Sargent: J. Opt. Soc. Am. B**2**, 8A (1985)
5.29 S.C. Mehendale, R.G. Harrison, A. Vass: Appl. Phys. Lett. **48**, 894 (1986)
5.30 X.G. Wu, P. Mandel: J. Opt. Soc. Am. B**3**, 724 (1986)
5.31 Note added in proof: The reader is also referred to the recent work:
   N.M. Lawandy, J.C. Ryan: "Limitations on coherently pumped molecular systems for studying two-level laser instabilities". Opt. Commun. (in press);
   C.O. Weiss, J. Brock: "Evidence for Lorentz-Type chaos in lasers". Phys. Rev. Lett. **57**, 2804 (1986);
   J.V. Moloney, J.S. Uppal, R.G. Harrison: "Origin of chaotic relaxation oscillations in an optically pumped molecular laser". Phys. Rev. Lett. (submitted)

6. Quantum Treatment
of Amplified Spontaneous Emission
in High-Gain Free-Electron Lasers

R. Bonifacio and F. Casagrande

The existence of a collective instability for an unbunched, monokinetic electron beam and no field excitation leads to an exponential growth of radiation in the free-electron laser (FEL) process. A high-gain FEL has been recently operated at Livermore in this amplified spontaneous emission regime. We have developed a fully quantum treatment which includes the initial stage of this regime so that we can describe the onset of the process, evaluate the effect of fluctuations on the build-up time and the power level of radiation, and give the photon statistics. In a proper limit of our treatment we recover the classical results in which the source of fluctuations is the electron shot-noise.

6.1 Introduction

In the basic configuration of a free-electron laser (FEL) [6.1] a beam of relativistic electrons interacts with a radiation field and a transverse, spatially periodic magnetostatic field in such a way that the radiation gets amplified while the particles are predominantly decelerated. This device can operate both as an amplifier in a single-pass process and as an oscillator if the interation takes place in and optical cavity. FEL can produce powerful coherent radiation from the microwave region up to the VUV and soft x-ray region. Broad tunability, high efficiency and possibility of operation at wavelengths not accessible to atomic and molecular lasers make this device a very attractive new laser source, with a wide spectrum of potential applications.

In the pioneering experiments by *Madey* and collaborators [6.1], like in the first FEL radiating in the visible [6.2], the FEL process can be described in terms of single-particle dynamics; the electrons radiate independently and the gain is due to an interference effect in which only a tiny fraction of the electron kinetic energy is transferred to the radiation field. On the other hand, a *high-gain* FEL has been recently operated [6.3], in which the particles are correlated by the common radiation field so that the emission of radiation is a cooperative effect.

The high-gain regime has been obtained in a single-pass process which starts from noise. Actually, the existence of a *collective instability* [6.4] for a monokinetic unbunched beam and no initial field excitation leads to an exponential growth of the emitted radiation before saturation sets in. This mode of operation is called *amplified spontaneous emission* (ASE) [6.5,6]. Though the first FEL working in the ASE regime radiates in the microwave region, it should be possible to scale this device to the visible [6.3]. Furthermore, this mode of operation should provide coherent radiation even in the VUV and soft x-ray regions [6.7,8], where it presents the definite advantage that it does not require mirrors, whose efficiency is rather poor at such short wavelengths. To this end, severe but not unattainable conditions must be imposed on the electron beam quality [6.7].

The classical theory of ASE is well understood [6.5–7]. We have recently developed a fully quantum treatment which can describe the initial stage of the exponential gain regime [6.9–11]. Quantum treatments have been applied to different FEL regimes [6.12]. In the ASE regime a quantum treatment is particularly relevant to describe the onset of the process and evaluate the effect of fluctuations on the build-up time and the power level of radiation. Furthermore, it allows the derivation of the photonstatistics of the field. On the other hand, the classical results can be recovered in a proper limit of our treatment.

We start by introducing the many-electron quantized Hamiltonian from which we derive the Heisenberg equations for the dynamics of the system (Sect. 6.2). In the classical approximation, discussed in Sect. 6.3, we recover the equations of the travelling-wave-tube type [6.5] and review some basic results of the classical analysis of the ASE regime. Next, in Sect. 6.4 we introduce electron collective operators [6.11] which allow a fully quantum description of the linear regime when the system is unstable. This is sufficient to our purposes, since the classical calculations [6.5] show the validity of the linearized equations nearly up to the first peak of the radiated intensity and therefore the possibility to evaluate the build-up (or delay) time by extrapolation of the linear dynamics.

In Sect. 6.5, assuming that each electron is initially in a minimum-uncertainty wave-packet and that the system operates on resonance, we obtain expressions of the mean photon number and of other relevant quantities which show the onset of the process due to quantum fluctuations of the electron positions and momenta. These expressions are discussed in two opposite limits.

First we define a "harmonic oscillator" limit in which initially each electron is well localized in a radiation wavelength. In this limit, discussed in Sect. 6.6, the dynamics of the system is that of three parametrically coupled harmonic oscillators. We derive the photon statistics of the field, obtaining a displaced Gaussian for the Glauber distribution $P(\alpha)$ which is typical of the superposition of a chaotic field with a coherent field. Hence, if the system starts from noise we obtain a Bose-Einstein photonstatistics, in agreement with the result first obtained by *Becker* and *McIver* [6.13] when the system operates in the stability region. Also, it follows that the radiation exhibits only first-order coherence, while there is no evidence of nonclassical effects such as antibunching or squeezing. On imposing that the build-up time does not exceed the interaction time, we give the threshold condition for ASE starting from quantum fluctuations. This condition turns out to be more severe than the corresponding classical condition.

In Sect. 6.7 we introduce an opposite, "classical" limit in which initially the particles are delocalized on the scale of a radiation wavelength. In this limit we succeed in deriving a fit formula for the build-up time of the field, obtained in the classical analysis [6.5] in which the source of fluctuations is the electron shot-noise, that is the randomness of electron positions in the beam. In this case, a classical treatment including shot noise appears to be fully adequate, in agreement with the conclusions of *Becker* and *McIver* [6.14]. Concluding remarks concern the connection of our results with those of [6.12–15] on the validity of a classical description.

6.2 The Quantum Hamiltonian Model

We consider a one-dimensional model of single-passage FEL described by the following N-electron, one-mode quantized Hamiltonian in a reference frame moving with the mean electron velocity:

$$H = \sum_{i=1}^{N} \left(\frac{\bar{p}_i^2}{2} \right) + iw \left[a^+ \sum_{J=1}^{N} \exp\left(-i\theta_J\right) - \text{h.c.} \right] - \Delta a^+ a , \qquad (6.1)$$

where the dimensionless variables and parameters are expressed in terms of the laboratory frame [Ref. 6.11, Appendix A] by

$$\theta_i = (K + K_0)z_i - \left(\frac{\omega}{2\omega_0}q + \Delta \right)\tau , \quad \bar{p}_i = \frac{p_i - \langle p \rangle_0}{\hbar(K + K_0)} ,$$

$$a = iE_0 \exp\left[i(\phi_0 + \Delta\tau)/[(4\pi/V)\hbar\omega]^{1/2}\right] ,$$

$$\tau = 2\frac{\omega_0}{q}t , \quad \Delta = \frac{\langle p \rangle_0 - p_R}{\hbar(K + K_0)}$$

$$q = \frac{\gamma_0 m_0 c}{\hbar(K + K_0)} , \quad w = \left(\frac{ec}{\omega_0} \right)^2 \frac{B_0}{2} \left(\frac{\pi}{V(\hbar\omega)^3} \right)^{1/2} \equiv \frac{(\varrho q)^{3/2}}{\sqrt{N}}$$

$$\varrho = \left(\frac{\mathbb{K}\,\Omega_{\text{p}}}{4\,\omega_0} \right)^{2/3} , \quad \mathbb{K} = \frac{eB_0}{m_0 c\omega_0} , \quad \Omega_{\text{p}} = \left(\frac{h\pi e^2 N}{m_0 V \gamma_0^3} \right)^{1/2} . \qquad (6.2)$$

Here, z_i and p_i are the position and momentum of the ith electron, γ_i its energy (in rest energy units); all electrons are assumed to have the same initial momentum, $\langle p_i \rangle_0 = \langle p \rangle_0$, and the same energy, $\gamma_i(0) = \gamma_0$; p_R and γ_R are the resonant momentum and energy parameter $\gamma_R = p_R/m_0 c = [(\omega/2\omega_0)(1 + \mathbb{K}^2)]^{1/2}$, where $\omega = ck$ is the radiation field frequency, $\omega_0 = ck_0 = 2\pi cN_0/L_0$ is the frequency corresponding with the periodicity of the wiggler of length L_0, number of periods N_0 and dimensionless magnetostatic amplitude $\mathbb{K}$; we assume that

$$\frac{p_i - \langle p \rangle_0}{\langle p \rangle_0} \ll 1 , \quad \frac{\langle p \rangle_0 - p_R}{\langle p \rangle_0} \ll 1 . \qquad (6.2')$$

Furthermore, a is the annihilation operator of the radiation field, such that $\hbar\omega a^+ a = |E_0|^2 V/4\pi$, with E_0 and ϕ_0 being the classical amplitude and phase of the field; Δ is the detuning parameter, w the coupling constant which depends on the "quantum" parameter q and on the strength or generalized Pierce parameter ϱ; Ω_p is the relativistic plasma frequency.

From the Hamiltonian (6.1), with the help of the commutation relations for the electron phases and momenta and for the field mode operators

$$[\theta_i, \bar{p}_J] = i\delta_{ij} , \quad [a, a^+] = 1 , \tag{6.3}$$

we derive the following Heisenberg evolution equations

$$
\begin{aligned}
d\theta_i/d\tau &= \bar{p}_i \quad (i = 1, \dots, N) , \\
d\bar{p}_i/d\tau &= -w[a\exp(i\theta_i) + \text{h.c.}] \quad (i = 1, \dots, N) , \\
da/d\tau &= w \sum_{J=1}^{N} \exp(-i\theta_J) + i\Delta a .
\end{aligned}
\tag{6.4}
$$

A constant of motion is

$$\sum_{i=1}^{N} \bar{p}_i + a^+ a = \text{const.} , \tag{6.4'}$$

namely the (dimensionless) total momentum. Equation (6.4') illustrates the basic mechanism of the FEL process in this model, that is the exchange of momentum between the electrons and the photons.

6.3 Classical Limit and Classical Treatment of Amplified Spontaneous Emission

In the classical approximation, that is when operators are replaced by c-number variables while fluctuations and correlations are neglected, (6.4) coincide with the well-known FEL equations of the travelling-wave-tube type [6.5] in the limit

$$(\gamma_i - \gamma_0)/\gamma_0 \ll 1 . \tag{6.5}$$

We can write the classical equations in the same form as (6.4), namely

$$
\begin{aligned}
d\theta_i/d\tilde{\tau} &= \eta_i , \\
d\eta_i/d\tilde{\tau} &= -\tilde{w}[\alpha\exp(i\theta_i) + \text{c.c.}] , \\
d\alpha/d\tilde{\tau} &= \tilde{w} \sum_{J=1}^{N} \exp(-i\theta_J) + i\tilde{\Delta}\alpha ,
\end{aligned}
\tag{6.6}
$$

where

$$\theta_i = (K + K_0)z_i - \left(\frac{\omega}{2\omega_0} + \tilde{\Delta}\right)\tilde{\tau} , \quad \eta_i = (\gamma_i - \gamma_0)/\gamma_0 \simeq (p_i - p_0)/p_0 ,$$

$$\alpha = iE_0 \exp\left[i(\phi_0 + \tilde{\Delta}\tilde{\tau})\right]/[(4\pi/V)\gamma_0 m_0 c^2]^{1/2} ,$$

$$\tilde{\tau} = 2\omega_0 t , \quad \tilde{\Delta} = (\gamma_0 - \gamma_R)/\gamma_0 , \quad \tilde{w} = \varrho^{3/2}/\sqrt{N} , \tag{6.7}$$

and the slowly-varying amplitude and phase approximation has been used to derive the field equation. Note that (6.6) depend only on the detuning parameter and the Pierce parameter (included in the coupling constant w), that is the very basic quantity in the model. Formally, the quantum equations (6.4) take the form (6.6) by the replacements

$$\bar{p}_i \to q\eta_i , \quad a \to \sqrt{q}\alpha ,$$

$$\tau \to \tilde{\tau}/q , \quad w \to q^{3/2}\tilde{w} , \quad \Delta \to q\tilde{\Delta} . \tag{6.8}$$

Note that the quantum-classical correspondence (6.8) is ruled by the parameter q, (6.2). This parameter, which contains the Planck constant h, is the inverse of the electron recoil in the emission of a photon; hence it is physically sound that it disappears from the equations in the classical approximation.

Equation (6.6) is the basic classical set of equations to describe amplified spontaneous emission. Actually, the initial condition with no field excitation and monokinetic unbunched electron beam

$$\alpha(0) = 0 , \quad \eta_i(0) = 0 , \quad \sum_{J=1}^{N} \exp\left(-i\theta_J^0\right) = 0 \tag{6.9}$$

is an equilibrium condition for the evolution equations (6.6). Let us introduce the scaled time $\bar{\tau}$ and the detuning parameter δ

$$\bar{\tau} = \varrho\tilde{\tau} = 2\omega_0\varrho t , \quad \delta = \tilde{\Delta}/\varrho \simeq (1/\varrho)(\gamma_0 - \gamma_R/\gamma_R) . \tag{6.10}$$

The linear stability analysis shows that the system is stable for values of the detuning parameter

$$\delta > \delta_T \equiv 3/2^{2/3} \simeq 1.89 \tag{6.11}$$

and unstable for $-\infty < \delta < \delta_T$ [6.16]. In the former case the FEL is "below threshold" because the radiated intensity is merely oscillating in time; in this regime, under proper approximations, one recovers the well-known small signal gain [6.1]. In the latter case, the FEL is "above threshold" and exhibits an exponential gain before saturation effects set in. One can describe these different regimes in terms of the *electron bunching parameter*

$$b = N^{-1} \sum_{J=1}^{N} \exp\left(-i\theta_J\right) \simeq N_\lambda^{-1} \sum_{J=1}^{N_\lambda} \exp\left(-i\theta_J\right) , \quad 0 \le |b| \le 1 , \tag{6.12}$$

where N_λ is the number of particles in one optical wavelength. The quantity (6.12) is the source term in the field equation and plays a role analogous to that of polarization in atomic lasers. In the small-signal regime the field does not correlate the particles, so that an initially unbunched electron beam ($b_0 = 0$), remains almost completely unbunched during the whole process. By contrast, in the collective unstable regime particles are correlated through the common radiation field and self-bunching occurs such that b reaches values close to one; this is the regime of *amplified spontaneous emission*. We recall that we have investigated, and obtained evidence of, the occurrence of Hamiltonian chaos in the classical nonlinear FEL equations when the system is unstable [6.17].

A crucial point is that in order to integrate (6.6) the system must be tilted from equilibrium, e.g. by introducing an initial value of the bunching parameter $b_0 \neq 0$. In other words, a classical simulation of noise must be introduced by hand. Now, if we define a threshold condition for ASE by imposing that the build-up time of the field $\bar{\tau}_D$, that is the time of the first peak of radiated intensity, does not exceed the transit time in the wiggler:

$$\bar{\tau}_D \lesssim 2\omega_0\varrho L_0/c = 4\pi\varrho N_0 , \tag{6.13}$$

this threshold depends critically on the initial noise level, since the numerical results show strong fluctuations of the delay time $\bar{\tau}_D$.

Another key point is that the time evolution of the field in the unstable regime that follows from numerical integration of the *linearized* equations fits nearly up to the first peak the evolution calculated from the full nonlinear equations. More precisely, if $|b(\bar{\tau})|$ diverges as $|b_0|\exp|\text{Im}\{\bar{\lambda}\}\bar{\tau}|$, where $\bar{\lambda}$ is one root of the cubic characteristic equation associated with the linear stability analysis around the equilibrium condition (6.8), one can evaluate the delay time simply on imposing $|b(\bar{\tau}_D)| \simeq 1$, obtaining $\bar{\tau}_D \simeq -(\text{Im}\{\bar{\lambda}\})^{-1}\ln|b_0|$. This evaluation turns out to be in agreement with the numerical solution of the *nonlinear* equations since a fit formula is [6.5]

$$\bar{\tau}_D \simeq -(\text{Im}\{\bar{\lambda}\})^{-1}\ln|b_0| + 1 , \tag{6.14}$$

for a very wide range of small values of b_0 which simulate noise. Hence the linear analysis is valid up to the first peak, giving the correct evaluation of the delay time. The problem is to evaluate b_0 which, according to (6.9), would be zero if θ_j^0 were given numbers. The simplest way of introducing noise is as follows: if we consider θ_j^0 as random quantities and define b_0 which appears in (6.14) as the square root of the ensemble average of

$$|b_0|^2 = N_\lambda^{-2}\sum_{i,j} \exp\left[\text{i}(\theta_i^0 - \theta_j^0)\right] ,$$

only the terms with $i = j$ will survive, leading to the shot-noise value

$$|b_0| = 1/\sqrt{N_\lambda} . \tag{6.15}$$

If, e.g., the system operates on resonance ($|\delta|\to 0$), $\text{Im}\{\bar{\lambda}\} = \sqrt{3}/2$ so that the delay time (6.14) and the threshold condition (6.13) become

$$\bar{\tau}_D \simeq (1/\sqrt{3})\ln N_\lambda + 1 \; , \tag{6.16a}$$

$$(1/\sqrt{3})\ln N_\lambda \lesssim 4\pi\varrho N_0 \; . \tag{6.16b}$$

In the next sections we derive (6.16) in a suitable classical limit of a general linear quantum treatment which, according to the classical results, is valid up to the first radiation peak allowing the evaluation of the delay time, threshold and photonstatistics.

6.4 The Basic Equations of the Linear Quantum Regime

Let us introduce the electron collective operators [6.11]

$$\Theta = i\left(\frac{\bar{\varrho}}{N_\lambda}\right)^{1/2} \sum_{J=1}^{N_\lambda} \exp\left(-i\theta_J\right) \; ,$$

$$P = \frac{1}{(\bar{\varrho}N_\lambda)^{1/2}} \sum_{J=1}^{N_\lambda} \exp\left(-i\theta_J\right)\bar{p}_J \tag{6.17}$$

where

$$\bar{\varrho} = \varrho q \; . \tag{6.18}$$

The quantum equations (6.4) can be written in terms of the collective operators (6.17) as follows:

$$d\Theta/d\bar{\tau} = P \; , \tag{6.19a}$$

$$dP/d\bar{\tau} = -a \; , \tag{6.19b}$$

$$da/d\bar{\tau} = -i\Theta + i\delta a \; , \tag{6.19c}$$

with $\bar{\tau}$ and δ defined in (6.10), where we consider a, $\bar{p}_i$ and $\sum_J \exp\left(-i\theta_J\right)$ as fluctuation operators, i.e. we consider initial states for the electrons and the field such that

$$\langle a\rangle_0 = \langle \bar{p}_i\rangle_0 = \sum_J \exp\left(-i\langle\theta_J\rangle_0\right) = 0 \; , \tag{6.20}$$

that is the quantum counterpart of (6.9), and where the higher-order quantities $a^+ \sum_J \exp\left(-2i\theta_J\right)$ and $\sum_J \exp\left(-i\theta_J\right)\bar{p}_J^2$ were neglected in (6.19b).

Looking for solutions of the linear system (6.19) of the form $\Theta(\bar\tau) = \exp(i\lambda\bar\tau)\Theta_0$, we find the same cubic characteristic equation of the classical analysis, namely

$$\lambda^3 - \delta\lambda^2 + 1 = 0 \ . \tag{6.21}$$

Once the three roots λ_1, λ_2, λ_3 of (6.21) are known, we can write, e.g., the field amplitude a and the collective operator θ in the form

$$a(\bar\tau) = f_1(\bar\tau)\Theta_0 + if_2(\bar\tau)P_0 - f_3(\bar\tau)a_0 \ , \tag{6.22}$$

$$\Theta(\bar\tau) = h_1(\bar\tau)\Theta_0 + ih_2(\bar\tau)P_0 + h_3(\bar\tau)a_0 \ . \tag{6.23}$$

The explicit expressions of f_i and h_i are given in Appendix 6.A, while the initial values $f_{1,2}(0) = 0$, $f_3(0) = -1$ and $h_1(0) = 1$, $h_{2,3}(0) = 0$ verify the initial conditions.

From (6.22) and its Hermitian conjugate equation we obtain the expression of the mean photon number

$$\langle a^+ a\rangle(\bar\tau) = |f_1(\bar\tau)|^2\langle\Theta^\dagger\Theta\rangle_0 + |f_2(\bar\tau)|^2\langle P^\dagger P\rangle_0$$
$$+ i[(f_1^* f_2)(\bar\tau)\langle\Theta^\dagger P\rangle_0 - \text{h.c.}] + |f_3(\bar\tau)|^2\langle a^+ a\rangle_0 \ , \tag{6.24}$$

where no correlation has been assumed initially between the electrons and the field. Equation (6.24) can be written with obvious notations

$$\langle n\rangle = \langle n\rangle_{\text{sp}} + \langle n\rangle_{\text{st}} \ , \tag{6.25a}$$

$$\langle n\rangle_{\text{sp}} = |f_1|^2\langle\Theta^\dagger\Theta\rangle_0 + |f_2|^2\langle P^\dagger P\rangle_0 + i(f_1^* f_2\langle\Theta^\dagger P\rangle_0 - \text{h.c.}) \ , \tag{6.25b}$$

$$\langle n\rangle_{\text{st}} = |f_3|^2 n_0 \ . \tag{6.25c}$$

Equation (6.25b) gives the mean photon number when the system starts radiating from the vacuum state, that is, it represents the *spontaneous emission* that arises from the electron fluctuations, which turns out to be composed of position fluctuations, momentum fluctuations and cross position-momentum fluctuations; this latter contribution to the start-up of the FEL process could not even be simulated in a classical treatment of FEL dynamics. Equation (6.25c) is the *stimulated* contribution to the radiated intensity, which is present (and eventually dominates) if the field does not evolve from vacuum. Note that if $n_0 = 0$ the radiation starts from the vacuum state since $\langle n\rangle_{\text{sp}}(0) = 0$.

From (6.22) and its h.c. equation one derives an expression for $\langle\Theta^\dagger\Theta\rangle(\bar\tau)$ which has the same form of (6.24), provided that the functions f_i are replaced by the functions h_i. Hence, if we define an *electron bunching operator*

$$\hat{b} = N^{-1} \sum_{J=1}^{N} \exp\left(-i\theta_J\right) \simeq N_\lambda^{-1} \sum_{J=1}^{N_\lambda} \exp\left(-i\theta_J\right) , \qquad (6.26)$$

the expectation value $\langle \hat{b}^\dagger \hat{b} \rangle$, that is the quantum analog of $|b|^2$, reads

$$
\begin{aligned}
\langle \hat{b}^\dagger \hat{b} \rangle(\bar{\tau}) &= (\bar{\varrho} N_\lambda)^{-1} \langle \Theta^\dagger \Theta \rangle(\bar{\tau}) \\
&= (\bar{\varrho} N_\lambda)^{-1} \{ |h_1(\bar{\tau})|^2 \langle \Theta^\dagger \Theta \rangle_0 + |h_2(\bar{\tau})|^2 \langle P^\dagger P \rangle_0 \\
&\quad + i[(h_1^* h_2)(\bar{\tau}) \langle \Theta^\dagger P \rangle_0 - \text{h.c.}] + |h_3(\bar{\tau})|^2 n_0 \} .
\end{aligned}
\qquad (6.27)
$$

Before evaluating the expectation values in the unstable regime of ASE, we notice that in the stable region, if we (i) neglect the spontaneous contribution to the field, so that $a(\bar{\tau}) = -f_3(\bar{\tau}) a_0$ and $\langle n \rangle(\bar{\tau}) = |f_3(\bar{\tau})|^2 n_0$, and (ii) take the limits $\delta \gg 1$ and $\delta/\bar{\tau} \ll 1$, we derive an expression of the radiated intensity $\langle n \rangle$ from which we obtain the well-known small-signal gain [6.1].

6.5 Analysis on Resonance and for Sufficiently Long Times

In this section we concentrate on the unstable region. First of all we consider the resonance limit $|\delta| \to 0$, in which the exponential growth rate is maximum [6.4,5]. In this case the explicit expressions of $f_i(\bar{\tau})$ and $h_i(\bar{\tau})$ are easily derived from the exact roots of the cubic equation (6.21) (the expressions of the relevant quantities which appear in (6.24,27) are reported in App. 6.A). In particular, if we keep only the divergent contribution for sufficiently long times $\bar{\tau} \gg 1$, we find the asymptotic expressions

$$
|f_i(\bar{\tau})|^2 \simeq |h_i(\bar{\tau})|^2 \simeq (1/9) \exp\left(\sqrt{3}\bar{\tau}\right) ,
$$
$$
f_1^*(\bar{\tau}) f_2(\bar{\tau}) \simeq h_1^*(\bar{\tau}) h_2(\bar{\tau}) \simeq -(1/18) \exp\left(\sqrt{3}\bar{\tau}\right) .
\qquad (6.28)
$$

Substitution of (6.28) into (6.24 and 27) gives (from now on we shall be always in the limits $|\delta| \to 0$, $\bar{\tau} \gg 1$):

$$
\langle n \rangle(\bar{\tau}) \simeq (1/9)[\langle \Theta^\dagger \Theta \rangle_0 + \langle P^\dagger P \rangle_0 - (i/2)
$$
$$
\times (\langle \Theta^\dagger P \rangle_0 - \text{h.c.}) + n_0] \exp\left(\sqrt{3}\bar{\tau}\right) ,
\qquad (6.29a)
$$

$$
\langle \Theta^\dagger \Theta \rangle(\bar{\tau}) = \langle n \rangle(\bar{\tau}) \to \langle \hat{b}^\dagger \hat{b} \rangle(\bar{\tau}) = (N_\lambda \bar{\varrho})^{-1} \langle n \rangle(\bar{\tau}) .
\qquad (6.29b)
$$

Equation (6.29b) can be easily understood from (6.19c and 21) for $\delta = 0$, since the cubic roots of unity, λ_i, are such that $|\lambda_i|^2 = 1$.

Next, we assume that at time $t = 0$ *each electron* is described by a *minimum uncertainty wave-packet*, so that

$$(\sigma_\theta)_i(0)(\sigma_{\bar{p}})_i(0) \equiv \sigma_\theta \sigma_{\bar{p}} = 1/2 \ . \tag{6.30}$$

In this case the expectation values in (6.29) are easily calculated and we get

$$\langle \hat{b}^\dagger \hat{b} \rangle (\bar{\tau}) \simeq (9N_\lambda)^{-1} \{ 1 - \exp(-\sigma_\theta^2) + \bar{\varrho}^{-2} [\sigma_{\bar{p}}^2 - (1/4)\exp(-\sigma_\theta^2)] $$
$$+ \bar{\varrho}^{-1} [\tfrac{1}{2}\exp(-\sigma_\theta^2) + n_0] \} \exp(\sqrt{3}\bar{\tau}) = (N_\lambda \bar{\varrho})^{-1} \langle n \rangle (\bar{\tau}) \ . \tag{6.31}$$

Note that in most cases $\bar{\varrho} \gg 1$ so that the first terms in (6.31) are expected to be dominant.

We stress that the result (6.31) could have been obtained, with the exception of the term $(1/2)\exp(-\sigma_\theta^2)$, by treating $\theta_i, \bar{p}_i$ (in Θ, P) as classical stochastic variables with independent Gaussian distributions with dispersions $\sigma_\theta, \sigma_{\bar{p}}$ not related by the uncertainty relation (6.30). Hence quantum mechanics comes in, in (6.31), if σ_θ and $\sigma_{\bar{p}}$ are related by the uncertainty relation (6.30).

We remark that at the peak of emitted intensity $\langle \hat{b}^\dagger \hat{b} \rangle_p \simeq |b|_p^2 \simeq 1$, so that from (6.29b) $\langle n \rangle_p \simeq \bar{\varrho} N_\lambda$. Hence the parameter $\bar{\varrho}$ represents the maximum number of photons that can be radiated per electron. Furthermore, the radiated peak power scales as $N_\lambda^{4/3}$, showing the *cooperative* nature of the radiation process. In the small-signal regime, in which the field does not vary appreciably from the initial level and the dynamics of the electrons is described in terms of weakly coupled pendula [6.18], the radiated peak scales simply as N_λ. We have recently suggested a novel, "superradiant" mode of operation in which the peak power should be proportional to N_λ^2 [6.9,19].

Now we discuss the result (6.31) in two opposite limits.

6.6 "Harmonic Oscillator" Limit

In the limit

$$\sigma_\theta \ll 1 \ , \tag{6.32}$$

i.e. $\sigma_z(0) \ll \lambda/2\pi$, (6.31) reduces to

$$\langle \hat{b}^\dagger \hat{b} \rangle (\bar{\tau}) \simeq (9N_\lambda)^{-1} [\sigma_\theta^2 + \bar{\varrho}^{-2}\sigma_{\bar{p}}^2 + \bar{\varrho}^{-1}(\tfrac{1}{2} + n_0)] \exp(\sqrt{3}\bar{\tau}) $$
$$= (N_\lambda \bar{\varrho})^{-1} \langle n \rangle (\bar{\tau}) \ . \tag{6.33}$$

This is the case in which at time $t = 0$ the particles in the beam are well localized on a wavelength scale. The limit (6.32) is a kind of harmonic oscillator limit, as it follows immediately by rewriting (6.33) in terms of energy

$$\hbar\omega\langle n\rangle(\bar{\tau}) \simeq (1/9)[\varrho\gamma_0 m_0\omega^2\sigma_z^2 + (\varrho\gamma_0 m_0)^{-1}\sigma_p^2$$
$$+ \hbar\omega(n_0 + 1/2)]\exp(\sqrt{3}\bar{\tau}) . \tag{6.34}$$

Note that the zero-point contribution $1/2$ in (6.34) comes just from the "purely quantum" terms due to the electron cross position-momentum fluctuations (6.24).

In this limit, we can expand the exponential in (6.17)

$$\sum_J \exp(-i\theta_J) = \sum_J \exp[-i(\langle\theta_J\rangle_0 + \delta\theta_J)]$$
$$= \sum_J \exp(-i\langle\theta_J\rangle_0)(1 - i\delta\theta_J + \ldots)$$

so that, with the help of (6.20), we can replace the collective operators Θ, P by the first-order-approximated operators

$$\tilde{\Theta} = (\bar{\varrho}/N_\lambda)^{1/2}\sum_J \exp(-i\langle\theta_J\rangle_0)\delta\theta_J ,$$
$$\tilde{P} = (\bar{\varrho}N_\lambda)^{-1/2}\sum_J \exp(-i\langle\theta_J\rangle_0)\bar{p}_J , \tag{6.35}$$

which obey the commutation rules

$$[\tilde{\Theta}, \tilde{P}] = [\tilde{\Theta}, \tilde{\Theta}^\dagger] = [\tilde{P}, \tilde{P}^\dagger] = 0 , \quad [\tilde{\Theta}, \tilde{P}^\dagger] = i . \tag{6.36}$$

By the ansatz

$$\sum_J \exp(-im\langle\theta_J\rangle_0) = 0 , \quad (m = 1, 2, \ldots) $$

the linear evolution equations for $\tilde{\Theta}$, $\tilde{P}$ and u take exactly the form (6.19). However, due to (6.16), they can be derived as a set of Heisenberg equations from the Hamiltonian

$$H = \tilde{P}^\dagger\tilde{P} + a\tilde{\Theta}^\dagger + a^+\tilde{\Theta} - \delta a^+a , \tag{6.37}$$

in which the time is $\bar{\tau}$ defined in (6.10).

Next, we can introduce two "collective electron harmonic oscillator operators" a_1, a_2

$$a_1 = \frac{1}{\sqrt{2}}\left(\frac{\chi}{\sqrt{\bar{\varrho}}}\tilde{\Theta} + i\frac{\sqrt{\bar{\varrho}}}{\chi}\tilde{P}\right) ,$$
$$a_2 = \frac{1}{\sqrt{2}}\left(\frac{\chi}{\sqrt{\bar{\varrho}}}\tilde{\Theta}^+ + i\frac{\sqrt{\bar{\varrho}}}{\chi}\tilde{P}^+\right) , \quad \chi = (\sigma_{\bar{p}}/\sigma_\theta)^{1/2} , \tag{6.38}$$

with the commutation rules $[a_1, a_1^+] = [a_2, a_2^+] = 1$, $[a_1, a_2] = 0$. These operators are defined so that the initial minimum-uncertainty state is the vacuum state of a_1 and a_2. On inverting (6.38), the Hamiltonian (6.37) can be written in the form

$$H = (\sigma_{\bar{p}}/\bar{\varrho})^2(a_2^+ a_2 + a_1^+ a_1 - a_1^+ a_2^+ - a_1 a_2)$$
$$+ \sqrt{\bar{\varrho}}\sigma_\theta[a^+(a_1 + a_2^+) + a(a_1^+ + a_2)] - \delta a^+ a \ . \tag{6.39}$$

Hence the dynamics of the system is that of *three parametrically coupled harmonic oscillators a, a_1, a_2, independently of the value of the electron parameter* N_λ. Note that this Hamiltonian has the constant of motion $a_1^+ a_1 - a_2^+ a_2 + a^+ a$, which compared with (6.4') shows that $a_1^+ a_1 - a_2^+ a_2$ corresponds to the total electron momentum in the linearized theory.

In the limit (6.32) we have derived the *photostatistics* of the field [6.9]. If the radiation mode is initially in a coherent state with an amplitude $\delta\alpha_0$, the Glauber quasiprobability distribution function is

$$P(\alpha, \bar{\tau}) = (\pi \langle n \rangle_{sp})^{-1} \exp\left[- |\alpha - \alpha(\bar{\tau})|^2 / \langle n \rangle_{sp}\right] \ , \tag{6.40}$$

where from (6.33)

$$\langle n \rangle_{sp} = (1/9)[\bar{\varrho}\sigma_\theta^2 + \bar{\varrho}^{-1}\sigma_{\bar{p}}^2 + 1/2] \exp(\sqrt{3}\bar{\tau}) \ ,$$
$$\langle n \rangle_{st} = |\alpha(\bar{\tau})|^2 = (1/9)|\delta\alpha_0|^2 \exp(\sqrt{3}\bar{\tau}) \ . \tag{6.41}$$

In the absence of initial field excitation ($\delta\alpha_0 = 0$) the photonstatistics is that of a chaotic field (as first obtained by *Becker* and *McIver* [6.13] below threshold). In the general case $\delta\alpha_0 \neq 0$ the distribution (6.40) is the displaced Gaussian typical of the *superposition of a chaotic (Gaussian) field with a coherent field* [6.20] with mean photon number, $\langle n \rangle = \langle n \rangle_{st} + \langle n \rangle_{sp}$, given in (6.41). It would approach a Poisson distribution only in the trivial limit of negligible spontaneous contribution; hence FEL radiation exhibits only *first-order coherence*. In particular, the photon number variance is

$$\sigma^2(n) = \langle n \rangle_{sp}(\langle n \rangle_{sp} + 1) + \langle n \rangle_{st} + 2\langle n \rangle_{st}\langle n \rangle_{sp} \ . \tag{6.42}$$

When $\langle n \rangle_{st} \gg 1$ and $\langle n \rangle_{sp} \ll 1$, $\sigma^2(n) \simeq \langle n \rangle_{st}$ (Poisson statistics); when $\langle n \rangle_{sp} \gg \langle n \rangle_{st}$, that is the case in which we are interested, $\sigma^2(n) \simeq \langle n \rangle_{sp}(\langle n \rangle_{sp} + 1)$ (Bose-Einstein statistics); if, e.g., $\langle n \rangle_{st} \gg \langle n \rangle_{sp} \gg 1$, $\sigma^2(n) \simeq 2\langle n \rangle_{st}\langle n \rangle_{sp}$, i.e., the interference term in (6.42) can be dominant over the other contributions.

One more comment on photon statistics is that our results exclude the occurrence of nonclassical effects in the statistical properties of radiation, namely photon antibunching and squeezing [6.21]. This should be due to the fact that in our quantum treatment (i) the many-particle effects are fully taken into account, (ii) the electron variables and the field variables are treated separately. If the system starts from noise, we do not expect the occurrence of these effects even when nonlinearities become dominant.

In analogy with the classical treatment, by extrapolation of the linear-stage results to the full nonlinear dynamics we give an estimate of the field delay time $\bar{\tau}_D$ simply on imposing that $\langle\hat{b}^\dagger\hat{b}\rangle(\bar{\tau}_D)\simeq 1$. Using the results (6.29b and 33) we obtain

$$\bar{\tau}_D = \frac{1}{\sqrt{3}}\ln\left(\frac{N_\lambda\bar{\varrho}}{\langle n\rangle(0)}\right) \equiv \frac{1}{\sqrt{3}}\ln N_c \ , \tag{6.43}$$

where N_c is an electron cooperation number. N_c is maximum when $\sigma_\theta^2 = 1/2$ and its maximum value is $(N_c)_{max} = 9N_\lambda\bar{\varrho}/(n_0+3/2)$. In the case of evolution from noise $(n_0 = 0)$, that is the most unfavorable case, the maximum delay time is

$$(\bar{\tau}_D)_{max} = (1/\sqrt{3})\ln(6N_\lambda\bar{\varrho}) \ . \tag{6.44}$$

Hence *the instability threshold for ASE starting from quantum fluctuations* is

$$(1/\sqrt{3})\ln(6N_\lambda\bar{\varrho})\lesssim 4\pi\varrho N_0 \ , \tag{6.45}$$

that is a much more severe condition than the classical one (6.16b). Further-more, by extrapolation of (6.33) backwards in time, and assuming that σ_θ^2 is dominant even in the limit (6.32), we obtain for the initial noise level

$$b_0 \equiv \langle\hat{b}^\dagger\hat{b}\rangle_0^{1/2} = (1/3)(\sigma_\theta/\sqrt{N_\lambda}) \ , \tag{6.46}$$

with a sensible reduction of fluctuations from the shot-noise estimate $b_0\simeq 1/\sqrt{N_\lambda}$.

6.7 "Classical" Limit

In the limit

$$\sigma_\theta\gg 1 \tag{6.47}$$

and keeping into account that $\bar{\varrho}\gg 1$, i.e. $\varrho\gamma_0 m_0 c^2\gg\hbar\omega$, (6.31) can be approxi-mated by the simple expression

$$\langle\hat{b}^\dagger\hat{b}\rangle(\bar{\tau})\simeq\frac{1}{9N_\lambda}\left(1+\frac{n_0}{\bar{\varrho}}\right)\exp(\sqrt{3}\bar{\tau}) = \frac{1}{N_\lambda\bar{\varrho}}\langle n\rangle(\bar{\tau}) \ . \tag{6.48}$$

Furthermore, if $\bar{\varrho}\gg n_0$ one obtaines $\langle\hat{b}^\dagger\hat{b}\rangle_0\simeq(1/9N_\lambda)$ which, in turn, implies $b_0 \equiv \langle\hat{b}^\dagger\hat{b}\rangle_0^{1/2} = 1/3\sqrt{N_\lambda}$. Comparing this result with (6.15), we have obtained the "shot-noise" expression of the bunching parameter within a factor 3, just taking the classical limit on (6.31) and neglecting the stimulated term n_0.

We call (6.47) "classical" limit because, contrary to the limit (6.32), it produces the disappearance from (6.31) of the purely quantum term $1/2$ and of the variances σ_θ^2, σ_p^2. Thus it is not surprising that we can reproduce the classical results where the single-particle fluctuations are due to shot-noise. However, the accuracy is rather surprising. Actually, if we repeat the steps from (6.43 to 45) using (6.48) instead of (6.33), we get for the delay time

$$\bar{\tau}_D = (1/\sqrt{3})\ln\left(9N_\lambda\bar{\varrho}/\bar{\varrho} + n_0\right) , \tag{6.49}$$

so that if we start from noise $(n_0 = 0)$

$$(\bar{\tau}_D)_{\max} = (1/\sqrt{3})\ln 9N_\lambda \simeq (1/\sqrt{3})\ln N_\lambda + 1 , \tag{6.49'}$$

namely we *derive* quantum-mechanically the classical fit formula (6.16a).

From (6.48) we see that spontaneous emission dominates over stimulated emission if

$$\bar{\varrho} \gg n_0 , \quad \text{i.e.} \quad \left(\hbar\omega\langle n\rangle_0/\gamma_0 m_0 c^2\right) \ll \varrho , \tag{6.50}$$

whereas stimulated emission is dominant in the opposite, "coherent" limit $n_0 \gg \bar{\varrho}$.

A final remark is in order on the connection of our results with the free-particle, small-recoil criterion [6.12,15] for the neglect of quantum effects. This criterion is obtained on imposing that (i) the spread of a (free-electron) wave packet is negligible with respect to a radiation wavelength, or that (ii) the electron recoil in the emission of a photon is negligible with respect to homogeneous broadening. It can be written

$$\varepsilon^{-1} \equiv \frac{1}{(4\pi)^2} \frac{\gamma_R^3 \lambda^2}{\lambdabar_C L_0} = \frac{\bar{\varrho}}{G} \gg 1 , \tag{6.51}$$

where $\lambdabar_C = \hbar m_0 c$ is the reduced Compton wavelength, $G = 4\pi\varrho N_0$ is the total gain (to within a factor on the order of 1), while ε is the free-particle quantum recoil parameter. Hence the condition $\bar{\varrho} \gg 1$ means

$$G/\varepsilon \gg 1 . \tag{6.52}$$

It follows that for low gain $(G \lesssim 1)$ condition (6.52) implies condition (6.51), whereas in the high-gain regime $(G \gg 1)$ it sets a less severe condition for the validity of a classical description, i.e., $\varepsilon < G$. This can be connected to the meaning of the parameter $\bar{\varrho}$ as the maximum number of photons per electron. Actually, if $\bar{\varrho} \lesssim 1$ the electrons will radiate practically in vacuo, so that quantum effects are expected to be relevant, whereas in the opposite limit the system will behave classically. However, to obtain the shot-noise initiation, the condition $\sigma_\theta \gg 1$ is necessary, in agreement with the recent results independently obtained by *Becker* and *McIver* [6.14].

6.A Appendix

If λ_1, λ_2, λ_3 are the roots of the cubic equation (6.21), the expression of the quantities f_1, f_2, f_3 which appear in (6.22) is

$$f_1(\bar\tau) = \sum_{i=1}^{3} f_{1i} \exp\left(i\lambda_i\bar\tau\right) ,$$

$$f_2(\bar\tau) = - \sum_{i=1}^{3} (f_{1i}/\lambda_i) \exp\left(i\lambda_i\bar\tau\right) ,$$

$$f_3(\bar\tau) = \sum_{i=1}^{3} \lambda_i f_{1i} \exp\left(i\lambda_i\bar\tau\right) ,$$

$$f_{1i} = \lambda_i [(\lambda_j - \lambda_i)(\lambda_i - \lambda_k)]^{-1} \quad (i \neq j \neq k = 1, 2, 3) , \tag{6.A.1}$$

while the expression of the quantities h_1, h_2, h_3 introduced in (6.23) is

$$h_1(\bar\tau) = \sum_{i=1}^{3} (h_{1i}/\lambda_i) \exp\left(i\lambda_i\bar\tau\right) ,$$

$$h_2(\bar\tau) = \sum_{i=1}^{3} (\lambda_i - \delta) h_{1i} \exp\left(i\lambda_i\bar\tau\right) ,$$

$$h_3(\bar\tau) = - \sum_{i=1}^{3} h_{1i} \exp\left(i(\lambda_i\bar\tau)\right) ,$$

$$h_{1i} = f_{1i}/\lambda_i . \tag{6.A.2}$$

In the limit $|\delta| \to 0$ the expressions of $|f_i(\bar\tau)|^2$ and $f_1^*(\bar\tau) f_2(\bar\tau)$ are

$$|f_{1,2}(\bar\tau)|^2 = \frac{2}{9}\left[2\cosh^2\left(\frac{\sqrt{3}}{2}\bar\tau\right) - 1 - \cos\left(\frac{3}{2}\bar\tau\right)\cosh\left(\frac{\sqrt{3}}{2}\bar\tau\right) \right.$$
$$\left. \pm\sqrt{3}\sin\left(\frac{3}{2}\bar\tau\right)\sinh\left(\frac{\sqrt{3}}{2}\bar\tau\right) \right]$$

$$|f_3(\bar\tau)|^2 = \frac{1}{9}\left[4\cosh^2\left(\frac{\sqrt{3}}{2}\bar\tau\right) + 4\cos\left(\frac{3}{2}\bar\tau\right)\cosh\left(\frac{\sqrt{3}}{2}\bar\tau\right) + 1 \right] , \tag{6.A.3}$$

$$f_1^*(\bar\tau) f_2(\bar\tau) = -\frac{2}{9}\left\{ \cosh^2\left(\frac{\sqrt{3}}{2}\bar\tau\right) + \cos\left(\frac{3}{2}\bar\tau\right)\cosh\left(\frac{\sqrt{3}}{2}\bar\tau\right) - 2 + i\sqrt{3} \right.$$
$$\left. \times \sinh\left(\frac{\sqrt{3}}{2}\bar\tau\right)\left[\cosh\left(\frac{\sqrt{3}}{2}\bar\tau\right) - \cos\left(\frac{3}{2}\bar\tau\right)\right] \right\} .$$

Furthermore, in the limit $|\delta| \to 0$

$$h_1(\bar\tau) = -f_3(\bar\tau) ,$$
$$h_2(\bar\tau) = f_1(\bar\tau) ,$$
$$h_3(\bar\tau) = f_2(\bar\tau) . \tag{6.A.4}$$

References

6.1 L.R. Elias, W.M. Fairbank, J.M.J. Madey, H.A. Schwettmann, T.I. Smith: Phys. Rev. Lett. **36**, 717 (1976);
   D.A.G. Deacon, L.R. Elias, J.M.J. Madey, G.J. Ramian, H.A. Schwettmann, T.I. Smith: Phys. Rev. Lett. **38**, 892 (1977)
6.2 M. Billardon, P. Elleaume, J.M. Ortega, C. Bazin, M. Bergher, M. Velghe, Y. Petroff, D.A.G. Deacon, K. Robinson, J.M.J. Madey: Phys. Rev. Lett. **51**, 1652 (1983)
6.3 T.J. Orzechowski, B. Anderson, W.M. Fawley, D. Prosnitz, E.T. Scharlemann, S. Yarema, D. Hopkins, A.C. Paul, A.M. Sessler, J. Wurtele: Phys. Rev. Lett. **54**, 889 (1985)
6.4 N.M. Kroll, W.A. McMullin: Phys. Rev. A**17**, 300 (1978);
   I.B. Bernstein, J.L. Hirschfield: Phys. Rev. A**20**, 1661 (1979);
   P. Sprangle, C.M. Tang, W.H. Manheimer: Phys. Rev. A**21**, 302 (1980);
   G. Dattoli, A. Marino, A. Renieri, F. Romanelli: IEEE J. QE-**17**, 1371 (1981);
   C.C. Shih, A. Yariv: ibid., 1387 (1981)
6.5 R. Bonifacio, C. Pellegrini, L.M. Narducci: Opt. Commun. **50**, 373 (1984)
6.6 R. Bonifacio, J. Murphy, C. Pellegrini: Opt. Commun., **53**, 197 (1985)
6.7 J. Murphy, C. Pellegrini: J. Opt. Soc. Am. B**2**, 259 (1985)
6.8 J.M.J. Madey, C. Pellegrini (eds.): *Free Electron Generation of Extreme Ultraviolet Coherent Radiation* (American Institute of Physics, New York 1984)
6.9 R. Bonifacio, F. Casagrande: Opt. Commun. **50**, 251 (1984)
6.10 R. Bonifacio, F. Casagrande: J. Opt. Soc. Am. B**2**, 250 (1985)
6.11 R. Bonifacio, F. Casagrande: Nucl. Instr. and Meth., A**237**, 168 (1985)
6.12 See e.g.: W. Becker, J.K. McIver: Phys. Rev. A**28**, 1838 (1983);
   A. Gover: In [Ref. 6.8, p. 144];
   J. Benson, J.M.J. Madey: In [Ref. 6.8, p. 173];
   A.T. Georges: Phys. Rev. A**28**, 3692 (1983);
   J. Gea-Banacloche: Phys. Rev. A**31**, 1607 (1985)
6.13 W. Becker, J.K. McIver: Phys. Rev. A**27**, 1030 (1983)
6.14 W. Becker, J.K. McIver: Opt. Commun. **53**, 39 (1985)
6.15 A. Renieri: In [Ref. 6.8, p. 1];
   W.B. Colson: In [Ref. 6.8, p. 260]
6.16 The inclusion of space-charge sets a lower limit to the instability region [6.6]
6.17 R. Bonifacio, F. Casagrande, G. Casati: Opt. Commun. **40**, 219 (1982); and in *Evolution of Order and Chaos*, ed. by H. Haken, Springer, Ser. Syn., Vol. 17 (Springer, Berlin, Heidelberg 1982) p. 248;
   R. Bonifacio, F. Casagrande, G. Casati, S. Celi: In *Coherence and Quantum Optics V*, ed. by L. Mandel, E. Wolf (Plenum, New York 1984) p. 801
6.18 W.B. Colson: Phys. Lett. **64A**, 190 (1977)
6.19 R. Bonifacio, F. Casagrande: In *Coherence and Collective Properties in the Interaction of Relativistic Electron Beams and Electromagnetic Radiation*, ed. by R. Bonifacio, F. Casagrande, C. Pellegrini (North-Holland, Amsterdam 1985) p. 36
6.20 R.J. Glauber: In *Laser Handbook*, ed. by F.T. Arecchi and E.O. Schulz-Dubois (North-Holland, Amsterdam 1972) p. 1
6.21 Review papers are: D.F. Walls: Nature **280**, 451 (1980); ibid. **306**, 141 (1983);
   R. Loudon: Rep. Prog. Phys. **43**, 913 (1980);
   H. Paul: Rev. Mod. Phys. **54**, 1061 (1982)

Additional References with Titles

Becker, W., Gea-Banacloche, J., Scully, M.O.: "Intrinsic linewidth of a free-electron laser", Phys. Rev. A**33**, 2174 (1986)
Bonifacio, R., Casagrande, F., Pellegrini, C.: "Hamiltonian model of a free electron laser", Opt. Commun. **61**, 55 (1987)
Gover, A., Amir, A., Elias, L.R.: "Laser line broadening due to classical and quantum noise and the free-electron laser linewidth", Phys. Rev. A**35**, 164 (1987)
Marshall, T.C.: *Free-Electron Laser* (Macmillan, New York, 1985)
Pike, E.R., Sarkar, S., (eds.): *Frontiers in Quantum Optics* (Adam Hilger, Bristol, 1986)
Scharlemann, E.T., Prosnitz, D., (eds.): *Free-Electron Laser* (North-Holland, Amsterdam, 1986)

7. Global Bifurcations and Turbulence in a Passive Optical Resonator

J.V. Moloney

With 22 Figures

Dynamical instabilities in an externally-pumped passive nonlinear optical ring resonator are reviewed. Starting with the simple plane-wave model, phase portraits are used to explain a new type of bifurcation associated with the formation of homoclinic orbits in the phase plane. The plane-wave map is shown to be more unstable to perturbations with a short-scale transverse structure than to plane-wave perturbations. This latter result has important ramifications, one of which is that period doubling cascades should be unlikely in high-finesse optical resonators. Instead, a modulational type of chaos is predicted to occur. Transverse effects are, in fact, inevitable and the study is extended to include pump beams with Gaussian spatial profiles in one and two transverse dimensions. The role of transverse solitons and solitary waves as steady states or as coherent spatial structures undergoing temporally chaotic oscillations is discussed. A self-focusing induced route to optical turbulence is identified and contrasted with plane-wave and self-defocusing models. The importance of transverse effects in determining the final asymptotic state of the optical field envelope will be emphasized.

7.1 Background Material

The transition to turbulence in physical systems is an old subject with much of the earlier focus being on fluid-dynamical systems. Understanding fully developed turbulence and its onset is still a major unsolved problem, although spectacular success has been made over the past few decades in understanding the onset of low-level turbulence in low-aspect ratio fluids [7.1]. Some very beautiful experiments on Rayleigh-Benard convection [7.2], and Taylor-Couette flow [7.3,4], have essentially confirmed the mathematical scenarios proposed as a result of the work of *Newhouse* et al [7.5], and *Feigenbaum* [7.6] and others [7.7]. The difficulty with these mathematical scenarios, however, is that they rely on topological ideas or on the study of the universal properties of few-dimensional maps. Bifurcation parameters appearing in these models cannot easily be identified with physical stress (control) parameters such as a Reynolds or Prandtl number for a fluid flow. An alternative approach is based on a Galerkin approximation, whereby the geometry of the system is exploited and the full fluid equations expanded in a Fourier series. Truncation of the Fourier series to a few spatial modes leads to systems of coupled nonlinear ordinary differential equations that can then be integrated directly. Such a procedure led *Lorenz* [7.8] to observe, for the first time, chaotic dynamics in a coupled system of three ordinary differential equations. However, it is not clear to what extent truncated-mode expansions mimic the true physical situation. While the models can exhibit complex bifurcation structure, the nature of the transition to chaos becomes sensitive to the level of truncation [7.9]. There is, in fact, evidence to show that, in some instances, the original partial differential equations from which these models are derived, exhibit no turbulence over the same parameter range.

Haken [7.10] in 1975 was the first to point out the close similarity between the Lorenz equations and the single-mode laser equations. The study of chaos in nonlinear optical systems took some years to develop since Haken's original paper. A number of chapters in this volume will review chaos in laser systems (Chap. 11). The study of chaos in passive optical cavities began with *Ikeda*'s paper in 1979 [7.11]. He showed that the optical field, in a plane-wave model of a bistable ring cavity, could go through period doubling cascades to chaos. Since then numerous nonlinear optical scenarios have been suggested as candidates for observing optical chaos [7.12]. The first experimental observation was in a hybrid bistable device [7.13]. This device and its later extensions to include a finite medium response time is be reviewed in Chap. 8. Recently, a number of experiments in all-optical bistable systems have shown clear evidence for the existence of transitions to chaos (Chap. 9) [7.14]. Most theoretical studies have invoked the plane-wave approximation, ignoring transverse spatial structure. Two notable early exceptions were numerical models of Gaussian beams in ring [7.15] and Fabry-Perot resonators [7.16]. One of these works noted that the transition sequences to chaos were closely analogous to recently observed experimental transitions in low-aspect ratio fluids [7.15]. Both works noted strong departures from the plane-wave predicitions.

In the present chapter we will concentrate on a single nonlinear optical model, namely, a passive optical ring resonator. This system will be studied at various levels of approximation starting with the simple plane-wave model. Our goal will be to understand the possible dynamical asymptotic states of the field in the resonator as one or more control parameters are varied. We will find that a much broader class of bifurcations is allowed when we move beyond the plane-wave assumption and include diffraction, a linear propagation effect, and self-focusing or self-defocusing nonlinearities. In fact, it will be shown that transverse spatial effects are inevitable even if we assume that the incident pump beam is a plane wave.

A central theme of this study, throughout, will be the use of geometric ideas from the modern theory of dynamical systems to track changes in the topology of the phase space. The phase space for the plane-wave model is just the complex plane (two degrees of freedom, amplitude and phase) and the dynamics of the intracavity optical field in this case reduces to the study of a nonlinear mapping of the plane [7.17]. Extending the study to include diffraction, self-focusing, or self-defocusing effects leads us to consider a nonlinear mapping in an infinite-dimensional space. Fortunately, our physical system is an infinite dimensional dissipative dynamical system; the dissipation of energy occurs primarily through mirror and diffraction losses. Under flow in forward time, or equivalently, under forward iteration of the map, the phase space volume contracts to zero and our expectation is that any dynamical asymptotic state for the optical field will lie on some few-dimensional attractor. Unfortunately, we do not as yet have any suitable nonlinear basis on which to project our infinite-dimensional system and we will have to rely on the idea of embedding of attractors from time series to investigate bifurcation of these attractors. In the plane-wave map we can explicitly construct global phase portraits in the plane (Sect. 7.4) and these will prove essential to uncovering a whole new range of nonlinear dynamical behavior for this apparently simple system.

In the next section we will briefly review the ring-resonator model pointing to approximations made in deriving the final systems of equations for each level of approximation. The problem will be seen to consist of two parts: a nonlinear evoluton of the laser-field envelope governed by the nonlinear wave equation and a feedback of the envelope onto the incident beam (resonator boundary conditions). Starting with the full three-dimensional problem (two transverse and one propagation dimension), we will motivate physically the various terms appearing in the equations and anticipate some of the results to be presented later. This three-dimensional problem is by far the most difficult to study computationally, and as a result there are relatively few concrete results. We will show, however, in Sect. 7.6 how two-dimensional transverse solitary waves and solitons can emerge as asymptotic states of the optical field [7.18]. We can retain essentially the same physics (diffraction, self-focusing, self-defocusing) but simplify the problem greatly if we drop one transverse spatial dimension. At this level the nonlinear evoluton equation can be identified with the nonlinear Schrödinger equation (NLS), an equation that arises in diverse areas of physics and is known to have soliton (solitary wave) solutions [7.19]. Consequently, we will find in Sect. 7.5 that coherent spatial structures can evolve even when the optical field may be undergoing temporally chaotic oscillations. Specifically, we will show that transverse solitary waves arise as fixed points of the infinite dimensional map [7.20], soliton-like structures appear at the onset of bifurcation from periodic to quasiperiodic motion giving rise to new routes to turbulence induced by self-focusing nonlinearities [7.21] and that a new type of modulational chaos can be inhibited by the presence of solitons [7.22]. In addition, we will find exotic bifurcations that depart from the usual well-known scenarios. Many of the bifurcation sequences to optical turbulence will be seen to be similar to experimentally observed transition sequences in low-aspect ratio fluids, consistent with the remarks at the beginning of this introduction. This last statement may appear surprising in view of the fact that the physical mechanisms giving rise to the instabilities and the governing equations in both cases are quite different.

7.2 Model and Theory

Our basic model is the optical ring resonator shown schematically in Fig. 7.1a. A CW laser beam (with either a constant (plane-wave) or Gaussian transverse spatial profile) is incident from the left, passes through the input mirror (intensity transmission coefficient $T = 1 - R$), and propagates through the nonlinear medium of length L_1. Part of the beam is monitored through the output mirror (transmission T) while the remainder is fed back around the resonator and added to the input beam. After an initial transient buildup (usually of the order of the resonator buildup time) the output beam reaches an asymptotic state. In some instances we shall see that the transient may last many hundreds of resonator round-trips before reaching steady state. It is well known that hysteresis (bistability) can be observed when the output is monitored against the input intensity (Fig. 7.1b). The lower (L) and upper (U) branches of the bistable loop may be stable or unstable, whereas the middle branch (M) is always unstable [7.23,11]. Our primary focus will be on unstable states of this system, although we will also discuss steady states (fixed points) where the transverse spatial profile possesses rich nonlinear structure (solitary waves).

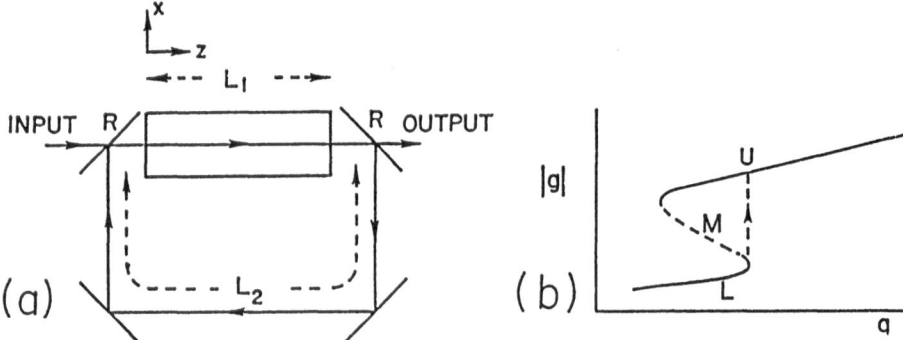

Fig. 7.1(a) Schematic of an optical ring resonator containing a nonlinear medium of length L_1. The total optical path length $L = L_1 + L_2$, L_2 corresponding to the free-space propagation length. In the soliton work (Sect. 7.5, 2 and 7.6) and the modulational instability study $L_2 = 0$. **(b)** Bistable loop when the output amplitude $|g|$ is monitored against the input amplitude a. The labels U, M, and L refer to the upper, middle, and lower bistable branches

The equations describing the time evoluton of the optical field envelope $\mathcal{E}$ in the ring resonator are collected in Table 7.1, for convenience. Rather than dwell at length on the derivation of these equations we shall instead summarize the important approximations made in deriving them and point to their important features. The Maxwell-Bloch equations have been treated in detail in *Allen* and *Eberly*'s book [7.24]. We have assumed that the nonlinear medium response is much faster than the time it takes the optical signal to circulate around the resonator. This allows us to adiabatically eliminate the medium variables from the Maxwell-Bloch equations and, furthermore, we assume the slowly varying envelope approximation for the field in the propagation (z) di-

Table 7.1

3D-Problem

$$\frac{\partial}{\partial(z/L_1)} \mathcal{E}_n(x,y,z) = \frac{\alpha_0 L_1}{2} \left(\frac{1+i\Delta}{1+\Delta^2+|\mathcal{E}_n|^2} - \frac{i\ln 2}{4\pi\alpha_0 L_1 F} \nabla_T^2 \right) \mathcal{E}_n(x,y,z) \qquad (7.1)$$

$$\mathcal{E}_n(x,y,0) = \sqrt{T}\mathcal{E}_{in}(x,y) + Re^{ikL}\mathcal{E}_{n-1}(x,y,L_1) \qquad (7.2)$$

Scale

$$\Big\downarrow \; [\Delta \gg 1]$$

3D Problem (Sect. 7.6)

$$2i\frac{\partial G_n}{\partial\varsigma} + \nabla_t^2 G_n - \frac{G_n}{1+2|G_n|^2} = 0 \qquad (7.3)$$

$$G_n(\eta,\varsigma,0) = a(\eta,\varsigma) + Re^{ikL}G_{n-1}(\eta,\varsigma,p) \qquad (7.4)$$

2D Problem (Sect. 7.5)

$$2i\frac{\partial G_n}{\partial\varsigma} + \frac{\partial^2}{\partial\eta^2}G_n - \frac{G_n}{1+2|G_n|^2} = 0 \qquad (7.5)$$

$$G_n(\eta,0) = a(\eta) + Re^{ikL}G_{n-1}(\eta,p) , \qquad (7.6)$$

$$G_0 = 0$$

Plane Wave (Sect. 7.4)

$$g_n = F(g_{n-1})$$

$$= a + R\exp\left[i\left(kL - \frac{p}{2(1+2|g_{n-1}|^2)} \right) \right] g_{n-1} \qquad (7.7)$$

Scaling:
α_0 : Linear absorption coefficient per unit length
L_1 : nonlinear-medium length
$L = L_1 + L_2$: total optical path length
$\Delta = (\omega - \omega_{ab})/\gamma_\perp$: laser two-level atom dimensionless detuning
$F = n_0 w_{1/2}^2/\lambda L_1 = n_0\ln 2w_0^2/\lambda L_1$: Fresnel number
$\varsigma = (\alpha_0/\Delta)z, \; p = \alpha_0 L_1/\Delta$
$(\eta,\varsigma) = \sqrt{f}(x,y)$
$f = \ln 2/4\pi Fp$

rection (i.e., $E(x,y,z) = \mathcal{E}(x,y,z)\exp[i(kz - \omega t)])$. The resulting map in Table 7.1 is a generalization of *Ikeda's* plane wave map [7.11] to include transverse effects. The map has been written explicitly for the full three-dimensional problem, a scaled nondimensional version of it (7.3,4), the two-dimensional scaled problem (7.5,6) and, finally, the plane wave case (7.7).

The significant parameters appearing in these equations are (i) $\alpha_0 L_1$, the linear absorption coefficient per resonator pass; (ii) $\Delta = (\omega_{ab} - \omega)/\gamma_\perp$, the dimensionless two-level atom frequency detuning where $\gamma_\perp$ is the dipole relaxation constant, $F = n_0 w_{1/2}^2/\lambda L$, the Fresnel number that measures the importance of diffraction in linear propagation, and $T = 1 - R$ the mirror intensity transmission coefficient. By assuming that we are detuned at least a few homogeneous line widths from line center, we can ignore saturable absorption ($\Delta \gg 1$) and the equations can be conveniently scaled to the dimensionless form shown in (7.3, 4). The number of parameters appearing in the problem is then reduced further with the "effective" nonlinear medium propagation length $pL_1 = \alpha_0 L_1/\Delta$. The scaled input field amplitude $a(\xi, \eta)$ can play a dual role both as a parameter or an initial condition (Sect. 7.4). In the nonlinear-medium propagation the effective Fresnel number Fp will measure the importance of diffraction.

The fact that the equations in Table 7.1 are written as a nonlinear mapping (a dynamical system in discrete time n) is crucial to our analysis of the asymptotic states of the optical field. In many instances it will be necessary to propagate the field about the resonator many thousands of times to get a clear picture of the topology of the attractor on which the final motion lies. This is true whether the motion is regular or chaotic.

Our study has been confined to high-finesse resonators where the input pump beam is a relatively small perturbation of the full intracavity field (we assume the $R = 90\%$ in most of our computations). In this situation we expect that the asymptotic states of the nonlinear evolution equation (Table 7.1) will be prevalent (solitons or solitary waves) as asymptotic states of the infinite dimensional map. What is particularly unique about the dynamical system is that while the asymptotic states may be solitons, for example, their heights and widths are determined, not by the initial data, but by the map itself (fixed points). This is in contrast to the usual beam propagation problem where the final state of the beam is a sensitive function of the initial data [7.25]. In low-finesse resonators (strong dissipation) we anticipate that the dynamics of the intracavity field will more closely mimic the plane-wave behavior and consequently, one-dimensional behavior, such as period-doubling cascades, should be more likely.

7.3 Bifurcation and Global Phase Portraits

Bifurcation occurs when one of the eigenvalues (λ) of the linearization of a map crosses the unit circle in the complex λ-plane. The three most common types of bifurcation (codimension one) for two-dimensional invertible maps are depicted in Fig. 7.2. In Fig. 7.2a, an eigenvalue pair lies on the positive real axis with the larger eigenvalue crossing through $+1$ as a parameter is varied; this is a saddle-node bifurcation. A period-doubling (or flip) bifurcation occurs in Fig. 7.2b when the largest eigenvalue (in magnitude) crosses through -1.

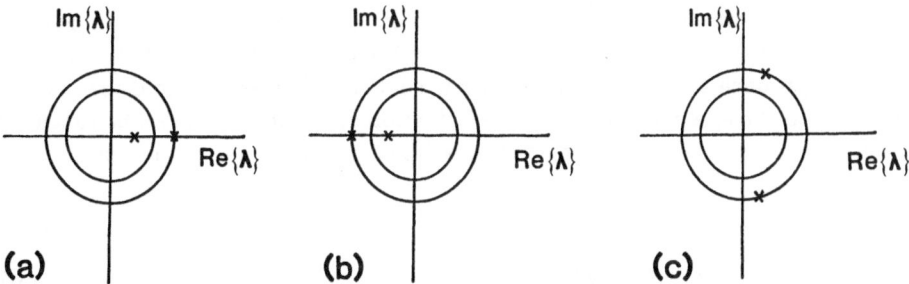

Fig. 7.2a–c. Eigenvalue behavior at bifurcation points. (a) Saddle-node bifurcation – the largest eigenvalue crosses the unit circle along the positive real axis in the complex λ-plane. (b) Period-doubling bifurcation – the eigenvalue with largest modulus crosses the unit circle along the negative real axis. (c) Hopf bifurcation – a complex conjugate eigenvalue pair cross the unit circle at an irrational angle off the real axis. The inner circle of radius R refers to the plane-wave map discussed in Sect. 7.4. The condition det $J = R^2 = \lambda_1\lambda_2 < 1$ forces eigenvalue pairs off the real axis to lie on this smaller circle and rules out a Hopf bifurcation

Note that the magnitude of the eigenvalue gives the rate of exponential growth ($|\lambda| > 1$) or contraction ($|\lambda| < 1$) of the discrete solution in the vicinity of a fixed point of a map [7.26]. Also, if the eigenvalue is negative, the solution (a discrete point) is flipped back and forth about the fixed point as it leaves ($|\lambda| > 1$) or approaches ($|\lambda| < 1$). The eigenvectors associated with these eigenvalues give the local direction of expansion ($|\lambda| > 1$) or contraction ($|\lambda| < 1$) near the fixed point in the phase plane. The final example in Fig. 7.2 is of a Hopf bifurcation where a complex conjugate pair of eigenvalues cross the unit circle off the real axis at a irrational angle; this bifurcation cannot occur for the plane-wave map (Sect. 7.4). Eigenvalue pairs crossing the unit circle at rational angles can lead to further complex bifurcation, an area of current active research in dynamical systems. Bifurcation in infinite-dimensional systems will often involve a single eigenvalue or a pair crossing the unit circle while all other eigenvalues remain stable ($|\lambda_i| < 1$). There exist more complex bifurcations in maps of dimension greater than two (codimension-two and higher) where, for example, *two* pairs of complex conjugate eigenvalues (a double Hopf bifurcation) or a complex conjugate pair plus one real eigenvalue (Hopf and saddle-node) simultaneously cross the unit circle. These latter behaviors are often termed nongeneric to distinguish them from the above. The recent book by *Guckenheimer* and *Holmes* [7.26], is an excellent reference on dynamical systems in general. In Sect. 7.5 we will find that such bifurcations are apparently of wide-spread occurrence in nonlinear optical resonators.

Bifurcation analysis provides us with a local picture near a fixed point (for example, an unstable saddle) in the phase plane. The picture can be extended to the entire phase plane by constructing global phase portraits using the plane-wave map itself. In fact, it is only by constructing such portraits that we can understand the origin of a whole new class of instabilities in the plane wave model. Figure 7.3 shows a schematic of the phase portraits for the middle branch (M) unstable saddle point in Fig. 7.1b. Under iteration of the map (7.7), points move away from the saddle (M) along the curve W^u (the unstable manifold) at an exponential rate whereas along the curve W^s (the stable manifold), points approach the saddle at an exponential rate. These manifolds are locally tangent to the eigenvectors of the linearization of the map and can be extended to the entire phase plane using the invariance property under the map,

$$F(W^u) = W^u = F^{-1}(W^u) \; , \tag{7.8}$$

$$F(W^s) = W^s = F^{-1}(W^s) \; . \tag{7.9}$$

Here the function F refers to the plane wave map in Sect. 7.4 and F^{-1} to its inverse. The inverse map is the discrete time analog of backward time integration in a differential equation (a flow). Applying F^{-1} to a segment of the stable manifold W^s will translate that segment away from the saddle point along W^s (backward time iteration). The following simple algorithm can be used to construct these manifolds globally over the phase plane: (i) locate the unstable saddle point from $g = F(g)$, the fixed-point equation for the plane-wave map in Table 7.1; (ii) determine the eigenvectors of the linearization of F; (iii) place a dense set of points in a line along the direction of the eigenvector v_u associated with the unstable eigenvalue $|\lambda_u| > 1$; (iv) apply the map F to this segment to translate it along W^u. The fact that the manifolds (W^u and W^s) are locally tangent to the eigenvectors (v_u and v_s) at the saddle point together with the invariance of the former under the map ensures that the mapped segment will trace out the appropriate manifold. The same procedure is used to construct W^s except that we replace v_u by v_s and F by F^{-1} in the above steps. In this manner the computer can be used as an analytic tool to study the changing topology of the phase plane. The above statements refer to an arbitrary composition of the map F, F^k, k an arbitrary integer, so that unstable saddles (fixed points) at any level of bifurcation can be treated as above. Examples of these constructions are given in Sect. 7.4. A detailed analysis of the plane-wave map using the above ideas appears elsewhere [7.17]. The exponential expansion or contraction associated with these manifolds (and the saddle point) is closely linked to the chaotic dynamics associated with the plane-wave map. The unstable manifold W^u and its limit points (for example, middle-branch saddle (M) and upper (U) and lower (L) stable fixed points) can be viewed as an attractor in the sense that remote or nearby points are smashed onto it and accelerated toward the stable fixed points. It is conjectured that the chaotic attractor appearing at the end of a period-doubling sequence is a subset of the unstable manifold. We provide direct evidence for this conjecture in [7.17]. The stable manifold W^s defines the attractor basin boundary separating two stable attractors; for example, the upper (U) and lower (L) branch fixed points in Fig. 7.3.

The concept of phase portraits and attractors in the phase space extends naturally to infinite dimensional systems. Even though we have no explicit few-dimensional coordinate system to deal with here, attractors can be embedded in phase space allowing us to differentiate between complicated periodic and

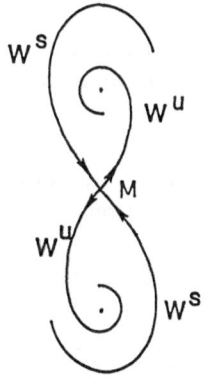

Fig. 7.3. Stable (W^s) and unstable (W^u) manifolds of the middle branch (M) saddle point in Fig. 7.1b. The basins of attraction of the upper and lower branch fixed points (*dots* in the figure) are separated by the curve W^s. Initial conditions on one side of W^s go to one fixed point while those on the other side go to the other fixed point. Points initially remote from W^u are smashed onto it and accelerated toward the respective fixed point (or attractor)

aperiodic motion. The idea of exponential expansion and contraction (usually called hyperbolicity) provides useful insights here also. In general, the manifolds may be surfaces in two or higher dimensions (for example, 2-tori or n-tori) instead of one-dimensional curves as in Fig. 7.3. Transients will provide very useful insights into the nature of these higher dimensional manifolds.

The well known routes to turbulence are made up of sequences of bifurcations involving either successive period doublings, or Hopf bifurcations, or combinations of these. The three most common scenarios are: (i) period doubling to chaos [7.6], which is less common in high-dimensional physical systems such as fluids; (ii) Ruelle-Takens scenario, which involves a stable solution going unstable to a periodic solution, a Hopf bifurcation to quasiperiodic motion followed by a rapid breakdown to turbulence (the final transition to turbulence may proceed directly or through a frequency locking followed by period doubling of the locked output); (iii) intermittancy [7.7], which is not well-characterized but is known to be associated with a saddle-node bifurcation in one-dimensional maps. We may (loosely!) refer to these scenarios as generic. All of these transition sequences occur in the present nonlinear optical model.

7.4 The Plane-Wave Map

Our philosophy in studying the plane-wave map is based on the observation that a significant quantitative change in the dynamical behavior of the optical field is associated with a profound change in the topology of the phase plane. The central idea then is to construct global phase portraits (associated with unstable saddle points) and track their behavior as one or more parameters are varied in the map. An important point to emerge from these studies is the multiparameter dependence of the map in Table 7.1, see (7.7), stressed earlier in [7.27]. Although unstable saddle points seldom appear explicitly in the dynamics of a physical system, we will show that much of the dynamical behavior of the system is organized about such fixed points and their stable and unstable manifolds. Many dynamical phenomena that might, under direct numerical iteration of the map, appear mysterious without any logical connection to other dynamical events can be easily understood. Inspection of phase portraits in the plane and their changes under variation of a parameter such as the input pump field amplitude will immediately convey a global picture of the dynamics even far away from the final attracting fixed points. This idea is important as we shall see that new dynamical behavior not directly associated with bifurcation of a fixed point (stable or unstable) but with transversal intersections of the manifolds are widespread in the plane-wave map.

As the stable (W^s) and unstable (W^u) manifolds approach homoclinic tangency and cross (a transversal intersection), new periodic sinks are abruptly created in the phase plane and they in turn are forced to period double to piecewise attractors as the full homoclinic orbit develops (Sect. 7.4.1). These events occur completely independently of the usual period doubling cascade of the lower branch (L) fixed point, for example. The end result is that the phase plane may be dotted with periodic sinks of arbitrary period, being abruptly created and period doubling to their own chaotic attractors. These sinks will have their own basins of attraction (i.e., W^s) interwoven in a complicated manner with the basins of attraction of coexisting attractors. The development of complicated attractor basin boundaries can be followed as they develop under variation of a parameter. Explosion (interior crises) and destruction (boundary crises) [7.28] of chaotic attractors are associated with the development of complicated basin boundaries. A few examples of the rich dynamical behavior of the plane wave map will be presented below. Full details can be found in [7.17].

Some useful properties of the plane-wave map that prove essential to the following discussion, (7.7), will now be reviewed. A fixed point of the map is computed by setting $g_n = g_{n-1} = g$ in (7.7) and solving the algebraic set of equations for g. Stability of the fixed point to small perturbations is investigated by setting $g_n = g + y_n$ and solving the resulting linearized system of equations for the perturbation y_n. This procedure provides the eigenvalues and eigenvectors referred to in the last section. The map (7.7) has the following properties: (1) it is two-dimensional invertible (g is complex), in contrast to the much discussed one dimensional logistic map studied by Feigenbaum; (2) the map is area contracting with the determinant of the Jacobian J, det $\{J\} = R^2 < 1$. The fact that det $\{J\} = \lambda_1 \lambda_2 = R^2$ is a constant less than one guarantees that the eigenvalues λ_i $(i = 1, 2)$ cannot cross the unit circle off the real axis. The only bifurcations possible for this map are therefore of the saddle-node or period-doubling type. Figure 7.4 shows schematics for these two types of bifurcation. One important point to note from Fig. 7.4a is that beyond the period-doubling accumulation point, only an infinite number of unstable saddle points remain (dashed line). It is also easy to see from Fig. 7.4b that the full hysteresis loop sketched in Fig. 7.1b constitutes a saddle-node (creation of M (saddle) and U (node)) and a reverse saddle-node (mutual annihilation of M and L) bifurcation. A saddle-node bifurcation can occur in any cmposition of the map F (F^k, k an arbitrary integer), and this accounts for the numer-

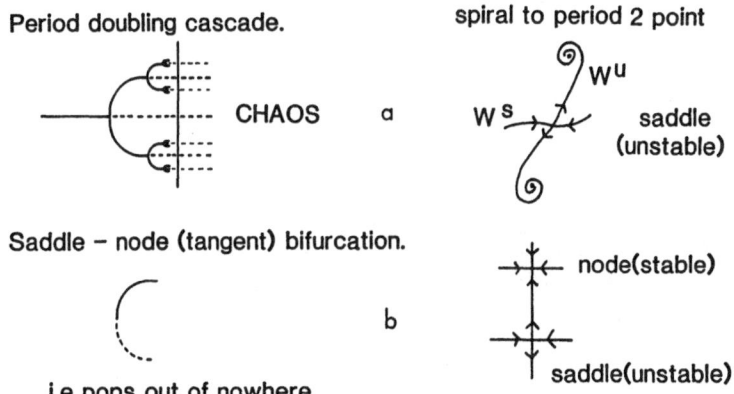

Fig. 7.4. (a) Period-doubling cascade under variation of a parameter. Beyond the accumulation point *(solid vertical line)* infinitely many unstable fixed points (saddles) exist. These are indicated by *dashed lines*. At the n-th bifurcation point $(n = 0, 1, 2 \ldots)$ in the period-doubling cascade, 2^n unstable saddle points appear in the phase plane. Each saddle has its own stable and unstable manifolds as indicated on the *right*. (b) Saddle-node bifurcation under variation of a parameter. A saddle (- - -) and a node (-) (see *right hand figure*) are simultaneously created in the phase plane

ical observation of hysteresis between perodic attractors on a single bistable branch. It is necessary therefore to discriminate between saddle-node bifurcations in the map itself that give rise to conventional bistability, and in its higher compositions that account for the creation of higher period sinks (Newhouse sinks) to be discussed below.

Figure 7.5 shows a bifurcation diagram in (p, a) parameter space for (7.7). This diagram shows the boundaries separating the one and three fixed point regions; the diagram should really be a surface but we can view the upper branch fixed point (labeled 3) as projected onto the (p, a) plane. The upper branch fixed point is stable over this entire parameter region. The remaining curves in this diagram (with the exception of MBH) refer to instabilities on the lower branch and the boundaries denoting bifurcation points in parameter space to higher period orbits in a period-doubling cascade are denoted $P2$, $P8$, etc. There are two separate period-doubling cascades to distinct chaotic attractors that originate in different regions of parameter space. This was noted earlier in [7.27], and Fig. 7.5 is similar to [Ref. 7.27, Fig. 1]. There are important differences, however. Besides the usual period-doubling cascades associated with bifurcation from the lower bistable branch fixed point L, a whole new class of dynamical phenomena arise as one approaches and crosses, under variation of the parameter a or p, the curves labeled LBH or MBH. These curves label points of homoclinic tangency between the stable (W^s) and unstable (W^u) manifolds to the lower (LBH) and middle (MBH) branch unstable saddle points; the lower branch fixed point must have at least undergone bifurcation to a period-2 orbit for LBH to exist (Fig. 7.4a). Having crossed these curves, a transversal intersection has occurred between the respective stable and unstable manifolds. Numerical iteration of the map (7.7) led to the accidental discovery of a coex-

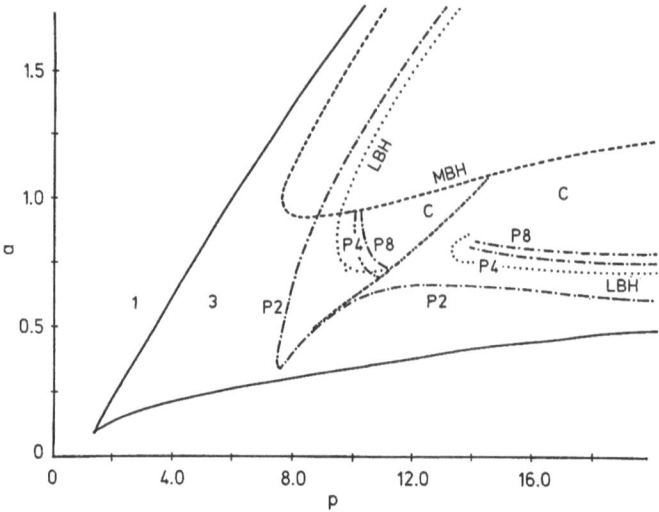

Fig. 7.5. Bifurcation diagram in (p, a) parameter space. The outer solid boundary separates regions of one and three fixed points. The inner boundaries denote the point of onset of period-doubling cascades to two distinct chaotic attractors. The curves labelled LBH (MBH) refer to points of homoclinic tangency between the stable (W^s) and unstable (W^u) manifolds of the unstable fixed point $L(M)$ [7.17]. A whole new sequence of bifurcations appear on crossing these curves and these are discussed in the text and in [7.17]. The line (-·-··-) denotes points of heteroclinic connection between the manifolds of distinct saddle points; explosion (interior crises) can occur along this curve [7.28]. ($R = 0.9$, $\phi = 0.4$, in this and the following figures unless otherwise stated)

isting period six attractor that underwent its own independent period-doubling cascade [7.27]. We shall see below that the abrupt appearance of such attractors in the phase plane is not an isolated anomaly but is of widespread occurrence as one proceeds through a period-doubling cascade; inspection of Fig. 7.5 shows that the line LBH is always crossed as the field goes through a period-doubling cascade to either chaotic attractor. As this dynamical phenomenon is probably unfamiliar, we shall discuss its origin with pictures below.

7.4.1 The Newhouse Sink Phenomena

One remarkable property of the stable (W^s) and unstable (W^u) manifolds is that they can intersect transversely forming a homoclinic orbit. In [7.17] we showed that the formation of such an orbit is intimately related to the development of the chaotic attractor at the end of a period doubling cascade. It can be proved that if the stable and unstable manifolds intersect once, they will have to do so an infinite number of times [7.26]. However, a manifold cannot intersect one of the same type; for example, the stable manifold cannot intersect itself or a stable manifold of another fixed point. As the manifolds approach tangency they are forced to oscillate wildly. Figure 7.6 shows a computer-generated picture of a homoclinic tangency between the stable and unstable manifolds of the lower branch fixed point; this picture was generated using the algorithm of the previous section at a point on the curve LBH. The longer fingerlike lobes

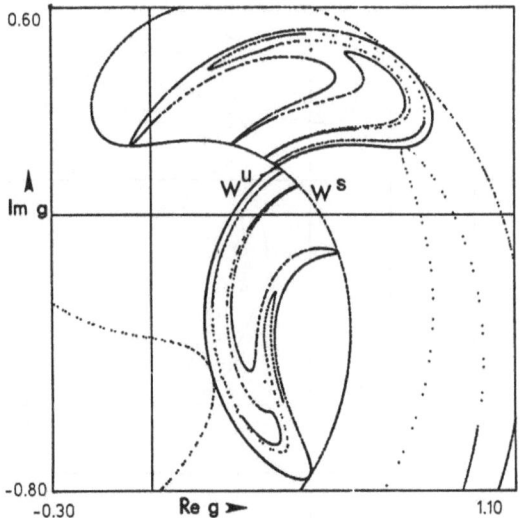

0.60

Im g

-0.80

-0.30　　　Re g ➤　　　1.10

W^u　　W^s

Fig. 7.6. A computer-generated homoclinic tangency between W^s and W^u of the lower branch unstable saddle point (L) at a point in (p, a) parameter space lying on the curve LBH in Fig. 7.5. The long fingerlike regions reflect the complicated nature of the attractor basin boundary (W^s), which at this point is a simple period-2 attractor (*two dots*). The fact that the manifolds must touch an infinite number of times and at the same time spiral into the attracting period-2 points leads to this complicated behavior

of W^u stretch out to touch W^s and vice versa. The area within each fingerlike lobe decreases by R^2 (areas contract by det $J = R^2$) as each finger begins to approach the saddle point L. If the map is iterated further, the lobes are forced to wind in a threadlike manner back into the fingers, with the result that very narrow fingerlike regions are mapped into fingers.

The period-6 attractor discovered in [7.27] has each attracting periodic point located within such a finger (Fig. 7.7). This period-6 cycle is created within the lobe via a saddle-node bifurcation in the sixth composition of the map (F^6) as the manifolds approach tangency. The six nodes (periodic points) are shown in the figure. The unstable saddles can easily be located [7.17], but are not shown here. As the manifolds pass through tangency and the full homoclinic orbit develops, the period-6 attractor is forced to period double. Note that a period-2 orbit coexists with the period-6 attractor, and is unaffected by the latter's period doubling; the period-2 orbit belongs to a period-doubling cascade from the lower branch fixed point. The period-doubling cascade of the period-6 attractor ends on a six-piece chaotic attractor that is abruptly destroyed via a boundary crisis (Sect. 7.4.2) on further increase of the input amplitude a. The period-6 attractor is not the only attractor contained within the lobes and, in fact, we have located a period-8 and a period-10 attractor as we move toward the tips of the fingers. These latter attractors are forced to undergo their own period-doubling cascades. The mathematical work of *Smale, Newhouse, Gavrilov* and *Silnikov*, and, more recently, *Yorke* and coworkers has established that infinitely many stable periodic attractors can be created within these lobes and all will be forced to period double [7.29]. Attractors of progressively higher period should be located closer to the tips of the lobes. Physically, only the lower period attractors will be observable, as the basins of attraction of the higher period ones become very small and noise in a physical

151

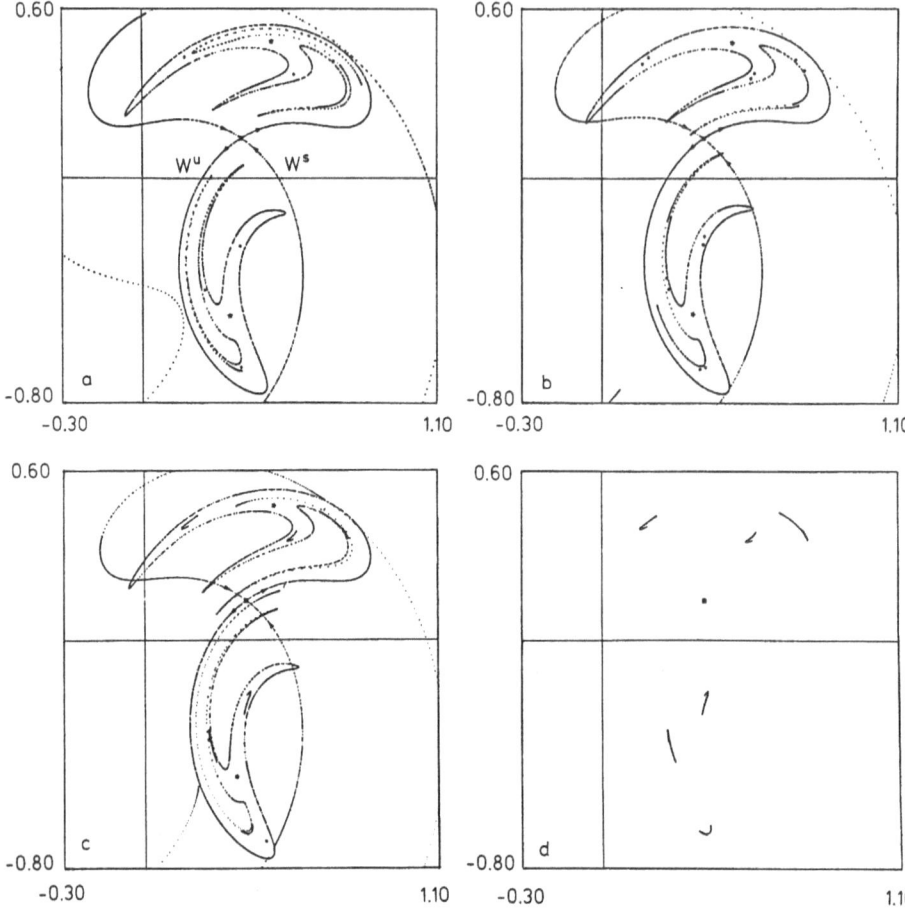

Fig. 7.7a–d. A period-6 attractor is spontaneously created within the finkgerlike lobes and is forced to period double to a six-piece chaotic attractor which is in turn destroyed via a boundary crisis. (a) Period-6 attractor created in lobes (*dots*) coexist with a period-2 orbit (*). (b) Period doubling to period 12. (c) Six-piece chaotic attractor at the end of the cascade. (d) The same 6-piece attractor with the lower branch manifolds deleted for clarity. Note that the period-2 orbit which has bifurcated from the lower branch fixed point L remains unaffected by these other dynamical events

system would preclude their observation. The mechanism for formation and forced period doubling of these attractors is associated with the study of a very special class of mappings in the plane – so-called protohorse shoe and horseshoe maps. The name derives from the fact that a rectangular region of the plane is mapped to a horseshoe region. Figure 7.8 shows that such a horseshoe mapping is associated with the creation of the period-6 cycle above.

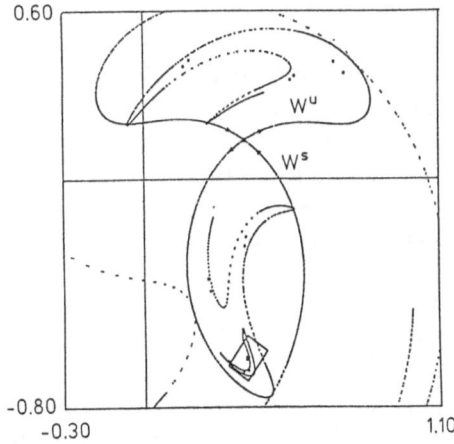

0.60

W^u

W^s

-0.80

-0.30   1.10

Fig. 7.8. Example of a protohorseshoe construction. The small rectangular box is mapped back onto itself as a horseshoe-shaped region under the sixth composition (F^6) of the map. Further appliation of F^6 shows that there is an invariant attracting region within the box in which a saddle and node pair have been created. In the picture the node has already period doubled, i.e., a period-12 orbit exists (*dots*)

7.4.2 Boundary Crisis

The role of the stable manifold W^s as an attractor basin boundary is clearly illustrated in Fig. 7.9. At the input amplitude $a<1.063$ ($p = 7$, $\phi = 0.4$), a chaotic attractor exists on the lower bistable branch. Under iteration, any initial condition on the attractor will stay on it for all time. At $a = a_c = 1.063$, a homoclinic tangency is created between the middle-branch stable and unstable manifolds. At this stage, the basin of attraction of the upper branch (U) fixed point (i.e., W^s) has barely touched the lower branch chaotic attractor. A slight increase in a beyond a_c causes the upper branch basin boundary to leak into

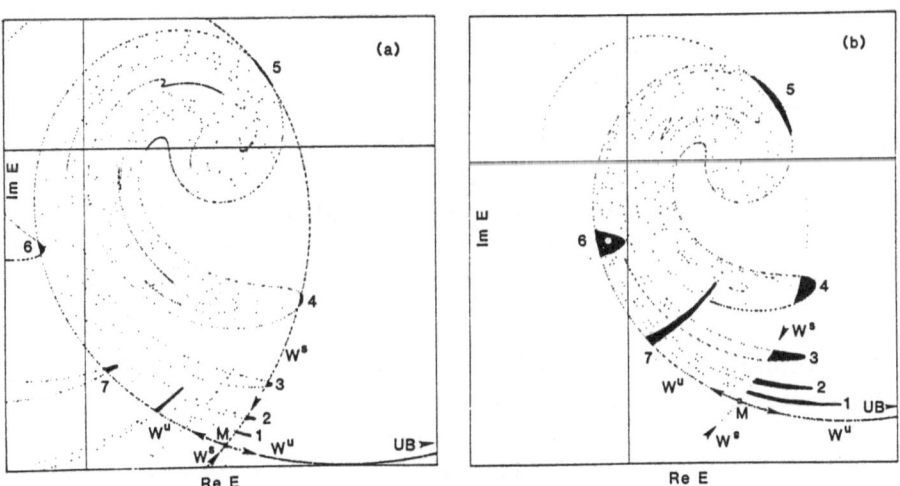

Fig. 7.9. A boundary crisis occurs on crossing the line MBH in Fig. 7.5. The attractor basin boundary (W^s) has leaked into the lower branch chaotic attractor causing the premature destruction of the attractor. The mean escape time of an arbitrary initial condition on the attractor scales as $\tau_{esc} \simeq (a - a_c)^{1/2}$ where $a_c = 1.063$. (a) $a = 1.07$ (b) $a = 1.12$

153

the chaotic attractor (Fig. 7.9a). There is now a finite probability that a point undergoing random excursions on the attractor can get trapped in one of the darkened buckets that lie in the basin of attraction of U. When this occurs, the point will be swept to the upper-branch fixed point U, and the attractor will be destroyed. Further increase in a reduces the average residence time of an initial condition on the attractor (Fig. 7.9b). The mean escape time of points from the attractor should scale as $\tau_{esc} \alpha (a - a_c)^{-1/2}$; in general $\tau_{esc} \alpha (a - a_c)^{-D/2}$, D being the topological dimension of the chaotic attractor.

7.5 Gaussian Beams (2D Problem)

In this section we discuss how spatio-temporal effects determine the asymptotic state of the field in the resonator. Solutions to the infinite dimensional map, (7.5,6) in Table 7.1, will be sought in a region of parameter space where we know that the plane-wave map exhibits simple behavior. The procedure for solving (7.5,6) is more involved, however, as we need to specify the initial data $a(\eta)$ (Gaussian spatial profile), solve the nonlinear evolution equation (7.5) to determine $G_1(\eta, p)$, and use this solution in (7.6) to determine new initial data $G_2(\eta, 0)$ for (7.5). The procedure is repeated until some asymptotic state is reached.

The inclusion of diffraction, self-focusing, or self-defocusing nonlinearities leads to significant departures from the plane-wave behavior. While the nonlinear phase shift encoded across the beam wavefront may be small on a single pass, the cumulative phase shift over multiple resonator passes may be very large, leading to large departures of the spatial profile from its original Gaussian shape; the evolution of transverse solitary waves (Sect. 7.5.2) is a case in point. Strong dissipation (in low-finesse resonators) may prevent the accumulation of large phase shifts and lead to a quasi plane-wave behavior. Also free-space propagation within the resonator will tend to smooth out soliton shapes; the analysis and results in the next two subsections assume a filled resonator (no free-space propagation). Most of the results presented in Sect. 7.5.3 on routes to turbulence assume a free-space propagation length equal to the nonlinear-medium length; the one exception will be the results on modulational chaos that are based on the analysis of Sect. 7.5.1.

7.5.1 Stability of the Plane-Wave Map to
Transverse Spatial Perturbations

In the previous section, we reviewed just a part of the very rich dynamical structure of the plane-wave map. We did show, however, that the only bifurcations allowed for the plane-wave optical field were of the saddle-node (hysteresis) and period-doubling (cascades to chaos) type. A question we now address is whether the plane-wave map itself is stable to perturbations with a transverse spatial dependence. To answer this question, we must study the stability of the plane-wave fixed point g in (7.5) to perturbations $y_n(x, z)$ with transverse x, ($\equiv (x, y)$) structure. Rewriting (7.5,6) in unscaled coordinates we get

$$2iG_{nz} + \gamma \nabla^2 G_n + pN(G_n G_n^*)G_n = 0 \ , \quad 0 < z < L \ , \tag{7.10}$$

$$G_n(x, 0) = \sqrt{T}\mathcal{E}_{in}(x) + Re^{i\phi}G_{n-1}(x, L) \ . \tag{7.11}$$

Writing $G_n(x, z) = [g + y_n(x, z)] \exp[ipN(I)z/2]$, $I = gg^*$ and using the plane wave map we find

$$y_n(x, z) = e^{\sigma z}(a_n e^{iK \cdot x} + b_n e^{-iK \cdot x}) + e^{-\sigma z}(c_n e^{iK \cdot x} + d_n e^{-iK \cdot x})$$

where $K^2 = K_x^2 + K_y^2$ ($K = 0$ corresponds to a plane wave). Here the nonlinear term $N(I) = -(1 + 2I)^{-1}$ for saturable and $N(I) = -1 + 2I$ for Kerr media. We investigate the stability of the plane wave fixed point on the lower branch so we assume $N(I) = -1 + 2I$. The growth rate of perturbations in the z-direction is given by σ and the coefficients a_n, b_n, c_n, and d_n are to be determined from the linearization. Rather than burden the reader with lengthy algebra, we will quote the final answer for the potentially unstable root ϱ ($\varrho = -\lambda = 1$ is the period-2 bifurcation point)

$$\varrho/R = b + \sqrt{b^2 - 1}$$

where if $\tau = \gamma K^2 > 4\mu$ ($\mu = pI$),

$$b(\mu, \tau) = \cos(\psi + \mu) \cos \nu + \frac{\tau - 2\mu}{2\nu} \sin(\psi + \mu) \sin \nu$$

and if $\tau < 4\mu$, the corresponding formula is

$$b(\mu, \tau) = \cos(\psi + \mu) \cosh \sigma + \frac{\tau - 2\mu}{2\sigma} \sin(\psi + \mu) \sinh \sigma \ .$$

In these equations $2\nu = \sqrt{\tau^2 - 4\mu\tau}$, $2\sigma = \sqrt{4\mu\tau - \tau^2}$, and $\psi = \pi + \phi - p/2$. It is a simple matter to see that the critical value of b at the period-doubling bifurcation point is $b_c = 1/2(R + R^{-1})$.

In Fig. 7.10 we plot $b(\mu, \tau)$ vs $\sqrt{\tau}$ for both the self-focusing and self-defocusing case. Note that in the self-focusing case at $p = 6$, $\phi = (kL) = 0.4$ the plane-wave fixed point g is stable. This is also evident in Fig. 7.10a where at $K = 0$ (the plane wave) $b(\mu, 0)$ is well below critical b_c. However, at $\mu > \mu_c = 0.11$, a finite band of short wavelength modes is already unstable! The mode with maximum growth [maximum $b(\mu, \tau)$] will actually grow. A similar instability occurs in the defocusing case at $p = -9$ (Fig. 7.10b).

In actual experiments, the laser beam is not a plane wave but has a Gaussian spatial profile. The analysis must be extended to allow for the fact that

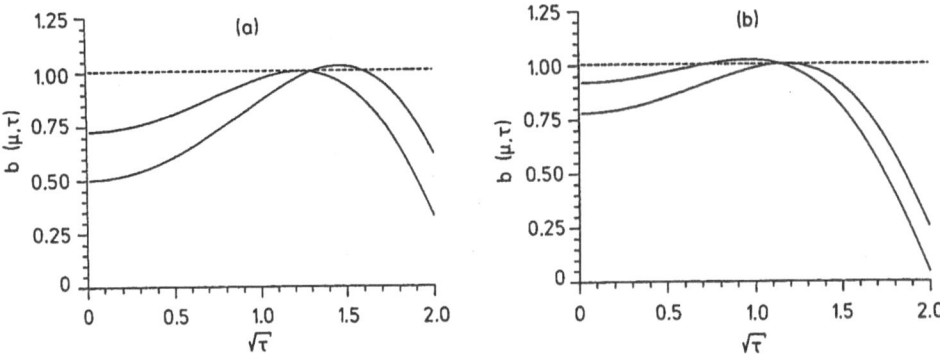

Fig. 7.10a,b. Plot of the function $b(\mu, \tau)$ against wavenumber $\sqrt{\tau}$ ($\tau = \gamma K^2$). The *dashed line* corresponds to the critical value of b for period doubling $[b_c = 1/2(R + R^{-1})]$. $K = 0$ corresponds to a plane wave. (a) Self-focusing ($p = 6$, $\phi = 0.4$). At critical $\mu = \mu_c = 0.11$ and above critical $\mu = \mu_c = 0.26$. (b) Self-focusing ($p = -9$, $\phi = -0.4$) at critical $\mu = \mu_c = -0.14$ and above critical $\mu = -0.25$

the growth rate of the instabilitiy must allow for energy flux from the center of the beam to the wings, in addition to the energy input to the fluctuation from the instability. Full details of the analysis have been given in [7.22]. Figure 7.11 shows how this instability manifests itself just beyond the period-doubling bifurcation point. The smooth field envelope breaks up into spatial rings that undergo a period-2 oscillation by phase shifting left and right in x by π. We will discuss briefly in Sect. 7.5.3 how this instability induces a new type of modulational chaos.

The above analysis shows that the plane-wave map cannot provide a correct description of the dynamics, at least in high finesse resonators. The above instability is reminiscent of the Benjamin-Feir or modulational instability that is so wide spread in physics. This latter instability causes high intensity laser beams to critically focus or filament as they propagate in a Kerr-like medium [7.25]. There are significant differences however as (i) the above instability occurs in the map (7.5,6) and is absent in the evolution equation alone, (ii) the unstable band of wavelengths can be outside the Benjamin-Feir band ($\tau \leq 4\mu$) (as in Fig. 7.10) and (iii) the instability occurs even in the defocusing case (Fig. 7.10b).

7.5.2 Solitary Waves as Fixed Points of the Infinite Dimensional Map

In [7.30], it was observed numerically that when a high-Fresnel-number beam switched to the high-transmission state of a bistable ring resonator, sharp gradients appeared at the edges of the on-spot. The subsequent dynamical evolution of the transverse-field envelope depended on whether the laser was tuned to the self-focusing or self-defocusing side of the atomic transition. Under self-focusing conditions, spatial rings appeared initially at the edges of the on-spot and slowly grew inwards toward beam center. The final asymptotic state of the transverse profile depends on the total energy (area) contained within the on-

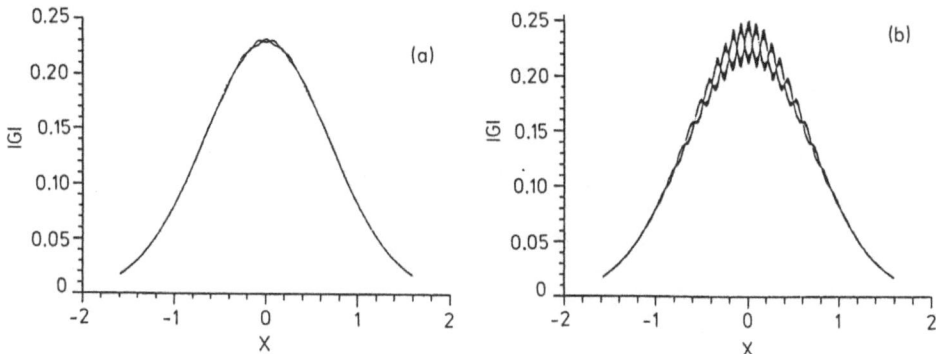

Fig. 7.11a,b. Spatial instability growth for a self-focusing (one transverse dimension) beam with an initial Gaussian spatial profile ($p = 6$, $\phi = 0.4$, $F = 33$, $|a(0)|^2 = 0.14$). (a) Initial growth of the instability from the weakly unstable fixed point ($K \simeq 0$) at resonator passes $n \simeq 150$. (b) Further growth and development of spatial rings ($n \simeq 180\text{--}190$). The rings keep growing and saturate at which point they undergo a stable period-2 oscillation by shifting left and right in phase by π

spot. The energy content of the on-spot in turn depends on how far the input Gaussian peak intensity exceeds threshold for switch-on of the bistable system. Figure 7.12 indicates the final asymptotic state of the field envelope at various locations along the upper branch ($p = 2$, $F = 200$, $kL = 0.4$). Observe that there exist finite bands in parameter space ($|a(0)|^2$) where 1, 3, 5, 7, ... spatial ring structures are stable. Between these bands the spatial rings undergo slow periodic oscillations in time while maintaining constant energy in the beam.

This phenomenon is quite different from the discussed in the preceding subsection. Here, large gradients at the edge of the on-spot make diffraction locally important so that $1/F\nabla_T^2$ is always of order unity, even though the Fresnel number itself may be large. Figure 7.13 shows three stages in the dynamical evolution of the beam profile toward its final seven stable ring state; the initial Gaussian shape on the first resonator pass, the on-spot at the 20th resonator pass (the cavity buildup time $\tau_c \simeq 30$ resonator passes), and the final asymptotic state of the field. In [7.20] it was shown that these transverse spatial rings were solitary waves of the nonlinear evolution equation [7.5] and

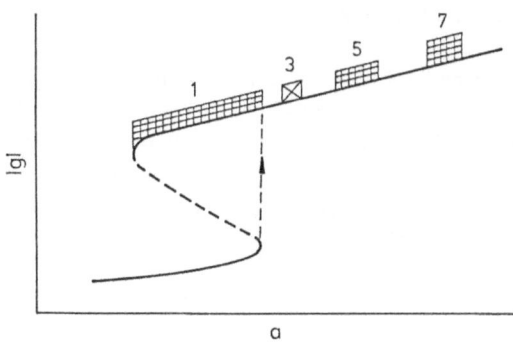

Fig. 7.12.
Schematic showing regions of stability of transverse n-solitary wavetrains ($n = 1, 3, 5, 7\ldots$) on the upper bistable branch

157

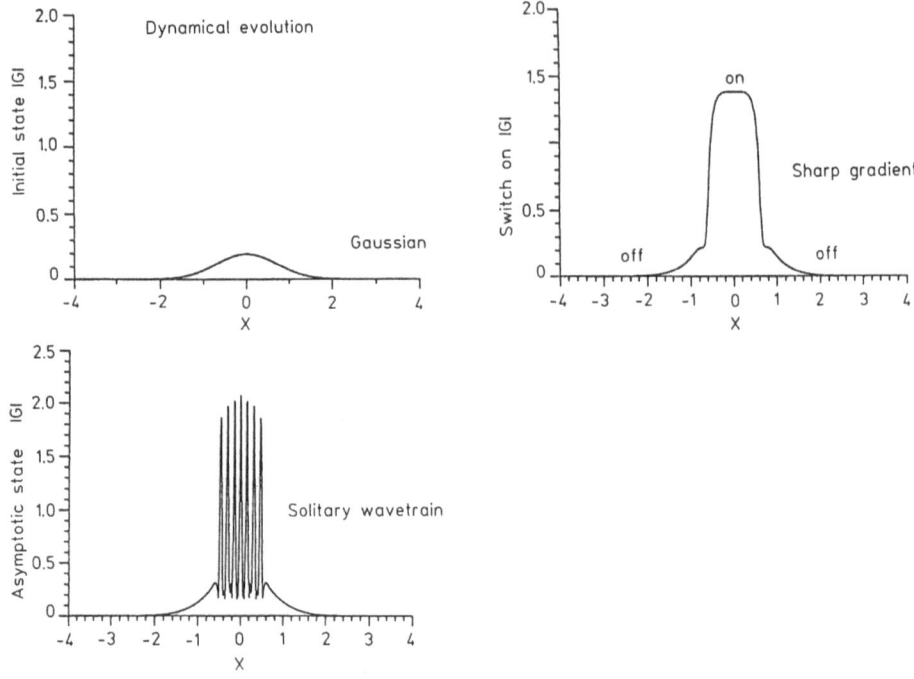

Fig. 7.13 Transient switching of the transverse profile to the upper branch. The peak amplitude $a(0)$ of the input Gaussian lies in the band in Fig. 7.12 where a seven solitary wavetrain is stable. The sharp gradient develops at about the 20th resonator pass and the profile reaches a stable asymptotic state after about 200 resonator passes ($p = 2$, $F = 200$, $\phi = 0.4$, $|a(0)|^2 = 0.0375$)

fixed points (in function space) of the infinite dimensional map. Solitary wave shapes computed from an independent theory were in excellent agreement with the numerically computed spatial rings. The central idea in the theory is that, in these high-finesse resonators, the shapes that emerge from the nonlinear medium are predominantly solitary waves of (7.5) (a nonintegrable nonlinear Schrödinger-type equation). Because the intracavity field is much larger than the broad input Gaussian, the latter can be treated as a small perturbation of the former. With these assumptions, the infinite dimensional map (7.5,6) can be reduced to a two-dimensional map in the solitary wave parameters (amplitude and phase).

Perhaps one of the most exciting results to emerge from this study is the possibility that a finite solitary wave or soliton nonlinear basis may capture the essential dynamics of these infinite dimensional systems at least over finite parameter ranges. Such a result could have far-reaching consequences beyond nonlinear optics. We shall see in the next subsection that there is compelling evidence for the existence of soliton-like spatial structures even when the beam is undergoing chaotic temporal motion.

7.5.3 Many Routes to Optical Turbulence

Evidence for complex dynamical behavior has already been provided in Sect. 7.4 where the simple plane-wave map was analyzed. Unlike the plane-wave map, we cannot now expect to readily locate unstable fixed points and their stable and unstable manifolds for the infinite-dimensional system. In the plane-wave map these manifolds were one-dimensional curves, whereas in the present case they may be two-dimensional surfaces (two tori, sheets, etc., in phase space) or even hypersurfaces (n-tori). The difficulties posed in analyzing the dynamical behavior of many dimensional systems must be faced by both experimentalists and theoreticians. Many of the questions that need to be addressed are the same; for example, what is the optimum means of detection of the output beam? Should one use a small aperture to record, say, the beam center intensity, or should one detect the total power output? We shall see shortly that many dynamical effects are intrinsically spatial in origin involving very small changes in the total power output from the resonator.

In the past few years, a number of techniques have been developed to analyze complex dynamical outputs from many-dimensional systems. Many of these techniques are extensions of the original ideas of Poincare (1880) and others, but made practically implementable with the constraints on experimentalists in mind. The idea of using a single experimental time series to embed an attractor in a higher-dimensional phase space has been successfully used to construct a chaotic attractor from the Belousov-Zhabotinski reaction [7.31]; a chemical reaction with up to 40 chemical constitutents. Recently this embedding construction has been used in studying instabilities in a laser [7.32]. Besides time series, power spectra are useful in identifying incommensurate or subharmonic frequencies in a time series. A very useful technique for studying orbits of maps is the construction of a Poincare surface of section (next amplitude map). This latter construction is also used in combination with the above embedding scheme to estimate the fractal dimension of chaotic attractors. These dimensions can be estimated by computing Liaponuv exponents that measure the average rate of exponential expansion and contraction on a chaotic attractor [in the plane wave map (Sect. 7.4) the Liaponuv exponents are simply the logarithms of the eigenvalues of the linearization of (7.7)] or by way of a correlation technique developed recently by *Grassberger* et al. [7.33].

We shall briefly review some numerical studies on (7.5,6) that have been carried out in a restricted region of parameter space. From the plane wave results we anticipate that the higher-dimensional phase space may contain many coexisting attractors of possibly high fractal dimension. As mentioned earlier, a much broader class of bifurcations is allowed, including Hopf bifurcations to invariant circles or tori. Interaction between coexisting attractors may lead to regular motion on higher-dimensional attractors that might otherwise be unstable [7.34]. Direct numerical evidence for the latter will be provided. Examples will be provided of bifurcation sequences that are closely analogous to experimentally observed transition sequences in low-aspect ratio fluids. New routes to optical turbulence induced by self-focusing nonlinearities will involve the active dynamical participation of solitonlike spatial structures. These transverse soliton structures appear to be the nonlinear optical analogs of small spatial scales appearing in fluids.

Referring to the plane-wave bifurcation diagram in Fig. 7.5, we study instabilities for an input beam with a transverse Gaussian spatial profile, where the plane wave itself is either stable or a simple period 2. We know from the discussion and results at the beginning of this subsection that the plane-wave map itself is unstable to short-wavelength transverse fluctuations. For a self-focusing beam, the boundaries denoting the period-2 bifurcation point are lowered in p relative to the plane-wave prediction. Also the transition to turbulence is significantly modified, leading to a modulational type of chaos to be discussed shortly.

As a first example we will discuss a transition sequence to optical turbulence that is closely analogous to experimentally observed transition sequences in low-aspect ratio fluids [7.15]. Rather than show time series (these were presented in [7.15]) we shall discuss the other techniques for analyzing the outputs. We assume strong diffractive coupling ($F = 0.54$) in a self-focusing beam. Other parameters used in the computation are $p = 10$, $\phi = 0.4$ and the range of input intensity $|a(0)|^2$ over which the beam is unstable (on the lower branch) is $0.43<|a(0)|^2<1.1$. The transition sequence involves stable – period 2 (P2) – quasiperiodic (QP2) – frequency locked (L6) outputs, as shown in Figs. 7.14 and 15. The beam profile oscillates between a Gaussian-like (TEM_{00}) and a doughnut-shaped (TEM_{01}) profile in both the quasi-periodic and frequency-locked regimes (Fig. 7.14). Strong diffraction prevents strong focusing of the beam. It is interesting to note in passing that if a confocal resonator were used (rather than plane mirrors as assumed here) a period-2 oscillation will, in fact, occur between TEM_{00}- and TEM_{01}-like transverse modes. Schematic frequency spectra of time series of the on-axis output are shown alongside the transverse outputs in this figure. These illustrate how a new incommensurate frequency (f_2) appears in the spectrum at the Hopf bifurcation point and how locking occurs when f_2 shifts to $1/6t_R$ ($f_3 = f_1 - f_2$ is just a combination frequency).

Finally we show in Fig. 7.15, a Poincaré surface of section (next amplitude map) as one proceeds through the bifurcation sequence. The Poincaré section can be viewed as a plane section through a limit cycle (continuous time) in three-dimensional space for the case of period 2. At the Hopf bifurcation point, the period-2 orbit goes unstable and bifurcates to an invariant circle. Motion on the invariant circle is quasi-periodic if points (iterates) cover it densely. Frequency locking occurs when points on the circles tend to accumulate locally in three regions on each circle. Finally the frequency-locked output L6 undergoes a period-doubling cascade to chaos (L12 is shown). We shall see shortly that the invariant circles (irrational tori) can break down directly to a chaotic attractor.

Figure 7.16 compares dynamical transverse beam profiles as the input Gaussian peak intensity $|a(0)|^2$ increases for self-defocusing ($p = -8$, $F = 10$, $\phi = -0.4$) and self-focusing beams ($p = 8$, $F = 10$, $\phi = 0.4$). Notice from

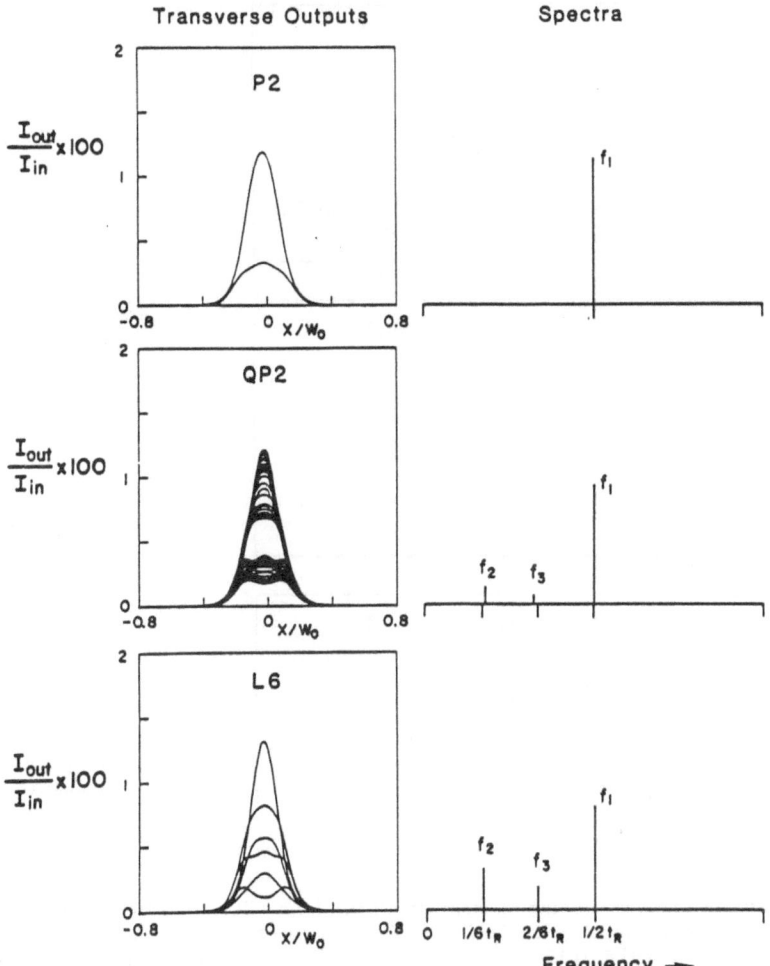

Fig. 7.14. Unstable transverse profiles corresponding to period 2 (P2), quasi-periodic (QP2), and frequency-locked (L6) oscillations on the low transmission branch ($p = 10$, $F = 0.54$, $\phi = 0.4$). On the right are schematics of spectra of the beam center times series [7.15] showing how frequency locking occurs

Fig. 7.5 that the plane-wave map goes from stable to period 2 and back to a stable fixed point in this parameter range. The behavior of the self-defocusing beam is qualitatively similar (Fig. 7.16a), except that now spatial effects are evident. With increasing input intensity, the center of the beam first returns to a stable fixed point while the wings maintain a period-2 oscillation. Further increase in intensity will cause the whole beam to return to a stable fixed point. The ring structures in the picture appear from a combination of the negative lensing effect of the defocusing nonlinearity and a simple interference effect over multiple resonator passes. The situation in Fig. 7.16b for self-focusing is very different. The focused beam bifurcates from a fixed point (in function space) to

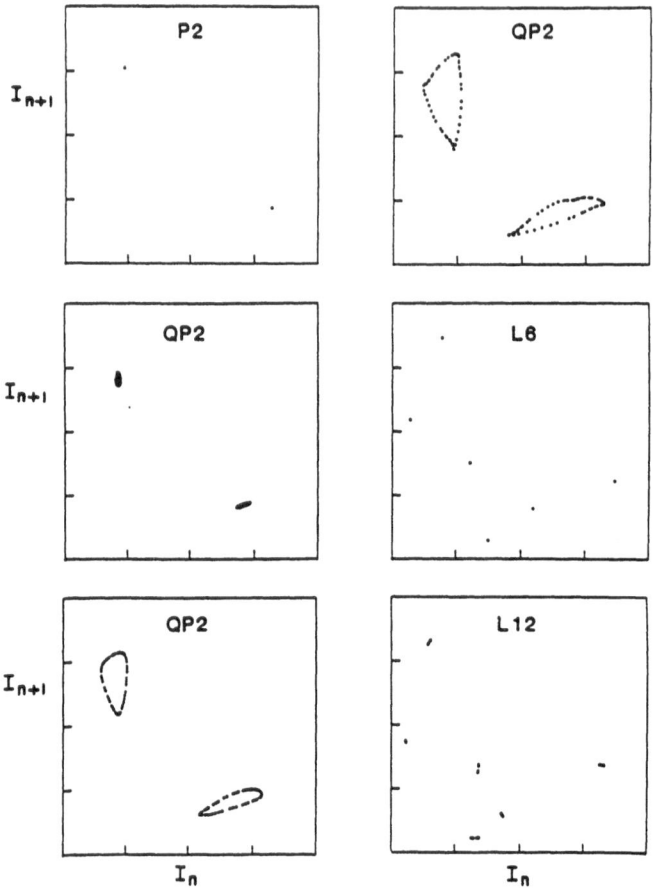

Fig. 7.15. Poincare sections showing in more detail the Hopf bifurcation from a period-2 solution to quasiperiodic motion (P2→QP2), the frequency locking to L6 and period doubling of the locked output (L6→L12). These sections are constructed by plotting the on-axis intensity from the time series at the $(n+1)$st resonator pass (I_{n+1}) against that at the nth pass (I_n).

a period-2 oscillation as before. However at a critical intensity $|a(0)|_c^2 = 0.17)$ the profile breaks up into transverse rings that then oscillate back and forth. Just beyond $|a(0)|_c^2$ the beam undergoes complicated but regular quasi-periodic oscillations. The transverse outputs however look incredibly complicated (each plot has 20 successive resonator outputs plotted). We see that self-focusing causes a breakup of the beam into transverse solitonlike structures and moreover, a further intensity increase causes the invariant circles (**QP2**) in Fig. 7.15 to break down directly to a chaotic attractor. The transition from period 2 (**P2**) to quasi-periodic motion (**QP2**) is an example of a Hopf bifurcation where a complex conjugate pair of eigenvalues from the linearization of the infinite dimensional map (7.5,6) cross the unit circle at an irrational angle.

In Fig. 7.17 we show a bifurcation diagram for the self-focusing case in (p, a) parameter space. Notice that this diagram would occupy a small square

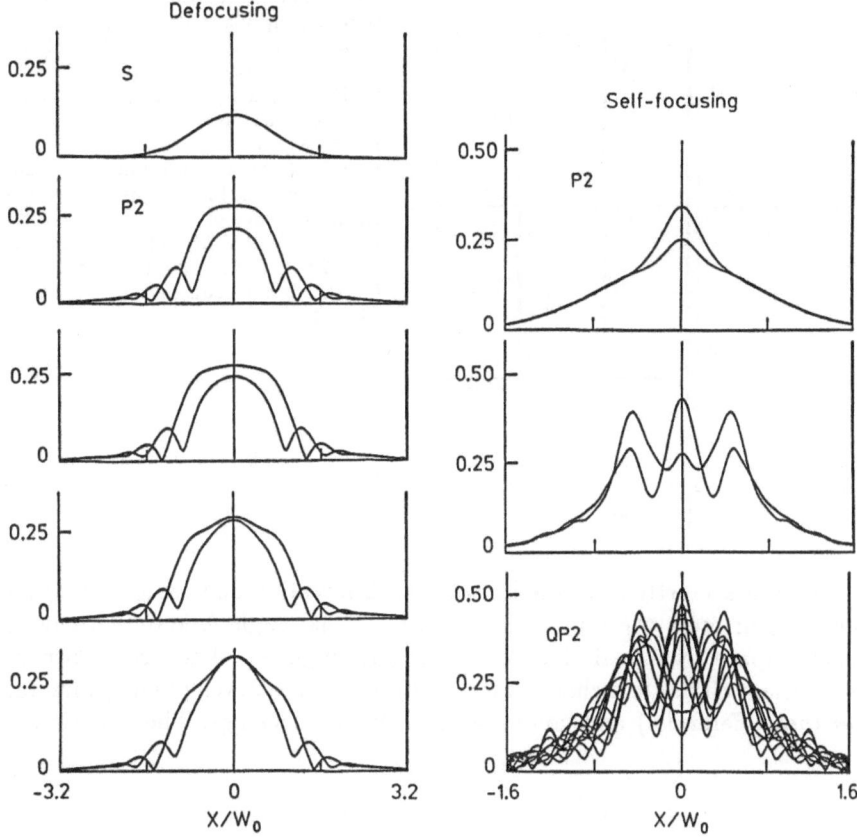

Fig. 7.16. Comparison of self-defocusing and self-focusing instabilities as the input Gaussian peak intensity $|a(0)|^2$ is increased from 0.1 to 0.5. The defocusing case is similar to the plane-wave behavior except that spatial effects are clearly evident with the beam center returning to a stable fixed point before the wings ($P = -8$, $F = 10$, $\phi = -0.4$, $R = 0.9$). Breakup of the profile into spatial rings signals the onset of quasi-periodic motion

region of the corresponding plane-wave bifurcation diagram (Fig. 7.5) where the latter is either stable or a simple period-2 orbit. The situation has now dramatically changed with bifurcation sequences to chaotic attractors occurring via a number of new routes. The notation in Fig. 7.17 is explained in the caption. Transition sequences involving a period-doubling bifurcation from a stable fixed point (S) to a period-2 orbit (T^1), followed by a Hopf bifurcation of the periodic orbit to invariant circles (T^2) are common. Again attractors coexist in the phase space as is evidenced by the appearance of different bifurcation sequences under variation of either $|a(0)|^2$ or p. Along the line $p = 8$, we observe a sequence involving stable – period 2 – intermittant chaos. Coexisting with this sequence is an unusual bifurcation involving period doubling of invariant circles. In Fig. 7.18 we show the attractors embedded in a three-dimensional phase space for this latter sequence. The period doubling of the invariant circles does not follow any of the usual bifurcation sequences. This

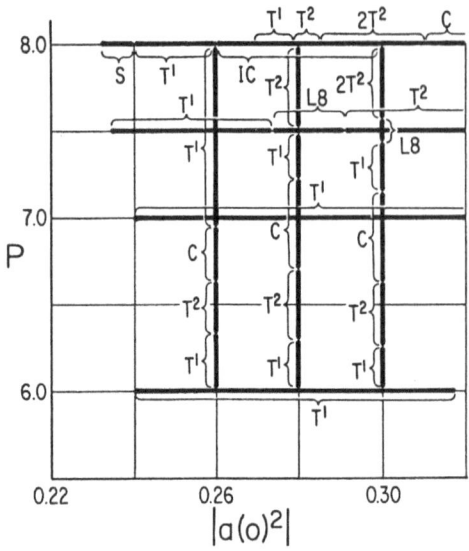

Fig. 7.17. Bifurcation diagram for a self-focusing Gaussian input beam in (p, a) parameter space. Typical sequences involve stable (S) – period 2 (T^1) – quasi-periodic (T^2) – chaos (C). Another sequence along the line $p = 8$ involves stable (S) – period 2 (T^1) to intermittant chaos (IC). On this line an unusual coexisting sequence of bifurcations involving period doubling of invariant circles $(T^2 \rightarrow 2T^2)$ occurs. Phase portraits for this latter sequence are shown in Fig. 7.18

doubling occurs shortly after the initial Hopf bifurcation and requires that an eigenvalue pair cross the unit circle at an irrational angle followed closely in parameter space by a real eigenvalue going through -1. The breakdown of these invariant circles to a chaotic attractor involves the development of fractal curves (nondifferential) and complicated frequency lockings. The chaotic at-

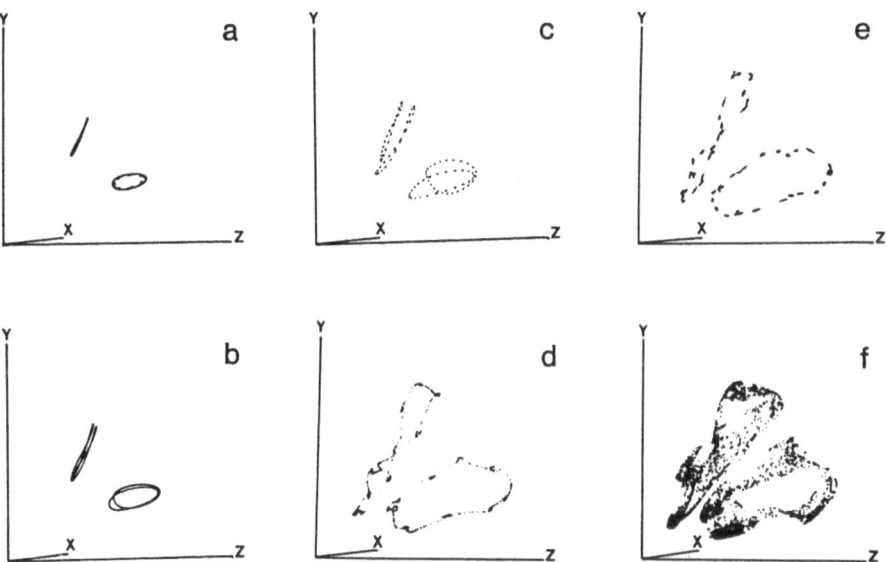

Fig. 7.18a–f. Attractors embedded in a three-dimensional phase space (I_{n+2}, I_{n+1}, I_n). **(a)** Invariant circles bifurcate from a period-2 orbit (not shown), **(b)** the circles period double, **(c)** frequency locking occurs to a high period orbit, **(d)** fractal (nondifferentiable) curves develop followed by **(e)** frequency locking, and **(f)** a two piece chaotic attractor

tractor consists of two pieces reflecting the underlying period-2 orbit (which is present even in the plane-wave map).

We end this subsection with an example of a new type of transition to chaos. At the beginning of the present subsection, we discussed how the plane-wave map itself is unstable to short-scale transverse fluctuations, and showed the initial instability growth from an initially smooth Gaussian spatial profile. There we alluded to the fact that the transverse instability induced a modulational-type chaos, thereby ruling out the possibility of period-doubling cascades to chaos even for plane waves. The instability growth depicted in Fig. 7.11 continues until it saturates, at which point there is a period-2 oscillation between finite K spatial rings. As the system is driven harder, energy begins to drift from the finite K spatial modes back toward the mean

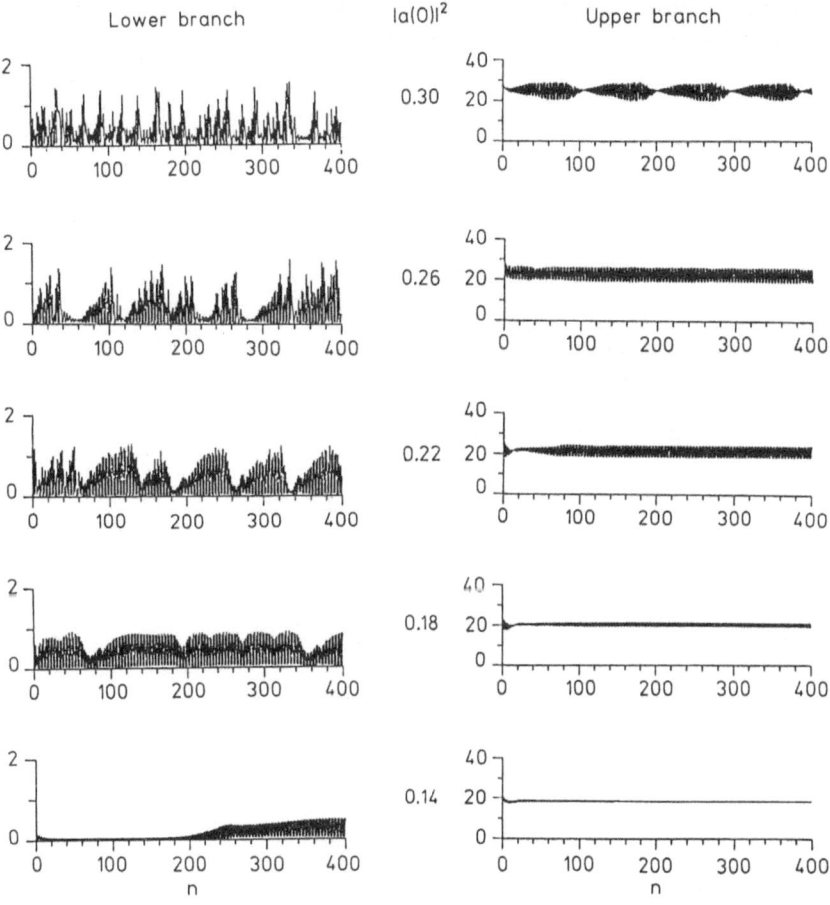

Fig. 7.19. Comparison of modulational chaos on the lower and upper bistable branches as the input peak intensity $|a(0)|^2$ varies. The time series (discrete n) correspond to beam center oscillations. The exponential growth of the lower branch solution from the weakly unstable fixed point is clearly evident at $|a(0)|^2 = 0.14$. (Compare with the full profiles in Fig. 7.11). ($p = 6$, $F = 33$, $\phi = 0.4$)

$(K = 0)$, which is an unstable saddle-fixed point; the mean is the original smooth Gaussian-like profile. This drifting of energy back and forth between different spatial scales destabilizes the period-2 oscillation. Figure 7.19 compares time series of the modulus of the beam center amplitude on the lower and upper bistable branches as the input intensity $|a(0)|^2$ is increased from 0.14 to 0.30 $(p = 6, F = 33, \phi = 0.4)$. The bottom trace on the left at $|a(0)|^2 = 0.14$ shows an initial transient that appears to have settled down to a stable fixed point (i.e., $K = 0$ smooth profile). However, by the 150th resonator pass, the smooth profile (Figs. 7.11a and b) begins to break up and the spatial rings grow exponentially and saturate by the 250th resonator pass. The final asymptotic state is a stable period-two oscillation between transverse spatial rings. At $|a(0)|^2 = 0.18$, the period-2 oscillation is destabilized and the system begins to drift back toward the mean, and we now observe a modulation of the period-two oscillation. As the system is driven harder, the modulation period decreases and the traces become progressively more chaotic.

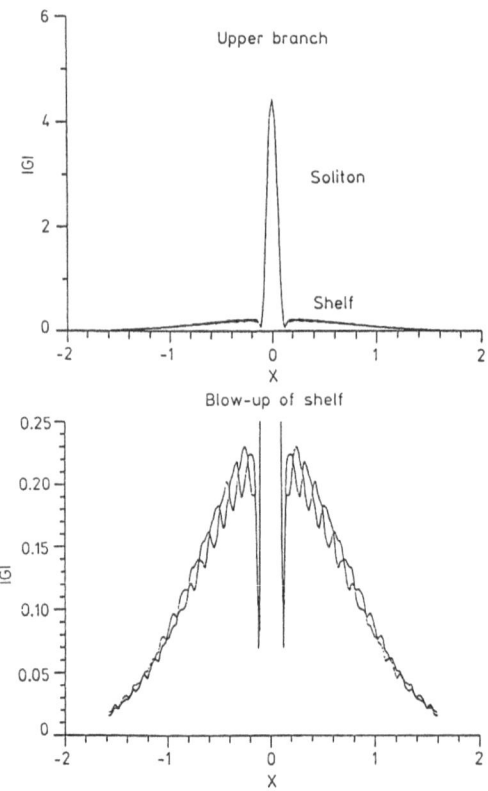

Fig. 7.20. Upper branch soliton sitting on a broad shelf at $|a(0)|^2 = 0.14$ $(p = 6, F = 33, \phi = 0.4)$. The blow-up of the shelf shows that it coincides with the lower branch solution (Compare with Fig. 7.11a). Now, however, the soliton in the center has frozen out the spatial rings preventing further growth as in Fig. 7.11b

A rather intriguing result is that the presence of a large solitary wave tends to stabilize the above chaotic behavior. The above time series on the left half of the figure refer to lower branch instabilities, but we know that at these parameter values, the upper bistable branch consists of a single solitary wave sitting on a broad low amplitude shelf (Fig. 7.20a). Figure 7.20b shows a blow-up of the shelf, and it is evident that this shelf is just the lower branch $K = 0$ solution of Fig. 7.11a; an overlay of the two figures shows that this is precisely the case. If we now begin to drive the system harder by increasing $|a(0)|^2$ and moving along the upper branch, we find it harder to destabilize the upper branch solution (right half of Fig. 7.19). At $|a(0)|^2 = 0.30$, for example, the lower branch solution is fully chaotic while the upper branch solution is a smoothly modulated period-2 type oscillation. Perhaps, even more interesting is the fact that if we drive the system with a strongly tapered input (apertured input beam) of the order of the upper branch solitary wave width, the upper branch solution is a stable fixed point! This latter result is consistent with the prediction of the plane-wave map where the upper branch fixed point is stable. We can see from this result that the lower branch instability drives the upper branch solution to chaos [7.22].

7.6 Gaussian Beams (3D Problem)

Relatively few concrete analytic results exist for the full three-dimensional non-linear evolution equation. Stability analysis shows that an initially smooth profile (or plane wave) will break up into spatial filaments if the nonlinearity is cubic: the so-called Benjamin-Feir or modulational instability. Mathematically, it can be shown that these filaments, once formed, will focus indefinitely; physically, diffusion or saturation effects will limit their growth, although in some instances material breakdown occurs first. As emphasized throughout the present chapter, our situation is quite different from conventional laser beam propagation [7.25]. Asymptotic states of the optical field envelope may be controlled by adjusting the degree of dissipation in the resonator. It may be even possible to control and stabilize filament growth by lowering the resonator finesse.

Our studies to date have been mainly confined to the case of a saturable nonlinearity as in (7.3). The dynamical evolution of the full two-dimensional beam profile has been tracked for the same case, as discussed in Sect. 7.5.2. We have found that the switching and initial growth of solitary waves is identical to the one-transverse dimensional problem. Figure 7.21 shows the time evolution of one quadrant of the two-dimensional profile $(|G_n(\eta, \xi)|)$ every 20th resonator pass $(n = 20, 40, \ldots)$. Solitary waves again grow from the outer edge of the cylindrical on-spot toward the center of the beam. By the 140th pass the outer two concentric rings appear to have stabilized; their heights and widths agree with the one transverse dimension predictions in Sect. 7.5.2. The center of the beam continues to oscillate and there appears to be a slow periodic recurrence of the full shape over a slow time scale $(n \simeq 400)$. The oscillation at beam center

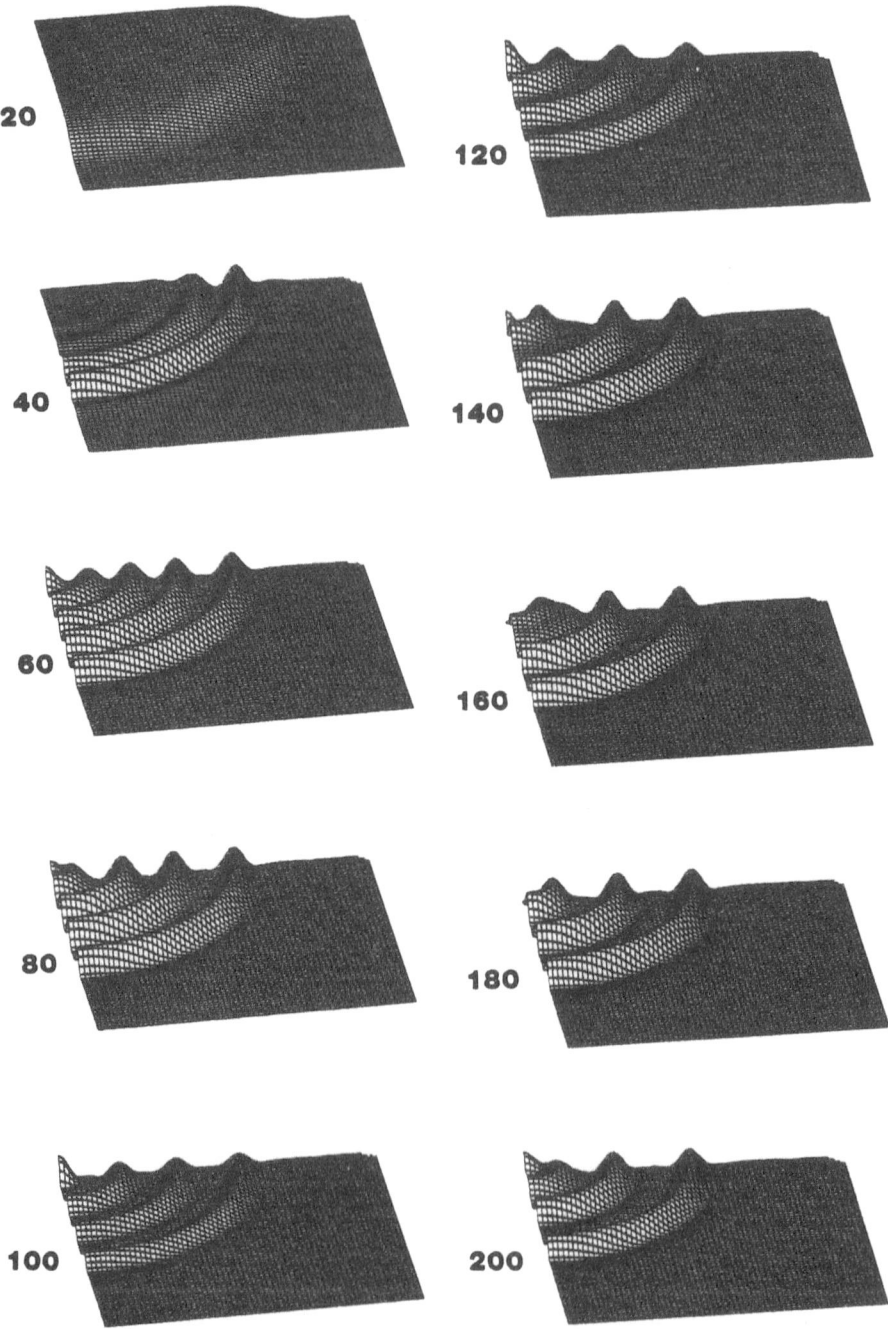

Fig. 7.21. Time evolution of one quadrant of the full two-dimensional beam profile ($p = 2$, $F = 200$, $\phi = 0.4$, $R = 0.9$). Sharp gradients on the cylindrical on-spot on the 20th resonator pass ($n = 20$) cause solitary waves to evolve toward beam center (compare with Fig. 7.13). For this case of saturable nonlinearity, the high contrast concentric rings are stable

appears to be due to the different nature of the solution near the origin where the radial nonlinear evolution equation has a singularity.

An important point to note here is that the outer concentric rings are stable and there is no evidence of filamentation. If, on the other hand, we used a Kerr nonlinearity the evolution of the beam would be quite different. Concentric rings would initially appear but these then start to break up into spatial filaments that then critically focus [7.18]. We have also found that filaments form in the saturable case if we choose the Gaussian peak intensity to just exceed threshold for switching to the upper bistable branch. In this case a low-intensity circular ring appears about the central intense soliton and then breaks up into four symmetrically disposed filaments. In contrast to the Kerr case, saturation stabilizes these filaments. The final figure (Fig. 7.22) shows how stronger dissipation ($R = 80\%$) can inhibit growth of high-contrast rings. The initial growth is similar to that of Fig. 7.21, but now instead of clean solitary waves, low-contrast circular ripples appear across the cylindrical beam profile. This latter figure is a nice example of how asymptotic states may be controlled by dissipation.

Clearly, there is much to be done on this three-dimensional problem. No attempt has been made to study this system in the turbulent regime along the lines discussed in Sect. 7.5.3. One particularly intriguing question is how the new spatial instability to short-wave transverse fluctuations in the map, predicted in [7.22] and discussed in Sect. 7.5.2, will manifest itself in higher dimensions. The two-dimensional degeneracy of the unstable modes (note that K^2 and *not* the direction K is chosen by dynamical considerations) makes it unclear as to what final asymptotic state will evolve. The answer to these problems will require access to an interactive computational facility with a dedicated high speed computer or array processor.

7.7 Summary

We have shown that a broad class of instabilities are associated with the intra-cavity field in a passive optical ring resonator. At the simple plane-wave level of approximation, quantitative information can be obtained by constructing global phase portraits [7.17]. A new type of bifurcation is associated with the transversal intersection (homoclinic orbit) of these phase portraits. The dynamical behavior, though complicated, can be completely understood in terms of the changing topology of the phase plane. The more realistic Gaussian beam problem extends significantly the allowable class of bifurcations. Indeed, the plane-wave map itself has been shown to be more unstable to perturbations with a short-scale transverse structure than to plane wave perturbations [7.22]. One important ramification of this latter result is that period-doubling cascades should be an unlikely scenario for the transition to optical chaos, at least in high finess resonators.

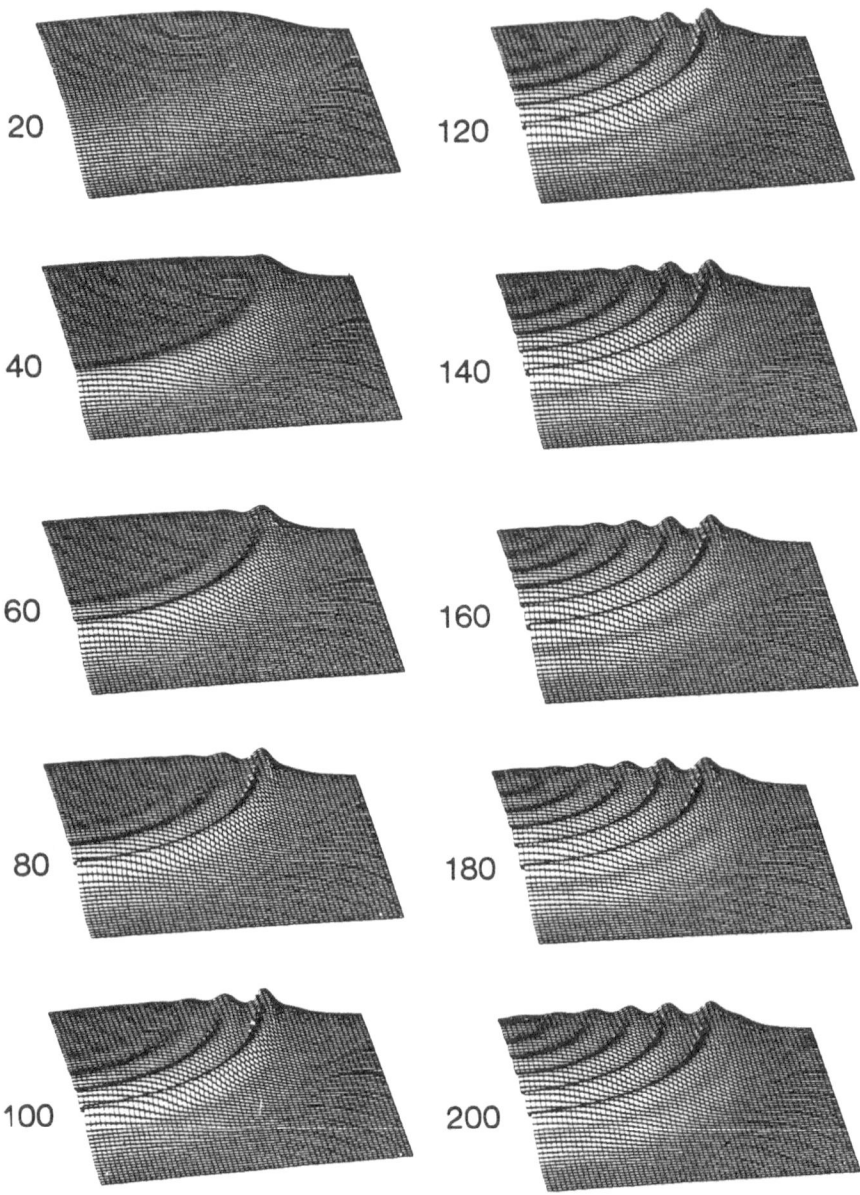

Fig. 7.22. Inhibition of solitary wave growth due to strong dissipation ($p = 2$, $F = 100$, $\phi = 0.4$, $R = 0.8$). The final asymptotic state consists of low-contrast circular ripples superimposed on a smooth envelope

Profound spatial changes can occur across the transverse beam profile for both self-focusing or self-defocusing nonlinearities. Self-focusing nonlinearities cause the profile to break up into transverse soliton or solitary wave structures. These latter nonlinear structures may exist as steady states (fixed points) of the field envelope or they may persist as spatially coherent structures that undergo regular or chaotic temporal motion. The chaotic state is analogous to experimentally-measured low level turbulent states in fluids. Coexistence of attractors leads to complex bifurcation involving, for example, period doubling of invariant circles [7.21]. Perliminary numerical evidence suggests that regular dynamical motion may occur on three- [7.35] or four-dimensional tori (three or four incommensurate frequencies in a power spectrum). This would violate the Newhouse-Ruelle-Takens scenario, but is consistent with a recent numerical study by *Grebogi* et al. [7.36].

The study of instabilities in infinite-dimensional systems is many orders of magnitude more difficult than the study of simple plane-wave models. These infinite-dimensional models capture the relevant physics and are closely related to actual experiments. It is evident from the results of Sect. 7.5 and 6 that experiments will have to be carefully designed to track and discriminate between various bifurcation sequences. In many instances, variation of more than one control parameter will be necessary; for example, both laser intensity and frequency. The reconstruction of phase portraits from experimental time series will prove a very powerful tool in analyzing the motion on higher dimensinal regular or chaotic attractors.

Many important theoretical questions remain to be addressed. The results reviewed here represent a very limtied exploration of parameter space. Strong dissipation may lead naturally to a one-dimensional behavior consistent with Feigenbaum period-doubling cascades. The full three-dimensional problem (Sect. 7.6) is only now being studied, and we anticipate further rich dynamical behavior for this system. Finally, structural stability of higher-dimensional attractors to external perturbations or noise will need to be investigated.

Acknowledgements. The work to be described in this chapter involved the direct collaboration of different groups of people. Much of the plane-wave work was carried out in collaboration with Steve Hammel and Chris Jones, the transverse soliton work with Dave McLaughlin and Alan Newell, and the earlier work with Hyatt Gibbs and Fred Hopf. The author is particularly indebted to Hyatt for his original suggestion of the need to study transverse effects and for his ongoing interest and active involvement. Finally, I would like to thank Maggie Whitney for her patience in typing and editing the manuscript. Various parts of the research were supported by the National Science Foundation, the Air Force Office for Scientific Research (AFOSR), the Office of Naval Research (ONR), and the Army Research Office (ARO).

References

7.1   A useful review of this topic may be found in "Nonlinear Dynamics", **357**, Annals of the New York Academy of Sciences, ed. by R.G. Helleman (1980)

7.2   A. Libchaber, S. Fauve, C. Laroche: Physica **7D**, 73 (1984) and references therein

7.3   H.L. Swinney: Physica **7D**, 3 (1984) and references therein

7.4   H.L. Swinney, J.P. Gollub (eds.): *Hydrodynamic Instabilities and the Transition to Turbulence*, 2nd. ed., Topics Appl. Phys., Vol. 45 (Springer, Berlin, Heidelberg 1985)

7.5   S.E. Newhouse, D. Ruelle, F. Takens: Commun. Math. Phys. **64**, 35 (1978)

7.6   M.J. Feigenbaum: J. Stat. Phys. **19**, 25 (1978)

7.7   Y. Pomeau, P. Manneville: Commun. Math. Phys. **74**, 189 (1980)
7.8   E.N. Lorenz: J. Atmos. Sci. **20**, 130 (1963)
7.9   A.J. Lichtenberg, M.A. Lieberman: *Regular and Stochastic Motion*, (Springer, Berlin, Heidelberg 1983) Chap. 7
7.10   H. Haken: Phys. Lett. **53A**, 77 (1975);
       Other contributions to this volume. See in particular the articles by N.B. Abraham, et al. and F.T. Arrechi (Chaps. 2 and 3)
7.11   K. Ikeda: Opt. Commun. **30**, 256 (1979)
7.12   Y. Silberberg, I. Bar Joseph: Phys. Rev. Lett. **48**, 1541 (1982);
       H.G. Winful, G.D. Cooperman: Appl. Phys. Lett. **40**, 298 (1982)
7.13   H.M. Gibbs, F.A. Hopf, D.L. Kaplan, R.L. Shoemaker: Phys. Rev. Lett. **46**, 474 (1981)
7.14   R.G. Harrison, W.J. Firth, C.A. Emshary, I.A. Al-Saidi: Phys. Rev. Lett., **51**, 562 (1983);
       R.G. Harrison, W.J. Firth, I.A. Al-Saidi: Phys. Rev. Lett. **53**, 258 (1984);
       M. LeBerre, E. Ressayre, A. Tallet, K. Tai, F.A. Hopf, H.M. Gibbs, J.V. Moloney: "Optical bistability and instabilities via diffration-free-encoding and a single feedback mirror", postdeadline presentation IQEC '84, Los Angeles. The first observation of period-2 and evidence for period-6 oscillations in a CW experiment was reported.
7.15   J.V. Moloney, F.A. Hopf, H.M. Gibbs: Phys. Rev. **A25**, 3442 (1982); and Appl. Phys. **B28**, 98 (1982);
       J.V. Moloney, F.A. Hopf: J. Opt. Soc. Am. **71**, 1634 (1981)
7.16   W.J. Firth, E.M. Wright: Phys. Lett. **92A**, 211 (1982);
       W.J. Firth, E. Abraham, E.M. Wright: Applied Phys. **B28**, 170 (1982); a number of analytic models that assume mode matched Fabry-Perot resonators are discussed and referenced in *Optical Bistability II*, ed. by C.M. Bowden, H.M. Gibbs, S.L. McCall (Plenum, New York 1984)
7.17   S.M. Hammel, C.K.R.T. Jones, J.V. Moloney: "Global Dynamical Behavior of the Optical Field in a Ring Cavity", J. Opt. Soc. Am. **B2**, 552 (1985);
       J.V. Moloney, S.M. Hammel, C.K.R.T. Jones: "Optical Bistability II", p. 87 (1984)
7.18   J.V. Moloney: "Solitons in Optical Bistability", Phil. Trans. Roy. Soc. **A313**, 429 (1984); J. Opt. Soc. Am. **B3**, 467 (1984);
       D.W. McLaughlin, J.V. Moloney, A.C. Newell: "Two-Dimensional Transverse Solitary Waves as Asymptotic States of the Field in an Optical Resonator", to be published
7.19   A.C. Newell: *Solitons in Mathematics and Physics*, CBMS Lecture Series, Vol. 48, SIAM, 1984
7.20   D.W. McLaughlin, J.V. Moloney, A.C. Newell: Phys. Rev. Lett. **51**, 75 (1983)
7.21   J.V. Moloney: Phys. Rev. Lett. **53**, 556 (1984)
7.22   D.W. McLaughlin, J.V. Moloney, A.C. Newell: "A New Class of Instabilities in Passive Optical Cavities", Phys. Rev. Lett. **55**, 168 (1985)
       D.W. McLaughlin, J.V. Moloney, A.C. Newell: Physica D (to appear)
7.23   A comprehensive review of Optical Bistability and Instabilities may be found in *Optical Bistability*, ed. by C.M. Bowden, M. Ciftan, and H.R. Robl (Plenum, New York, 1981); "Optical Bistability II", ed. by C.M. Bowden, H.M. Gibbs, and S.L. McCall (Plenum, New York, 1984);
       R. Bonifaccio, L.A. Lugiato: Lett. Nuovo Cimento **21**, 510 (1978)
7.24   L. Allen, J.H. Eberly: *Optical Resonance and Two Level Atoms* (Wiley-Interscience, 1975)
7.25   R.Y. Chiao, E. Garmire, C.H. Townes: Phys. Rev. Lett. **13**, 479 (1964);
       E. Garmire, R.Y. Chiao, C.H. Townes: Phys. Rev. Lett. **16**, 347 (1966);
       A.J. Campillo, S.L. Shapiro, B.R. Suydam: Appl. Phys. Lett. **23**, 628 (1973)
7.26   J. Guckenheimer, P. Holmes: *Nonlinear Oscillations, Dynamical Systems, and Bifurcations of Vector Fields* (Springer, New York 1983)
7.27   J.V. Moloney: Opt. Commun. **48**, 435 (1984);
       J.V. Moloney, H.M. Gibbs, F.A. Hopf: *Optical Bistability II*, ed. by C.M. Bowden, H.M. Gibbs, and S.L. McCall (Plenum, New York 1984)
7.28   J.V. Moloney, S.M. Hammel, C.K.R.T. Jones: J. Opt. Soc. Am. **B3**, 499 (1984);
       S.M. Hammel, C.K.R.T. Jones, J.V. Moloney: to be published

7.29 S. Smale: Bull. Amer. Math. Soc. **73**, 747 (1967);
S.E. Newhouse: "Progress in Mathematics", No. 8 (Birkhauser-Boston 1980) pp. 1–114;
N.K. Gavrilov, L.P. Silnikov: Math. USSR S6, **88**, 467 (1972);
J.A. Yorke: private communication
7.30 J.V. Moloney, H.M. Gibbs: Phys. Rev. Lett. **58**, 1607 (1982)
7.31 J.C. Roux: Physica **7D**, 57 (1983)
7.32 R.S. Gioggia, N.B. Abraham: Phys. Rev. Lett. **51**, 650 (1983);
R.S. Gioggia, A.M. Albano, C.M. Searle, T.H. Bhyba, N.B. Abraham: J. Opt. Soc. Am. B1, 499 (1984)
7.33 P. Grassberger, I. Procaccia: Phys. Rev. Lett. **50**, 346 (1983); Physica **9D**, 189 (1983)
7.34 G. Iooss, W.F. Langford: p. 489 in Ref. 7.1
7.35 P. Davis, K. Ikeda: Phys. Lett. **100A**, 455 (1984)
7.36 C. Grebogi, E. Ott, J.A. Yorke: "Attractors on an N-Torus: Quasiperiodicity versus chaos", preprint

8. Experimental Verification of Regenerative Pulsations and Chaos

M.W. Derstine, J.L. Jewell, H.M. Gibbs, F.A. Hopf, M.C. Rushford, L.D. Sanders, and K. Tai

With 19 Figures

Experiments are described which verify the basic principles of regenerative pulsations and the Ikeda instability. The regenerative pulsations of an intrinsic optical bistable device resemble those we obtain from a noise-free computer simulation. The agreement improves when the pulsation frequency is much lower than that of the laser noise. Bifurcations up to period-8 and various paths to chaos are seen in a hybrid optical experiment, consistent with the theoretical predictions including real noise levels. A test for distinguishing a chaotic system from one that is simply noisy is outlined. Finally, we describe the apparent observation of instabilities and chaos in an intrinsic cavity-less bistable device.

8.1 Regenerative Pulsations

The earliest prediction of an instability in a passive optical device was made by *McCall* [8.1]. His theory and his experiment using a hybrid optical bistable device showed that relaxation oscillations can occur when the nonlinearity has two contributions of opposite sign and different time constants. McCall further outlined an intrinsic pulsation process in which switching is due to a fast electronic effect, but a slower thermal effect prevents either state from being stable. We reported observation of this intrinsic process in a 4.2 μm-thick bulk GaAs etalon at $\simeq 80$ K, with the free-exciton resonance providing the electronic nonlinearity [8.2]. Subsequent room-temperature observations using the same sample as well as bulk GaAs and multiple-quantum-well (MQW) structures with thicknesses 1.5–2.0 μm show very similar results.

The hysteresis loop of Fig. 8.1a shows bistability in our GaAs device using the decrease in refractive index from saturation of the free-exciton absorption below resonance [8.3]. Optical bistability due to the increase in optical path length with temperature [8.3] has also been seen in GaAs. Figure 8.1b illustrates the competition between excitonic and thermal effects, with the result that the switch-down intensity is higher than the switch-up intensity for the particular input pulse used. In this case, significant tuning of the Fabry-Perot transmission peak relative to the laser frequency occurs during the 40-μs "on" time. Figure 8.2a shows that during a long ($>100\,\mu$s) square-top input pulse, sufficient heating occurs to cause excitonic switch-down followed by cooling and

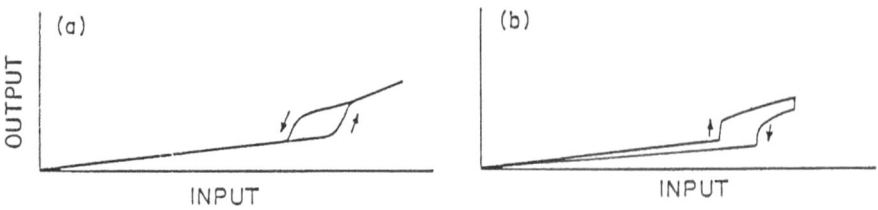

Fig. 8.1. Experimental hysteresis loops: (**a**) "normal" bistability with 1-μs triangular input pulse, (**b**) "backwards" hysteresis with 40-μs flat-topped input

excitonic switch-up, etc. The experimental data gives a peak transmission in the on state of 15–20%.

Theory predicts periodic oscillations, but they are found to be somewhat random in occurrence, mainly due to noise from the input laser. Random oscillations can be obtained from a single nonlinear effect if the input intensity fluctuations exceed the difference between switch-up and switch-down intensities. This case is shown in Fig. 8.2b. The difference in appearance from Fig. 8.2a is readily apparent. The switching times in Fig. 8.2a are about 100 ns or less, consistent with excitonic switching [8.3]. Throughput decreases as a result of heating in the on state and increases during cooling in the "off" state (Fig. 8.2a).

The regenerative pulsations are interpreted as follows. Denote the Fabry-Perot peak transmission frequency as ν_{FP}, the laser frequency as ν_L, and the exciton frequency as ν_{EX}. Initially $\nu_{FP} < \nu_L < \nu_{EX}$. The input intensity shifts ν_{FP} sufficiently for excitonic switch-on, yielding $\nu_L \simeq \nu_{FP} < \nu_{EX}$. Heating then decreases ν_{EX} toward ν_L resulting in increased intracavity absorption and thus lower etalon transmission and finesse. The heating also decreases ν_{FP} toward

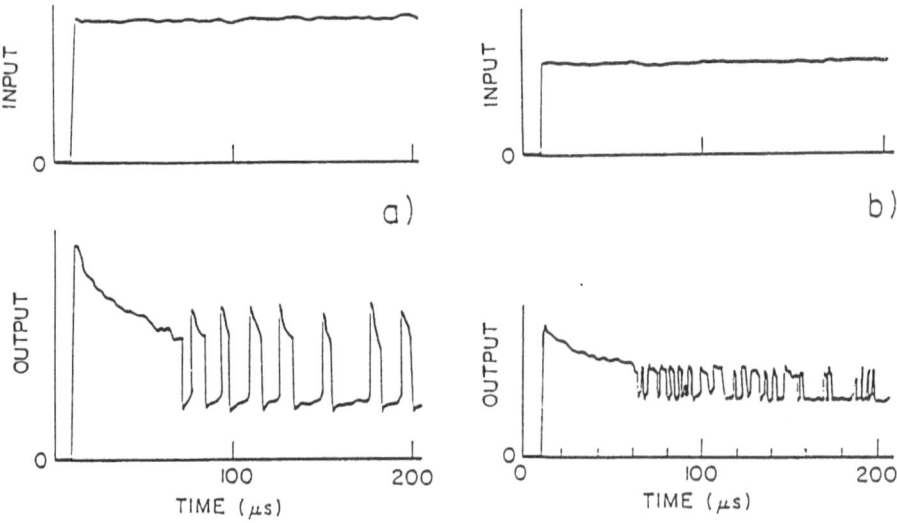

Fig. 8.2. Output vs time at 20 μs/div: (**a**) regenerative pulsations, (**b**) random-noise switching

ν_L until it is slightly less than ν_L. Rapid excitonic switch-down then occurs resulting in a jump of ν_{FP} to a value less than its original value. In the off state, cooling brings ν_{FP} back toward ν_L and decreases intracavity absorption until excitonic switch-up occurs again. In principle, the process repeats indefinitely. In practice, our sample lacked a heat sink and pulsations were not observed for pulses longer than 100 ms. This model is the basis for the computer simulation discussed below.

Figure 8.3 shows the experimental setup. The input laser was a Coherent Radiation 590 dye laser containing oxazine 750 perchlorate dye pumped by the 647 and 676 nm 5-W output of a Spectra Physics 171 Kr$^+$ laser. The dye laser output was about 1 Å wide at 0.4 W and tunable from 770 to 870 nm. Triangular and square-top optical pulses were obtained by controlling an acousto-optic modulator with a Tektronix PG 508 pulse generator. A feedback loop from the input detector to the modulator driver reduced input noise to about 3 %. The diffracted beam was focused to about a 10-μm diameter 100-mW spot on the sample. The GaAs sample was grown by molecular beam epitaxy. It consisted of 4.2 μm of GaAs sandwiched between 0.21-μm Al$_{0.42}$Ga$_{0.58}$As layers. A 1–2mm diam. hole was etched through the 150-μm GaAs substrate to provide optical access. Multilayer coatings increased the surface reflectivity to 0.9. Silicon photodiodes with 0.6-ns FWHM pulse response and a fast oscilloscope were used to observe the input and output intensities.

A computer program was written to simulate the effects previously discussed. Assumptions of transverse spatial uniformity in the beam and no diffraction or scattering cause this model to more closely resemble a waveguide device than our setup. The etalon transmittance and reflectance are determined by iteration of steady-state formulae over time intervals Δt which are smaller than thermal conduction times but larger than the medium response time and are much larger than the cavity round-trip time. For each Δt, the temperature change is calculated from the difference between the energy absorbed (product of absorbed power and Δt) and that dissipated or stord. The absorbed power is calculated by subtracting from the input power the sum of the reflected and transmitted powers. The excitons store a small amount of the absorbed energy, and dissipation is due mostly to thermal conduction through the GaAs. Since transverse effects are ignored, a uniform temperature T_0 is assumed in the focal region. Heat conduction is calculated across the boundary between the focal

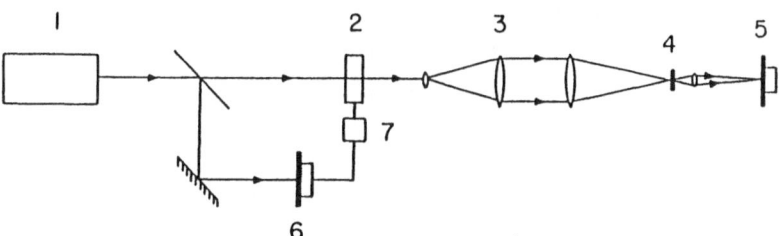

Fig. 8.3. Experimental set-up: *(1)* tunable laser, *(2)* acousto-optic modulator, *(3)* focusing optics, *(4)* GaAs etalon, *(5)* output detector, *(6)* input monitor, *(7)* feedback noise reducer

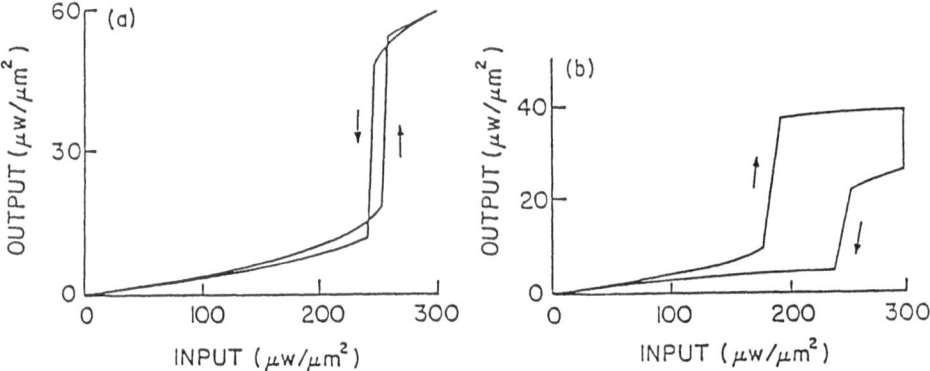

Fig. 8.4. Computer simulated hysteresis loops: **(a)** "forward" loop in a 1-μs pulse, **(b)** "backwards" loop from a 6-μs pulse

region and an annular ring around it of equal area and of uniform temperature T_1. Areas outside this ring are assumed to be at 77 K, and conduction across the outer boundary of the T_1 region is also calculated. The values of ν_{EX}, ν_{FP}, absorptance and finesse are updated every Δt yielding the transmission. A saturable two-level homogeneously broadened model was used for the exciton absorption feature.

Good agreement between the structure and behavior of the simulated results and the experimental data was achieved. Figure 8.4 shows simulation results corresponding to Fig. 8.1, and regenerative pulsations computed from a noise-free input are displayed in Fig. 8.5. With the sample in the on state, heating increases absorption in the sample. This lowers the peak transmission, but at the same time ν_{FP} moves closer to ν_L. Thus it is not qualitatively obvious whether throughput should increase or decrease with heating. The experiment shows a sharp decrease (Fig. 8.2). The simulation revealed this structure to be very sensitive to the slope of the absorption edge of GaAs. An exponential band-tail model was used for the band-band absorption (assumed to be

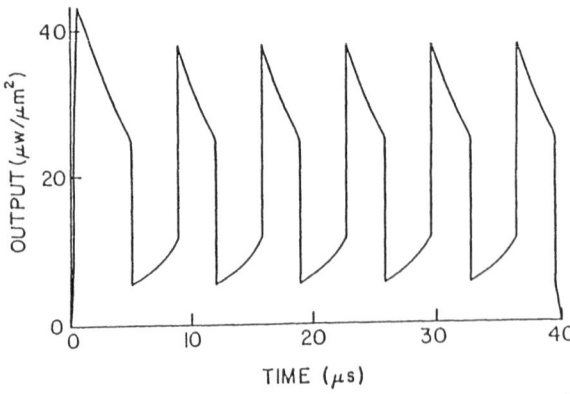

Fig. 8.5. Computer simulated regenerative pulsations

nonsaturating at our intensities) and the slope, determined from Sturge's data [8.4], yielded a structure similar to the experiment (Fig. 8.5). Slopes less than about half this value resulted in increased transmission with temperature. In both experiment and simulation, raising the input intensity sufficiently results in the device remaining on. Although the simulation employs a rather simplified model, its qualitative agreement with experiment indicates that it contains the essential theoretical aspects of the process. Quantitative agreement in areas such as power requirements, time scales, and throughput would require accurate detailed knowledge of our sample characteristics (e.g., band-edge structure).

Our room-temperature observations were made on a variety of samples having GaAs thicknesses 1.5–10 μm and quantum well thicknesses 53, 152, 299, and 336 Å as well as bulk GaAs. Frequencies of several MHz with the pulsations resembling Fig. 8.2a in appearance were achieved in 1.5–2 μm devices. Pulsations closely resembling the noise-free simulation of Fig. 8.5, but on a ms time scale, are shown in Fig. 8.6. The device consisted of three "flakes" of the 336 Å sample stacked and cemented between dielectric mirrors to form an etalon with ~6 μm GaAs thickness and ~15 μm total thickness. The increased volume of the device produced a longer thermal response time. In this case most of the laser noise was at frequencies much higher than the thermal frequency response, thus the effects due to the noise were much less noticeable than in Fig. 8.2a.

Much faster regenerative pulsations, perhaps approaching the ns time scale of GaAs carrier relaxation, might be obtained by reducing the device size and hence the thermal response time. Alternatively a device with two opposing nonthermal mechanisms might be extremely fast. Other possible optical square-wave generators for non-gain media include ring-cavity self-pulsing resulting from mode competition in absorptive bistability [8.5], and the Ikeda instability [8.6,7] discussed below.

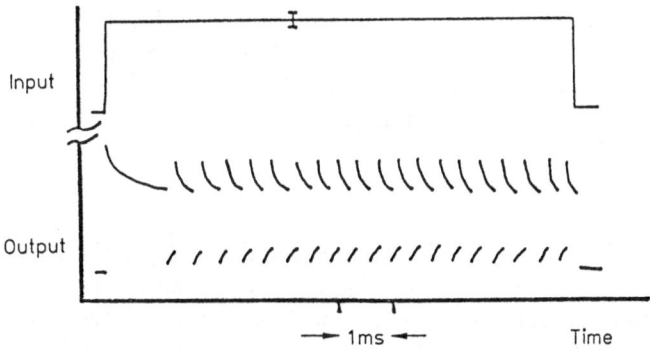

Fig. 8.6. Regenerative pulsations when the thermal response time is much slower than the input noise. The error bar on the input trace shows the approximate width of the trace on the original photograph. Switching times were too fast to be seen with the oscilloscope on the 1-ms time scale

8.2 Ikeda Instabilities

In 1979 *Ikeda* showed that previously neglected time-retardation effects in a ring cavity could lead to instabilities in the output vs. time under conditions of dispersive optical bistability [8.6]. The requirement for such oscillations is that the medium response time τ must be shorter than the cavity round-trip time t_R. In 1980 *Ikeda* et al. further analyzed the ring cavity and pointed out that in one limit the dynamical equations of the cavity reduce to those for a hybrid bistable optical device in which a delay of t_R is introduced in the feedback circuit [8.7]. They analyzed the hybrid dynamics and found periodic outputs as well as chaotic or turbulent outputs.

Chaos is distinguished from noise in that it occurs in a purely deterministic noise-free system. Chaos is distinguished from periodic oscillations in that the smallest change in the initial conditions will result in exponentially diverging outputs in the chaotic case. The chaos and the path to chaos are exciting areas of study not only because of their possible importance in the design and operation of bistable devices but even more because of their importance in many diverse fields including mathematics, population dynamics, physiological control systems, hydrodynamics, etc.

8.2.1 Experimental Setup

Shortly after the predictions of *Ikeda* et al., the present authors reported the observation of periodic oscillations and chaos in a hybrid bistable optical device employing a transparent piezoelectric crystal [8.8]. The experiment described here is similar except that the intra-cavity phase shift is produced with an electro-optical modulator and the delay is produced by an optical fiber [8.9,10].

Figure 8.7 shows the experimental set-up. A helium-neon laser beam passes through a Glans prism polarizer, then through a KDP crystal four times before coming back through the polarizer where it is coupled into the optical fiber. The light emerging from the fiber is detected with an RCA 8852 photomultiplier, and the amplified electrical signal is applied to the KDP crystal.

The equation that describes our system is [8.9]

$$\tau \dot{X}(t) + X(t) = \mu\pi\{1 - \xi \cos [X(t - t_R) + X_b]\} \tag{8.1}$$

where $X = \pi V/V_h$, V is the voltage applied to the modulator, V_h is the half-wave voltage of the modulator, and $X_b = \pi V_b/V_h$ is a variable bias. The parameter μ is the bifurcation parameter. The value that it takes determines the character of the output of the system. In our experiment μ is proportional to the product of the input laser intensity and the amplifier gain. We measure μ directly in the apparatus by procedures discussed below. We have found [8.9,10] that we get the same results whether we vary μ by changing the laser power or by changing the gain. We prefer the latter procedure, since changing the laser power changes the noise level of the device, which is determined by

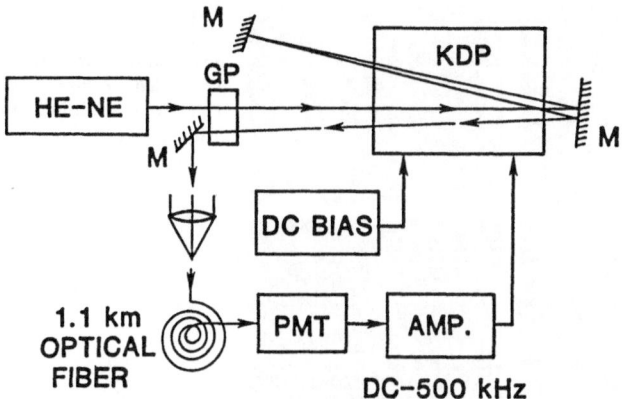

Fig. 8.7. Experimental layout of the hybrid device: He-Ne laser; GP, Glans prism; KDP, crystal; M, mirrors; PMT, photomultiplier

shot noise in the detector. The ability of the system to achieve extinction is measured by ξ, which in our experiment is 0.98 ± 0.01. We have explored the ranges of $-\pi<X_b<-\pi/2$. All of the data that are shown in detail were taken for $X_b = -\pi$ which is representative of the other cases. This choice of bias maximizes the domains of μ over which individual waveforms are stable and thus minimizes the consequences of the 1 % drift in the laser power, which is the major uncertainty in measuring μ. The range of μ that we can test at this bias lies below the "upper branches" of the bistable device. The procedure to measure μ is first to break the feedback loop between the final amplifier and the modulator. The input to the modulator is then shorted to insure a zero voltage input (the bias voltage is not changed), and the voltage V at the output of the final amplifier is measured. This determines μ as $\mu = V/2V_h$.

Ikeda showed that for $\tau\ll t_R$ the representation of the experiment could be simplified to the difference equation [8.7]

$$X_{n+1} = \mu\pi[1 - \xi \cos (X_n + X_b)] \; , \qquad (8.2)$$

where $X_n = X(nt_R)$ is the value of the amplitude at one instant in time within a time interval of duration t_R. Using the parameters for the current experiment, $X_b = -\pi$, and $\xi\approx1$, (8.2) reads

$$X_{n+1} = \pi\mu(1 + \cos X_n) \; . \qquad (8.3)$$

For low values of μ the output of the device is constant. As μ is increased the system bifurcates and a periodic output appears. In Fig. 8.8a is shown the time trace $X(t)$ and the power spectrum $S(\nu)$ of the first periodic waveform. As μ is increased further the system again bifurcates and the waveform and spectra in Fig. 8.8b appear. Further increase produces the balance of the waveforms and spectra shown. This route to chaos through the waveforms shown in Fig. 8.8a–c is known as the period doubling sequence.

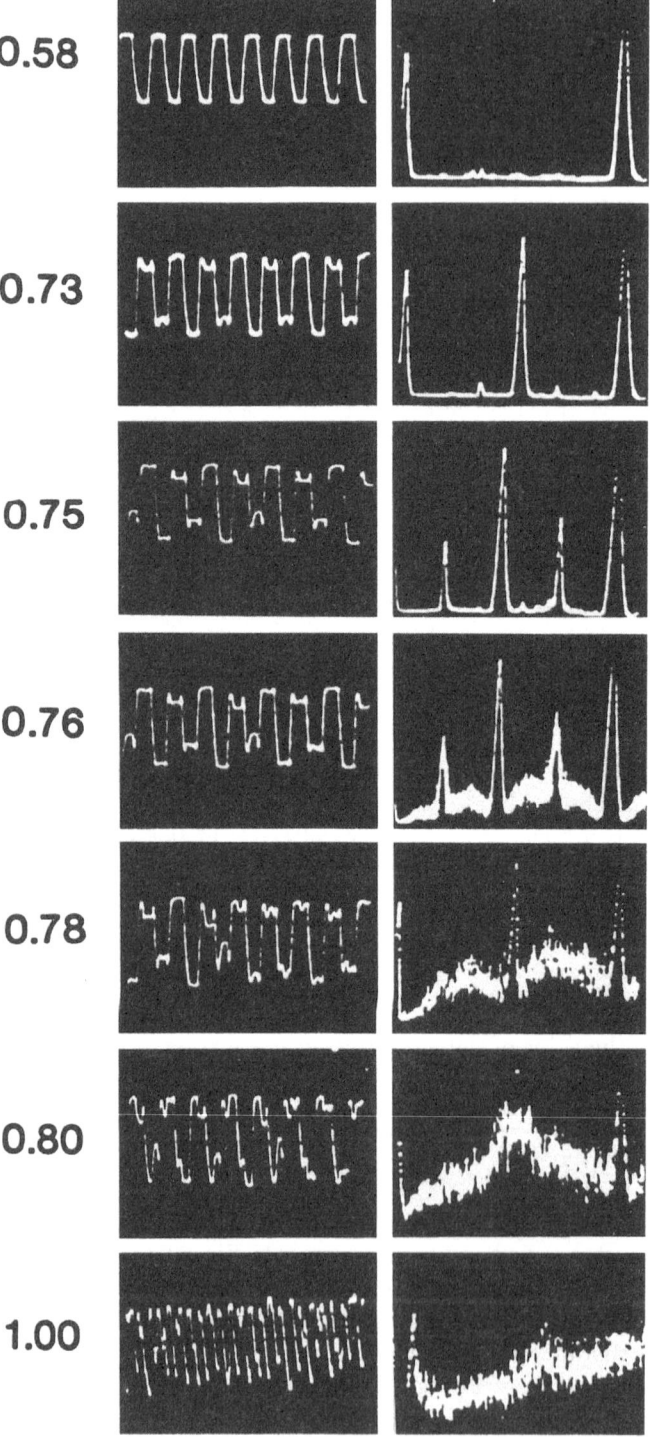

0.58

0.73

0.75

0.76

0.78

0.80

1.00

Fig. 8.8.
Waveforms *(right column)* and power spectra *(left column)* of the device shown in Fig. 8.7 as a function of increasing μ

8.2.2 Review of Period Doubling

Period doubling was first seriously investigated by *Lorenz* in 1963 [8.11], but it failed to attract much attention until the work of *Feigenbaum* [8.12] (discussed below).

As we have seen period doubling occurs as some parameter, called the bifurcation parameter, increases. As noted before the bifurcation parameter in our system is μ. If we consider our system to be ideal and modeled exactly by (8.3), then for small values of μ the output intensity of the system is constant. As μ is increased the output changes to oscilliatory with a period $2t_R$ (Fig. 8.8, $\mu = 0.58$). The power spectrum of this waveform naturally shows a strong peak at $1/2t_R$ as well as higher harmonics. As μ is increased further, the waveform's structure changes to have a period of $4t_R$ and the power spectrum now adds a peak at $1/4t_R$ (Fig. 8.8, $\mu = 0.73$). This doubling of the period of the waveform (and adding of peaks in the power spectrum) continues ad infinitum, until the period is infinite. In Fig. 8.8 $\mu = 0.75$ is the next period doubling. As we shall see in Sect. 8.2.3, this process takes place within a finite range of μ. When the system has period doubled as far as possible (an infinite number of times in an ideal system) an increase in the bifurcation parameter causes the system to become chaotic, that is, show irreproducible, seemingly random oscillations.

In the chaotic regime a reverse cascade takes place. The power spectrum of waveforms in the chaotic regime appear to have sharp peaks with a noisy background. The time traces, of course, look almost periodic, but never reproduce exactly (Fig. 8.8, $\mu = 0.76$–1.00). As the bifurcation parameter is increased the background rises up and washes out the peaks. This process is known as a reverse cascade since the basic periodicity halves as the bifurcation parameter is increased. In Fig. 8.8 $\mu = 0.76$ is shown a chaotic period 8. As μ is increased the output changes to that in Fig. 8.8 $\mu = 0.78$, a chaotic period 4, and then to that in Fig. 8.8 $\mu = 0.80$, a chaotic period 2. If periodic waveforms are denoted by P_p where the period of the waveform is pt_R, and the chaotic waveforms denoted by N_μ, the route to chaos can be described by

$$\text{steady} \rightarrow P_2 \rightarrow P_4 \rightarrow P_8 \rightarrow \ldots \rightarrow P_\infty \rightarrow N_\infty \rightarrow \ldots \rightarrow N_8 \rightarrow N_4 \rightarrow N_2 \rightarrow N_1 \ .$$

This scenerio applies to ideal systems. Our system is not ideal, with implications discussed in Sect. 8.2.5. For a more rigorous definition of period doubling, see Chap. 7.

8.2.3 Universality

In 1979 *Feigenbaum* discovered that some period-doubling systems behaved in qualitatively and quantitatively similar waves [8.12]. This similarity is known as universality. Some of the features that universality predicts for period-doubling systems are the range of μ over which specific waveforms exist and the heights of the peaks of the power spectrum.

If the value of μ that causes the system to switch from a period of pt_R to $2pt_R$ is denoted by μ_p, then the ratio $\delta_n \equiv (\mu_{2n} - \mu_n)/(\mu_{4n} - \mu_{2n})$ reaches a constant value δ for large n. *Feigenbaum* found that $\delta = 4.6692016...$ in a large class of situations. This ratio has been measured as a function of n in real systems as well [8.13]. One important point that this result shows is that the domains of existence of the waveforms will become geometrically smaller and reach μ_∞ at a finite μ. In our system we are able to measure a rough approximation to this number, but because we do not see bifurcations at $n>8$ we cannot be sure if it converges to δ.

As the system period doubles, Feigenbaum predicted that the height of the peaks in the power spectrum of successive subharmonics differ by 8.2 db. Since we are able to take spectra in real time, we find, in fact, that the subharmonics are about 8 db down, from the previous one.

8.2.4 Nature of Transitions

Another prediction of Feigenbaum is that the transitions between waveforms are like second-order phase transitions. This means that the waveforms change continuously. To determine experimentally the character of the transitions, we measured the power in the spectral peaks and plotted the results as a function of μ. The results of these measurements are shown in Fig. 8.9. The numerals label the experimental curves: 2 for P_2, 4 for P_4, and 8 for P_8. The height of the continuous background (labeled CHAOS) was used to determine the nature of the transition to chaos. The experimental curves show the proper shape at the bifurcations between the periodic waveforms to be analogous to second-order phase transition. The transition to chaos is not significantly different (to within the 1 % uncertainty in μ) than the transitions among the periodic waveforms. Within the error of experiment we can say that the transition to chaos is like a second-order phase transition. We show below cases that are very different from this.

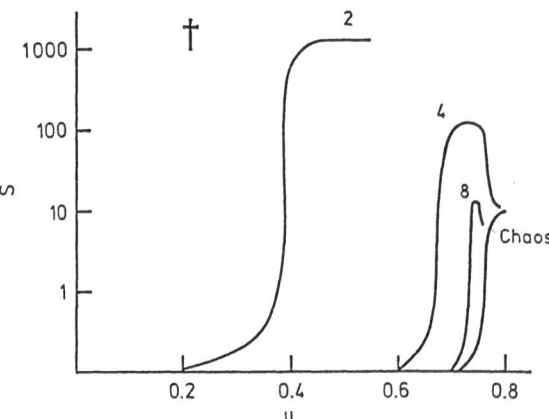

Fig. 8.9. Measurement of the character of the bifurcations for the transitions from P_1 to P_2, P_2 to P_4, P_4 to P_8 and from P_8 to chaos

8.2.5 Effects of Noise

By numerically studying a forced anharmonic oscillator and *Feigenbaum*'s quadratic map [8.12], *Crutchfield* and *Huberman* predicted that the addition of Gaussian noise to a system would cause the number of period doublings to be finite [8.14]. They called this a "bifurcation gap" since, instead of period doubling ad infinitum, some waveform of finite period bifurcates to the chaotic waveform of the same periodicity, leaving a gap in the sequence of observed waveforms. We observed such a gap after two or three period doublings [8.8–10]. Others have also observed a gap in the period-doubling sequence [8.15].

In contrast, it was found in a hydrodynamic experiment that noise has no significant effect on the transition to chaos [8.16]. Moreover, noise is not the only potential cause of a bifurcation gap. *Chow* [8.17] has predicted that a gap should be observed in our system in the absence of noise, if the ratio of the response time τ to the delay time t_R is small enough. With the time scales of our present system we would not expect to see this type of truncation. We showed that the location of the truncation in the bifurcation sequence moves monotonically toward fewer period doublings with increasing noise. If the truncation was deterministic, we would expect the bifurcation sequence to be independent of the noise level.

The major source of noise in the experiment was the shot noise from the photomultiplier. We increased the noise level in our system in the following manner: First we attenuated the light reaching the photomultiplier tube using the polarizer placed in front of the fiber; then we increased the voltage on the dynodes of the tube to return the signal to its previous level. Note that the spectral width of the noise at the photomultiplier tube is much larger than the band width of the rest of the device. Other noise sources (the amplifiers, dark current, etc.) contribute about half of the noise signal at the lowest noise levels used in the experiment. We investigated noise levels from 0.3 to 10 rms (measured with the modulator at maximum transmission).

To determine the bifurcation points in the presence of noise we located the rapid rise in the peaks of the power spectrum like Fig. 8.9. We measured the rise of the spectral power of the background to locate the bifurcation to chaos.

In Fig. 8.10 the solid curves show the bifurcation points as a function of noise. The numbers indicate the waveform present in a domain. (For clarity, only the period of the waveform is shown on the graph. A starred number indicates a chaotic waveform.) Figure 8.10 shows that more noise is needed to eliminate a chaotic waveform than is needed to eliminate the corresponding periodic waveform. The bifurcation gap presented by *Crutchfield* and *Huberman* showed the elimination of waveform and the corresponding chaotic waveform at the same noise level [8.14]. One possible explanation for the elimination of the chaotic waveforms at different noise levels in the two systems is the differing characteristics of the noise. The noise in our system is intensity dependent, but Crutchfield and Huberman used intensity-independent Gaussian noise. Another possible explanation lies in the difference of the systems. Our device is a

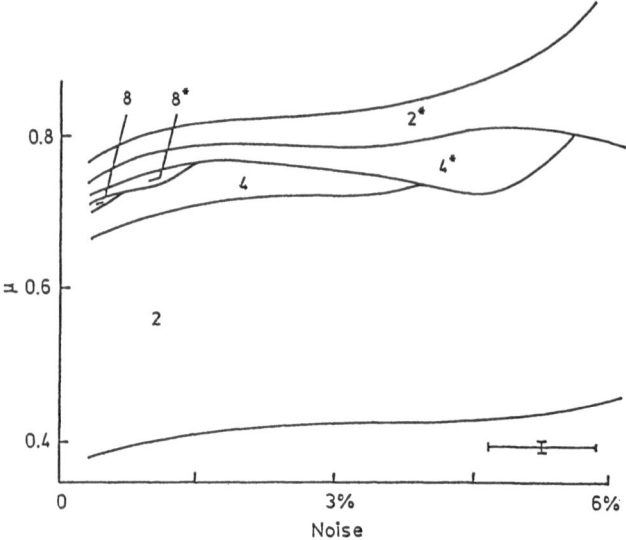

Fig. 8.10. Bifurcation points in the hybrid device as a function of increasing noise levels. The numerals 2, 4, and 8 indicate the period of the waveform in units of t_R. The numerals subscripted with an asterix indicate chaotic waveforms

differential delay system while the systems that they studied involve mappings or nonlinear differential equations.

8.2.6 Other Waveforms / Alternate Paths

In this subsection we describe alternate paths to chaos that occur in our device. We found that transitions between waveforms along these paths show hysteresis and other phenomena characteristic of first-order phase transitions. In a strict sense they are not first-order transitions, but occur as a result of complicated stability characteristics of the solutions. We show here that the solutions are built up from the many different ways that continuous-time waveforms can be constructed from the normal period-doubling sequence of a discrete-time map [8.18].

Equation (8.3) predicts what will happen one delay time later, but it does not specify what happens in the intervening time interval. For that reason, the association of a particular solution of (8.3) with a solution of (8.1) is not unique. To see how to construct potential solutions of (8.1), let us denote the solutions of (8.3) schematically using integers [we denote as "1" smallest amplitude in the iteration of (8.3)]. The P_1 is then $1{\rightarrow}1{\rightarrow}1\ldots$, P_2 is $1{\rightarrow}2{\rightarrow}1{\rightarrow}2$, P_4 is $1{\rightarrow}2{\rightarrow}3{\rightarrow}4{\rightarrow}1$, etc. Let us divide one delay-time interval of (8.1) into n equal-size sub-intervals, and let us take $n = 3$ for specificity. In constructing a possible solution of (8.1) from (8.3) in the P_2 regime, we can start off with any three combinations of 1 and 2. From all of the possible initial combinations, only two, the 111 and the 121, have different waveforms (the rest are phase shifted versions of these two). The iteration of the first sequence reads

$111 \rightarrow 222 \rightarrow 111 \ldots$, and is shown in Fig. 8.11a. Since each triplet lasts one t_R, this waveform has a period $2t_R$ and is just the usual perod two. The iterations of the second sequence is $121 \rightarrow 212 \rightarrow 121 \ldots$, and is shown in Fig. 8.11b. This sequence repeats every two thirds of an interval, and hence has a period $2t_R/3$, which is one third of the period two. The spectrum of the waveform in Fig. 8.11b has its strongest spectral component at a frequency $f = 3/2t_R$ which is the third harmonic of the fundamental component of the period-two spectrum. It is conventional in hydrodynamics [8.16] to refer to waveforms with

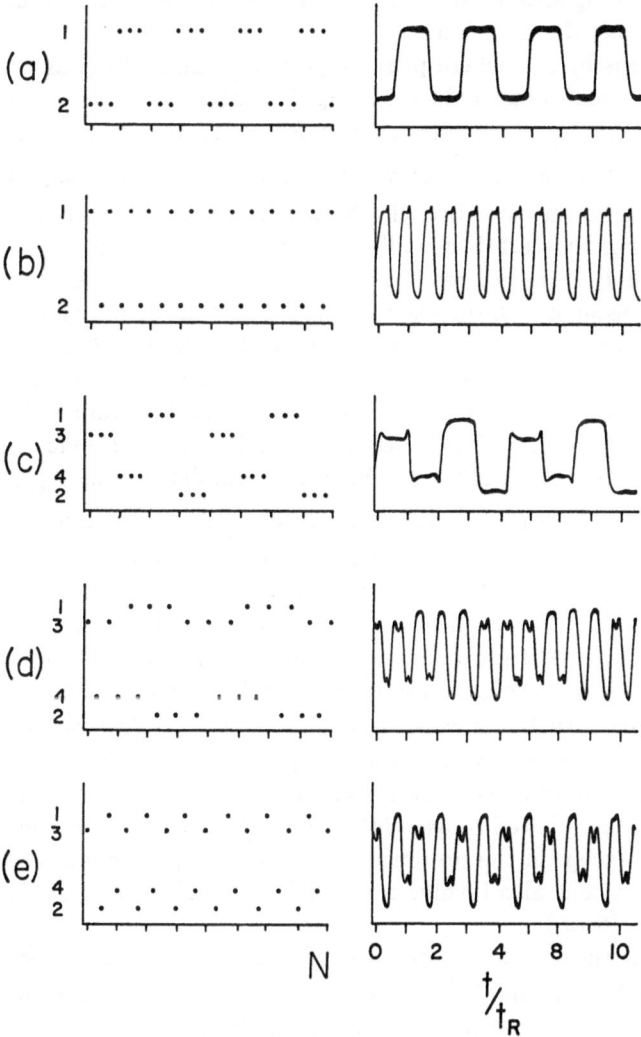

Fig. 8.11a–e. Illustration of how three interspersed maps can model the frequency locked behavior: left column; interspersed map, in which each point determines a point three spaces to the right via (8.3). *Right column;* resulting waveforms in continuous time. *Rows:* (a) P_2, (b) $L^3_{2/3}$, (c) P_4, (d) L^3_4, (e) $L^3_{1/3}$. In both columns, increasing signal is plotted downward. In all cases the modulation depth is nearly 100 %

such spectra as "frequency locked" when they are found on the path to chaos, and we use that terminology here, even though the waveform is constructed from the subharmonic sequence.

The division of the interval t_R into three subintervals is arbitrary, and any other division would do as well. *Ikeda* et al. [8.19] have determined that there is a regularity criterion that is a necessary criterion for stability that applies to all waveforms other than those of the period-doubling sequence. This criterion implies that the same or nearby amplitudes cannot appear twice in a row in the sequence. Divisions into even multiples do not meet this criterion, and in practice never occur, so we ignore them from now on. We confine our attention to cases of odd n that meet the criterion.

We denote waveforms by L_p^n, where p is the period in units of t_R, and n denotes the strongest harmonic of period two. Using this notation the waveform in Fig. 8.11b is denoted by $L_{2/3}^3$, the waveform in Fig. 8.11d by L_4^3, and the Fig. 8.11e by $L_{4/3}^3$. We will denote the set of waveforms with the same strongest harmonic as all belonging to the same branch. We will denote a branch by L^n, where n is harmonic. Thus Figs. 8.11b, d, e all belong to the L^3 branch.

In our notation, the P_4 sequence reads $1 \to 2 \to 3 \to 4 \to 1 \ldots$, where $X_1 < X_3 < X_4 < X_2$. For this case, using $n = 3$, the starting conditions that do not violate *Ikeda* et al.'s stability criterion in an obvious manner are 111, 123, 124, 134, and 143. Figure 8.11c shows the $111 \to 222 \to 333 \to 444 \to 111 \ldots$, a normal P_4 (in the figure the sequence is phase shifted and reads $333 \to 444 \to 111 \ldots$). Figure 8.11d will be discussed below. Figure 8.11e shows the sequence $143 \to 214 \to 321 \to 432 \to 143 \ldots$ (phase-shifted to $321 \to 432 \to 143 \ldots$). This waveform, denoted by $L_{4/3}^3$, has a period $4t_R/3$, and is the period-doubled version of $L_{2/3}^3$. The distinction among the other three waveforms is more subtle. Let us compare the sequences $123 \to 234 \to 341 \to 412 \to 123 \ldots$ and $124 \to 231 \to 342 \to 413 \to 124 \ldots$. To see how a solution occurs it is necessary to recognize that in the bifurcation $P_2 \to P_4$ the amplitude 4 grows out of 2, and 3 grows out of 1. Thus a simple procedure to determine the character of the waveform is to take the limit $4 \to 2$, $3 \to 1$. In that case the sequence starting with 123 goes to $L_{2/3}^3$, and those starting with 124 and 134 go to P_2. While all three waveforms have a perod $4t_R$, the one starting with 123 is observed to have its strongest spectral component at $3/2t_R$. It is shown in Fig. 8.11d (phase-shifted to $34 \to 341 \to 412 \to 123 \to 2 \ldots$). The other two, which are unstable except in the presence of external locking signals, are variants on P_4 (these solutions also involve some difficulties in meeting *Ikeda* et al.'s stability criterion [8.19]).

In Fig. 8.12 we show the waveforms denoted $L_{2/n}^n$ and their power spectra for $n = 1, 3, 5$, and 7. The conditions under which they were found are described later. The case $n = 1$ is the P_2 waveform. It has, as expected for a square-wave waveform, strong odd-harmonic peaks in its spectrum at frequencies $f = n/2t_R$, $n = 1, 3, \ldots$. Each spectrum of the frequency-locked waveforms shows a dominant spectral peak at the appropriate odd harmonic.

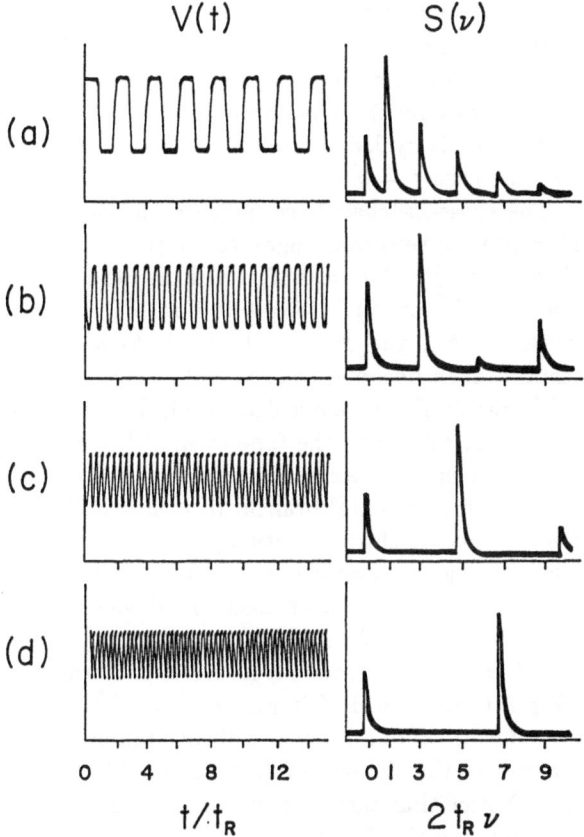

V(t) S(ν)

(a)

(b)

(c)

(d)

0   4   8   12   0 I 3   5   7   9

t/t_R $2 t_R \nu$

Fig. 8.12. Waveforms of the type $L^n_{2/n}$ (period $2t_R/n$). The rows show the cases $n = 1, 3, 5,$ and 7, respectively. The *left column* is the waveform (voltage vs. time). the *right column* is the power spectrum (power vs. frequency)

The spectra still have the dominant spectral peak at $f = n/2t_R$, $n = 1, 3, \ldots$. This readily separates waveforms of the type L^n_4, $n>1$, from P_4, which otherwise have the same spectral components. In waveforms of the type $L^n_{4/n}$, the spectral components associated with the fundamental branch are mostly absent, and new components, e.g., at $f = n/4t_R$, $n = 3, 5, \ldots$, are observed which do not occur in the spectra of P_p.

The transitions between branches that increase n occur via crisis [8.20]. A crisis manifests itself experimentally in the following manner. Suppose the system is producing a chaotic output at a fixed value of the bifurcation parameter. As the output wanders the system may "find" a periodic waveform that is more stable (e.g., with respect to noise) than the present chaotic output. At this point the crisis will occur. That is, the system will change to the periodic waveform in a short time. It has been shown [8.19] that this type of transition may show hysteresis. Since it may take some time for the system to "find" the stable waveform the chaos may persist as a long-lived transient, as documented in [8.9]. In our present system the long-lived transients persist for no more than milliseconds so it appears to be instantaneous. We, therefore, consider only the asymptotic waveforms in the following discussion.

In Fig. 8.13 we show a diagram of how the waveforms jump from branch to branch as the gain of the amplifier is cycled. Figure 8.13a shows the case $\tau = 0.50\,\mu s$. Figure 8.13b shows the case $\tau = 0.38\,\mu s$. The branch of observed waveforms are plotted vs. μ. The cross-hatched lines indicate chaotic waveforms. As μ is increased from zero, the waveform changes from a steady voltage to an oscillating waveform of the type P_2 at $\mu = 0.37$, which has the fundamental as the strongest component (L^2 or P). Along this line, P_2, P_4, and the corresponding chaotic waveforms N_4, and N_2 are observed. At $\mu = 0.74$ the waveform on the fundamental branch becomes chaotic and at $\mu = 0.76$ the system abruptly jumps to the L^3 branch. If the gain is decreased, the waveform stays on the L^3 branch rather than returning to the fundamental branch and forms a hysteresis loop. However, when μ is decreased below $\mu = 0.73$, the device abruptly jumps back onto the fundamental branch. If, at $\mu = 0.76$, μ is increased, rather than cycled back, the waveform eventually becomes chaotic. Then, when μ is increased further, the system jumps to the L^5 branch, becomes chaotic and jumps to the L^7 branch. When μ is decreased, the device follows the same path it went up, until it reaches the L^3 branch.

If a smaller value of τ is used ($\tau = 0.38\,\mu s$), the structure of the branches changes to the one shown in Fig. 8.13b. Instead of jumping to the L^3 branch the device jumps, again at $\mu = 0.76$ (for $\tau \ll t_R$, the device always jumps at this value of μ, which is the high-μ end of the domain of N_2), to the L^7 branch, skipping the L^3 and L^5 branches. Notice that there are two observable hysteresis loops, one between the L^7 branch and the L^9 branch, and another between the L^7 and the fundamental branch.

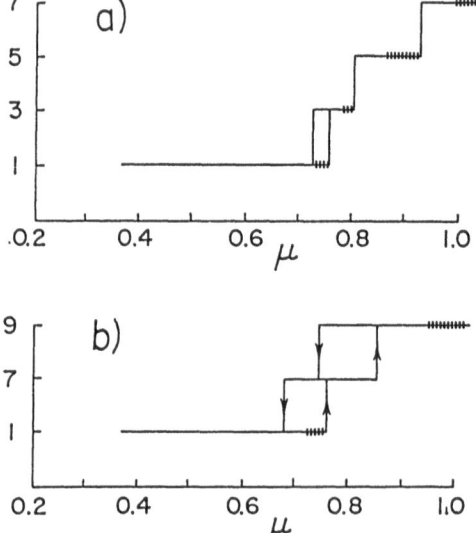

Fig. 8.13a,b. Bifurcation structure between major subharmonics. Vertical axis denotes observed strongest harmonic. Horizontal axis is the bifurcation parametre μ. (a) $\tau = 0.50\,\mu s$; the observed harmonics are the fundamental branch ($n = 1$) and $n = 3, 5$. (b) $\tau = 0.38\,\mu s$; the observed harmonics are the fundamental branch and $n = 7, 9$

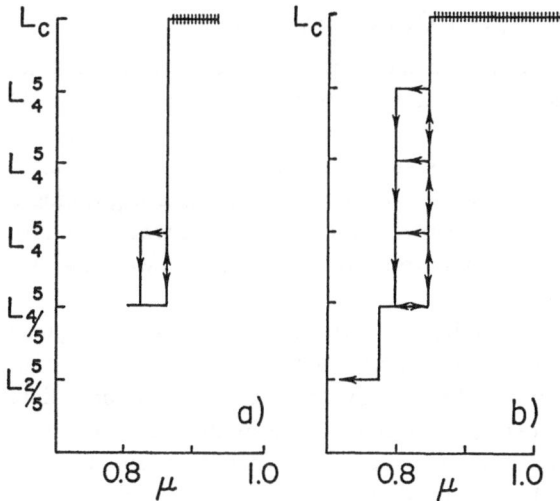

Fig. 8.14a,b. Bifurcation structure of the waveforms within the branch $n = 5$; (a) without locking signal, (b) with a locking signal of frequency $f_P = 5/2t_R$. The vertical axis denotes different waveforms of the type L_P^5 (period pt_R). There are three distinguishable waveforms of the type $L_{4/5}^5$, which can be observed with the locking signal. For simplicity, we have chosen not to develop a notation to distinguish between them

In Fig. 8.14a we show, in detail, the existence of various waveforms of the L^5 branch of Fig. 8.13a. The vertical scale shows the period of the waveform, the horizontal axis is μ. Along the L^5 branch, only two periodic waveforms are stable, the $L_{4/5}^5$ and only one of the three possible waveforms denoted L_4^5. When the system jumps onto the L^5 branch at $\mu = 0.81$, the waveform is $L_{4/5}^5$. (This is different from the L^3 branch where the first observed waveform is L_4^3.) When μ is increased, the waveform remains $L_{4/5}^5$, up to the value $\mu = 0.87$, at which point the output becomes chaotic. When μ is subsequently decreased, the system comes out of chaos statistically, sometimes with the waveform L_4^5, more often with the same $L_{4/5}^5$ waveform it had on the way up.

In Fig. 8.14b, we show the bifurcation structure of the L^5 branch when a small fifth-harmonic locking signal is applied to the bias of the device. The primary consequence of the locking signal is to stabilize the normal unstable $L_{2/5}^5$ waveform, which is then observable at low μ. Since we are able to produce this waveform with a small perturbation we conclude that it is a possible output, but not a stable one. The bifurcation structures with and without a locking signal are similar, except that there are several L_4^5 waveforms, all of which can be observed with the locking signal, while only one is observed without it.

As shown by the cross hatched lines in Figs. 8.13 and 14, different kinds of chaos exist. Ikeda has shown that different types of chaos may coexist at the same μ and we have observed many qualitatively different spectra in this regime. In the next subsection we have been careful to avoid regions that may show different chaoses, but a more sophisticated version may be able to distinguish between the chaoses better than the use of the power spectrum.

191

8.2.7 Test for Chaos in High-Speed Optical Systems

A great deal of recent work has addressed the question of how one determines whether a system is erratic because of chaos or because of noise [8.21]. The techniques that have been developed are dependent on devices being drift free and require that a long time trace of the signal be digitized. However, most optical system do drift, and most oscillate so rapidly as to make digitizinig a waveform a difficult and expensive proposition. Another method which we have proposed [8.22] involves the digitization of data, but at a much slower rate and by much simpler equipment.

This method exploits the sensitivity to initial conditions to distinguish between chaos and noise. The basic technique used is: turn the device off, either by breaking the feedback in a hybrid, or by turning off the incident field in an intrinsic device. Then, quickly (in relation to the oscillation frequency) turn the device on. If this process is done repeatedly and is observed on an oscilloscope, pictures like Fig. 8.15, are produced. In Figs. 8.15a and c are P_2 and P_4, respectively. Notice how they reproduce exactly, whereas the chaotic waveforms in Figs. 8.15b and d start out the together but "fuzz out" in time.

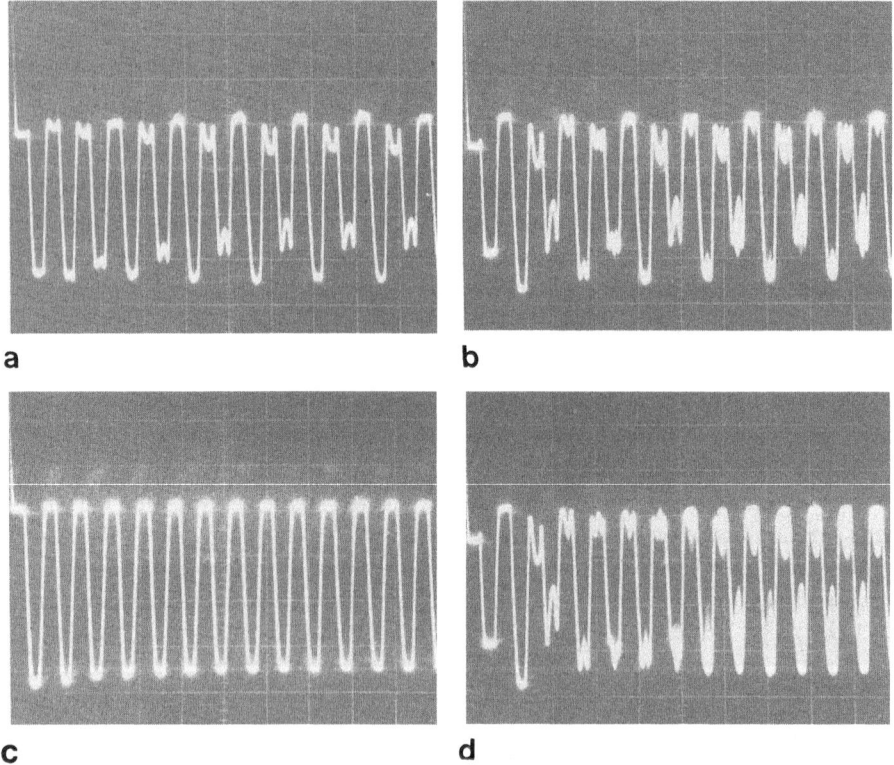

a   b

c   d

Fig. 8.15a–d. Many overlayed traces of X after analog switch closed. Time scale is approximately 15 ms/div. **(a)** Period four, **(b)** period-4 chaos, **(c)** Period two, **(d)** period-2 chaos. Notice the sharp onset of fuzziness, after a finite delay, in the chaotic traces

A noisy system would show waveforms that are "fuzzy" as soon as the device is turned on. We now quantify this "fuzzing out".

Figure 8.16a shows the log of the variance of the traces as a function of time. It is obtained by turning the device on, waiting a delay time t and then sampling the waveform with an analog-to-digital converter. This process is repeated ten times for each t, and the variance is computed. Notice that for the chaotic system the rise is linear, for a time, and then saturates at some maximum value. This linear rise (on the log plot) shows that the divergence of nearby initial points is exponential, just as predicted by the *Ruelle* and *Takens* [8.23]. Since the data in Fig. 8.16a are, by themselves, unconvincing, we have taken the same number of data points in a shorter time, these data are shown in Fig. 8.16b. Notice that the rise is indeed linear. If a noisy (random-walk) process were at work, the divergence would have a $\sqrt{t}$ dependence, the shape of the curves superimposed on the data in Fig. 8.16c. Notice that they do not fit. This difference in the character of the rise allows chaos and noise to be distinguished. The slope that is measured on a chaotic graph is also very much like another theoretical measure of chaos, the metric entropy [8.24]. We have not yet, however, made an unambiguous connection between the two. If this quantity is the metric entropy then theoretical results say that if the slope

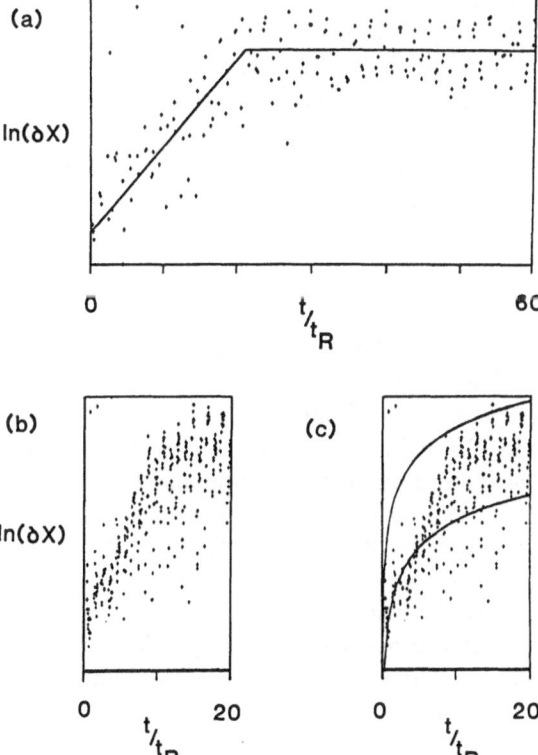

(a)

$\ln(\delta X)$

0
t/t_R
60

(b)

$\ln(\delta X)$

0
t/t_R
20

(c)

0
t/t_R
20

Fig. 8.16. (a) A plot of the log of the standard deviation of the points as a function of time, measured from the instant the system was switched from an non-oscillating steady state to parameters which lead eventually to period two chaos. Notice the linear initial rise (the lines are fit by eye). (b) The details of the initial rise to saturation. (c) data in (b) fitted to noise curves 1/2 Int

measured is positive and finite the system is chaotic; if it is unfinite, (as if the erratic behavior starts immediately) the system is noisy, and if it is negative the system is periodic.

8.3 Instabilities in Sodium Vapor

The observation of possible chaos in an all-optical system has been reported in two laser systems [8.25,26], and four bistable systems [8.27–30]. In general, lasers are poorer candidates for studying chaos than bistable devices, since they are much noisier. The first all-optical passive system was the experiment described in the thick sample case (Sect. 8.3.1) below [8.27]. The other systems sought to demonstrate the Ikeda $2t_R$ instability. They include a fiber-optic cavity configured as a ring pumped with a pulsed, mode-locked YAG [8.28], and two with an ammonia cell in a ring cavity [8.29] and a Fabry Perot [8.30], both excited by a pulsed CO_2 laser (Chap. 9). They all demonstrate the Ikeda instability, but as pulsed experiments they do not permit detailed investigation of the bifurcation structure.

The frequency stability requirements of the laser and of the passive cavity make CW experiments with the Ikeda instability difficult. By ridding the experiment of the passive cavity the stability problems diminish. To this end we have investigated variants on the self-focussing device described by *Bjorkholm* et al. [8.31]. It has the advantage of a large nonlinearity from an atomic vapor, and, of course, no cavity. We have used a modified version of their device as illustrated in Fig. 8.17. The changes in the optical beam are radial in this case, and the fluctuations are in the power density, rather than in the total power. The pinhole assures detection of a small portion of the beam. We have operated this device in two configurations: with a long, optically thick medium, and with a short, optically thin medium.

8.3.1 Optically Thick

The apparatus can be operated in several different configurations. The earliest one we tried had a long cell and was operated close to resonance so that the optical thickness was large. It operated very much like the original device [8.31]. The round-trip time of the feedback (time from the sodium to the mirror and back to the sodium) was short compared to the response time of the medium. This differs from the operation of the hybrid and from the normal Ikeda instability. Recent theoretical work by *Winful* [8.32] and others [8.33] has shown that chaos can occur in this regime of operation. A major difference in the experimental results is that, in the hybrid, chaos and bistability are two independent phenomena (all data shown above were taken in regimes where there are no upper branches), while in the device in Fig. 8.17, instabilities are observed only on the upper branches of bistable loops. In Fig. 8.18, a sequence of waveforms and power spectra are shown.

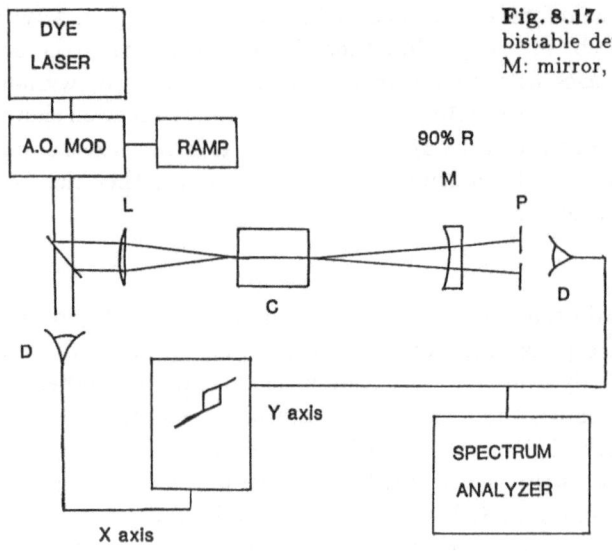

Fig. 8.17. Schematic of the all-optical bistable device. (C: Sodium cell, L: lens, M: mirror, P: pinhole, D: detector)

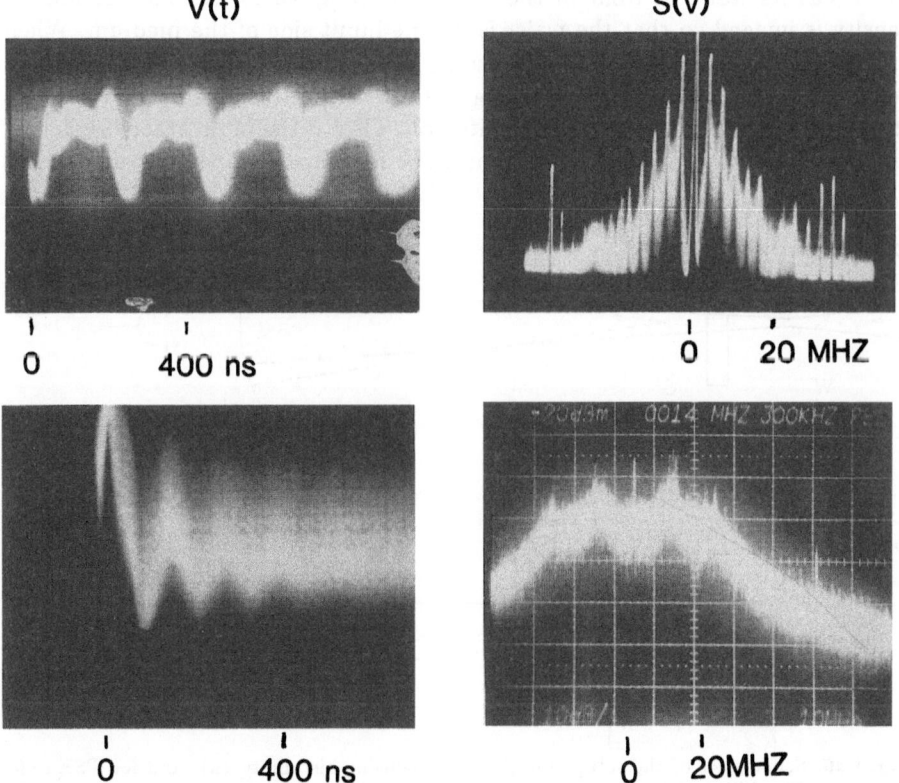

Fig. 8.18. Waveforms *(left column)* and power spectra of the waveform *(right column)* from the device in Fig. 8.17

The progression of waveforms and spectra in Fig. 8.18 are qualitatively similar to those in the hybrid, except that there is no bifurcation structure to the periodic regime. Cases without bifurcation sequences (with or without noise) are known in hydrodynamics [8.16], as well as the noise truncation shown in Fig. 8.10. We have no experimental way, at present, of knowing whether our device has a sequence that has been truncated by noise, or whether this is a case in which there is no sequence.

8.3.2 Optically Thin

It is also possible to operate the device in a regime where the round-trip time is long compared to the response time of the medium (like the hybrid), and where the medium is optically thin. This is accomplished by adding a buffer gas (argon) to collisionally broaden the sodium line, and using a shorter sodium cell. In this configuration the light does not self-trap, rather, bistability results from a change of the phase of the wavefront. We call this phase encoding. The physics of thin-sample-encoding (TSE) self-lensing bistability resides in an enhancement of the intensity in the medium via feedback from past nonlinear self-lensing (i.e., self-focusing or self-defocusing). The basic idea is illustrated in Fig. 8.19. In the self-focusing case (Fig. 8.19a) the diverging Gaussian input beam has its waist in front of the short cell. Single-pass feedback at low intensity is imaged so that the waist is on the input side of the medium. When the intensity is increased the self-lensing of the medium moves the waist toward the medium, thereby increasing the intensity in the medium. With this increased intensity the device can switch on. Then if the input intensity is lowered, the combination of the pump and the feedback keep the beam focused,

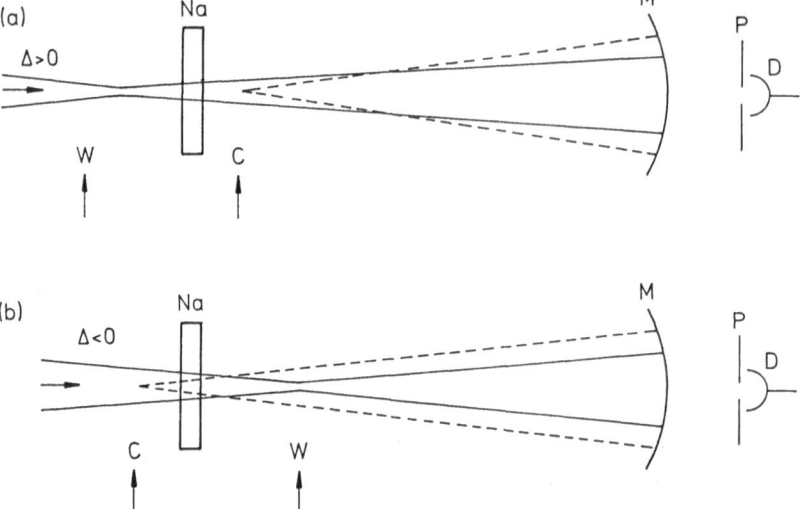

Fig. 8.19. Schematics of the setups for TSE self-focusing bistability (**a**), and for TSE self-defocusing bistability (**b**) using a short sodium cell. (W: input Gaussian beam waist, C: center of curvature, M: spherical mirror, P: pinhole, D: detector)

i.e., the device stays in the on state; it is bistable. When the round-trip time is made long compared to the medium response time, this device shows an Ikeda $2t_R$ instability. The configuration on the self-defocusing side (Fig. 8.18b) is the same except that the imaging point for the waist has been moved to the other side of the sample to compensate for the opposite sign of the nonlinearity. An advantage of TSE is that the diffraction and absorption can be ignored inside the medium, which allows straight-forward theoretical modeling.

This device has been shown to work on both the self-focusing and self-defocusing sides of resonance. Although the device has not been fully investigated, it has shown oscillations with a period slightly longer than twice the feedback round-trip time, and oscillations with a much longer period like the oscillations of the optically thick configuration, shown in Fig. 8.18. The $2t_R$ oscillations have a period that is largely insensitive to the transverse alignment and the detuning. The oscillations of longer periodicity are strongly affected by the transverse alignment and the detuning, and are qualitatively the same as those seen in the thick-medium device. No period-doubling structure has been seen for either type of oscillation. Both types of instability can be seen independently of the other, that is, one or the other may be observed, or they may be observed simultaneously. When observed together a quasi-periodic and a P_6 waveform are sometimes observed. Preliminary theoretical calculations [8.34] show the existence of an oscillation with a period close to $2t_R$, period doubling bifurcations $(P_2 \rightarrow P_4)$ and P_6. The observation of two independent frequencies obscures the interpretation of the observed P_6.

8.4 Issues for Fast Chaotic Systems

Optical systems present a unique opportunity to study chaos. Since experiments with features much like those seen in hydrodynamics can be performed in a matter of minutes (milliseconds!) optical systems offer the possibility to more fully study a system. Since these experiments occur quickly the experimental parameters can be varied adiabatically. This keeps transients from complicating the experiment. Without this adiabatic following we would have been unable to untangle the alternate paths seen in Sect. 8.2.6.

Another benefit of speed is the ability to take power spectra in real time. It enables measurements like those used in Sect. 8.2.5 to distinguish between waveforms, and locate the transition between waveforms.

At the present time most of the systems used to investigate chaos have been systems that are slow compared to optical systems. This difference has important consequences for the distinguishing of chaos from noise.

In an optical system the bandwidth is high, often as high as 100 MHz. The noise band width is naturally the same, and since noise $\propto$ (band width)$^{1/2}$ optical systems will tend to have more noise than slower systems. As such, noise will play an important role in the behavior of such systems, and cannot be ignored. Lasers may be at a particular disadvantage in this regard because of spontaneous emission, although work on optical chaos in laser systems is extensive [8.36] (Chaps. 2–6).

Present techniques [8.21] for distinguishing chaos from noise require that a time trace be digitized with several points per period over many periods. After the data are collected a significant, although tractable, amount of computation is required. We have discussed alternatives in this chapter. In the future we envision chaotic systems made with a nonlinear semiconductor material replacing the sodium in the optically thin configuration. Such a system's speed would surpass the sampling rate of a state-of-the-art transient digitizer. The test that we have proposed could be applied to these cases.

Acknowledgements. We acknowledge helpful discussions with Drs. M. Le Berre, E. Ressayre, and A. Tallet. We are grateful for support by NSF Phy-810492, Phy-8216191, and Int-8313178 and NATO 882/83.

References

8.1 S.L. McCall: Appl. Phys. Lett. **32**, 285 (1978)
8.2 J.L. Jewell, H.M. Gibbs, S.S. Tarng, A.C. Gossard, W. Wiegmann: Appl. Phys. Lett. **40**, 291 (1982)
8.3 H.M. Gibbs, S.L. McCall, T.N.C. Venkatesan, A.C. Gossard, A. Passner, W. Wiegmann:CLEA 1979; Appl. Phys. Lett. **35**, 451 (1979);
     H. Walther, K.W. Rothe (eds.): *Laser Spectroscopy IV*, Springer Ser. Opt. Sci., Vol. 21 (Springer, Berlin, Heidelberg 1979) p. 441
8.4 M.D. Sturge: Phys. Rev. **127**, 768 (1962)
8.5 Predicted in R. Bonifacio, L.A. Lugiato: Lett. Nuovo Cimento **21**, 510 (1978)
8.6 K. Ikeda: Opt. Commun. **30**, 257 (1979)
8.7 K. Ikeda, H. Diado, O. Akimoto: Phys. Rev. Lett. **45**, 709 (1980)
8.8 H.M. Gibbs, F.A. Hopf, D.L. Kaplan, R.L. Shoemaker: Phys. Rev. Lett. **46**, 474 (1981)
8.9 F.A. Hopf, D.L. Kaplan, H.M. Gibbs, R.L. Shoemaker: Phys. Rev. A**25**, 2172 (1982). Throughout the present chapter we simplify the discussion by taking the period to be precisely a multiple of t_R. The real period is always somewhat longer, and its magnitude is discussed here [8.8].
8.10 M.W. Derstine, H.M. Gibbs, F.A. Hopf, D.L. Kaplan: Phys. Rev. A (Rapid commun.) **26**, 3720 (1982)
8.11 E.N. Lorenz: J. Atmos. Sci. **20**, 130 (1963)
8.12 M.J. Feigenbaum: Los Alamos Science **1**, 4 (1980)
8.13 P.S. Lindsay: Phys. Rev. Lett. **47**, 1349 (1981);
     J. Testa, J. Perez, C. Jeffries: Phys. Rev. Lett. **48**, 714 (1982)
8.14 J.P. Crutchfield, B.A. Huberman: Phys. Lett. **77A**, 407 (1980)
8.15 W. Lauterborn, E. Cramer: Phys.Rev. Lett. **47**, 1445 (1981);
     J. Chrostowski: Phys. Rev. A**26**, 3023 (1982);
     R.W. Rollins, E.R. Hunt: Phys. Rev. Lett. **49**, 1295 (1982)
8.16 J.P. Gollub, S.V. Benson: J. Fluid Mech. **100**, 449 (1980);
     H.L. Swinney, J.P. Gollub (eds.): *Hydrodynamic Instabilities and the Transition to Turbulence*, 2nd ed., Topics Appl. Phys., Vol. 45 (Springer, Berlin, Heidelberg 1985)
8.17 S.N. Chow: Talk presented at the workshop on Coupled Nonlinear Oscillators, Los Alamos, NM (1981)
8.18 M.W. Derstine, H.M. Gibbs, F.A. Hopf, D.L. Kaplan: Phys. Rev. A**26**, 3720 (1982)
8.19 K. Ikeda, K. Kondo, O. Akimoto: Phys.Rev. Lett. **49**, 1467 (1982)
8.20 E. Ott, J. York: Phys. Rev. Lett. **48**, 1507 (1982)

8.21 A. Brandstater, J. Swift, H.L. Swinney, A. Wolf, J.D. Farmer, E. Jen, J.P. Crutchfield: Phys. Rev. Lett. **51**, 1442 (1982);
P. Grassberger, I. Procaccia: Phys. Rev. Lett. **50**, 346 (1983);
P. Grassberger, I. Procaccia: Phys. Rev. **A28**, 2591 (1983)

8.22 M.W. Derstine, H.M. Gibbs, F.A. Hopf, L.D. Sanders: J. Opt. Soc. Am. **B1**, 464 (1984) and Phys. Rev. Lett., submitted

8.23 D. Ruelle, F. Takens: Commun. Math. Phys. **20**, 167 (1971)

8.24 R. Shaw: Z. Naturforsch. **36A**, 80 (1981);
J.D. Farmer: Physica **4D**, 566 (1982);
J.P. Crutchfield: In *Evolution of Order and Chaos*, ed. by H. Haken, Springer Ser. Syn., Vol. 17 (Springer, Berlin, Heidelberg 1982) p. 215

8.25 L.T. Arecchi, R. Meucci, G. Puccioni, J. Tredicce: Phys. Rev. Lett. **49**, 1217 (1982)

8.26 N.B. Abraham, M.D. Coleman, M. Maeda, J.C. Wesson: App. Phys. **B28**, 169 (1982);
M. Maeda, N.B. Abraham: Phys. Rev. **A20**, 3395 (1982)

8.27 F.A. Hopf, M.W. Derstine, H.M. Gibbs, M.C. Rushford: In *Optical Bistability II*, ed. by C.M. Bowden, H.M. Gibbs, and S.L. McCall (Plenum, New York 1984) p. 68

8.28 H. Nakatsuka, S. Asaka, H. Itoh, K. Ikeda, M. Matsuoka: Phys. Rev. Lett. **50**, 109 (1983)

8.29 R.G. Harrison, W.S. Firth, C.A. Emshary, I.A. Al-Saidi: Phys. Rev. Lett. **51**, 562 (1983)

8.30 R.G. Harrison, W.S. Firth, I.A. Al-Saidi: Phys. Rev. Lett. **53**, 258 (1984)

8.31 J.E. Bjorkholm, P.W. Smith, W.J. Tomlinson, A.E. Kaplan: Opt. Lett. **6**, 345 (1981). Our experiment differs from the device described t here only by the location of the pinhole. The pinhole in our device is located before the detector rather than between the mirror and the sodium cell.

8.32 H.G. Winful, G. Cooperman: Appl. Phys. Lett. **40**, 29 (1982)

8.33 K. Ikeda, O. Akimoto: Phys. Rev. Lett. **48**, 617 (1982);
Y. Silberberg, I. Bar Joseph: Phys. Rev. Lett. **48**, 1541 (1982);
J.A. Goldstone, E.A. Garmire: IEEE J. QE-**19**, 208 (1983)

8.34 M. Le Berre, E. Ressayre, A. Tallet, K. Tai, F.A. Hopf, H.M. Gibbs, J.V. Moloney: Postdeadliine Paper PDD5, IQEC (1984); and in preparation

8.35 N.B. Abraham: Laser Focus, 73 (May 1983). For a overview of chaos in a variety of systems see N.B. Abraham, J.P. Gollub, Harry L. Swinney: Physica **11D**, 252 (1984)

8.36 R.S. Gioggia, A. Albano, R.M. Searle, T. Chyba, N.B. Abraham: J. Opt. Soc. Am. **B1**, 499 (1984)
R.G. Harrison, D.J. Biswas: Prog. Quantum Electron. **10**, 147 (1985)
J.R. Ackerhalt, P.W. Milonni, M.L. Smith: Phys. Rep. **128**, 205 (1985)

9. Instabilities and Routes to Chaos in Passive All-Optical Resonators Containing Molecular Gases

R.G. Harrison, W.J. Firth, and I.A. Al-Saidi

With 20 Figures

Dispersive optical bistability and instabilities leading to period doubling, higher harmonics and chaos in an all-optical passive quantum system are discussed theoretically and experimentally. These phenomena have been observed for a wide range of operating conditions in ring and Fabry-Perot resonators containing molecular gases as the nonlinear medium.

Laser input intensity, gas pressure and cavity tuning are the main control parameters in observation of these effects. The experimental results are well modelled by our generalisation of standard two-level system theory to include standing-wave, reservoir and transverse effects.

9.1 Background

Nonlinear optical resonators display a wide range of phenomena of fundamental interest. In passive resonators, optical bistability [9.1] and related phenomena have attracted much attention. We are here concerned with an outgrowth of optical bistability which has wide interest in its own right, namely instabilities and chaos in the output of a (passive) nonlinear resonator. This class of phenomena, currently undergoing intense cross-disciplinary investigation [9.2] are illustrated in a particularly interesting manner in optics, since simple behaviour characteristic of low-dimensional systems can be obtained [9.1,3], while there is the interesting possibility of a quantum description of the phenomena, particularly when the nonlinearity arises from the saturation of two-level atoms [9.4].

Ikeda first described an instability with period $2t_R$, where t_R is the cavity round-trip time, in a ring resonator containing a two-level medium [9.4], with a period-doubling cascade leading to chaotic output with a conitnuous wave input [9.3]. Since then similar phenomena, including all the "universal" routes to chaos, have been predicted in a wide class of passive optical systems with feedback [9.1]. Experimental evidence, however, was slow in arriving, and remains patchy. The first demonstration was in a hybrid system with electronic feedback [9.5], which provided an enormous stimulus despite its only quasi-optical nature (Chap. 8). Only in 1983 did the first all-optical demonstration of the Ikeda instability appear in the literature [9.6], employing glass fibre as both waveguide and nonlinear medium.

The main reason for the Ikeda instability being much harder to observe than optical bistablity itself is the requirement that the response time of the nonlinear medium, τ, be short enough to allow $2t_R$ oscillation, i.e., a "good", or at least fairly good, cavity is required, in order that the medium band width is able to cover two or more adjacent longitudinal cavity modes. There is no such requirement for bistability, e.g., $\tau \geq 10^4 t_R$ is typical for InSb [9.7], in which bistability is observable at very low powers, because of the resonant nature of the nonlinearity. A non-resonant nonlinearity, on the other hand, will be fast enough for the Ikeda instability, but will require high powers, and there will be no discrimination against competing nonlinear processes – both problems in the fibre experiment [9.6], where picosecond pulse excitation was necessary to avoid stimulated Brillouin scattering.

Ideally then, one seeks a medium with a nonlinearity resonantly enhanced sufficiently to outweigh competing processes and reduce power requirements, but with a response time of a few nanoseconds to allow compact resonator design.

Considerations such as these point to gases as optimal media, and we were led to look at molecular gases where there are many absorption lines near to resonance with CO_2 laser lines, while the response time is readily adjustable by pressure variation to obtain the best resonance response-time trade-off. In particular, ammonia has well-documented coincidences with CO_2 laser lines [9.8], and energy levels sufficiently widely-spaced to approximate reasonably closely to an ideal two-level system. We recently reported $2t_R$ oscillation in a ring resonator containing NH_3 gas [9.9], and subsequently observed bifurcations and chaos in a compact Fabry-Perot resonator, again containing ammonia [9.10]. In other work on sulphur hexafluoride we have also reported evidence of optical bistability [9.11].

The purpose of this chapter is to collate and expand our theoretical models and experimental observations of OB and instabilities leading to chaos in ring and Fabry-Perot optical resonators. We first review in Sect. 9.2, the theoretical picture for OB and instabilities in ring and Fabry-Perot resonators. In Sect. 9.3 we describe the nonlinear absorption characteristics of ammonia and sulphur hexafluoride and their molecular energy-level-schemes. We then describe in Sect. 9.4 the experimental arrangement and experimental results in comparison with the theoretical predictions. Finally, in Sect. 9.5, we summarise our experimental and theoretical results and discuss possible future developments.

9.2 Theory of Instabilities in Passive Nonlinear Resonators

In this section we develop and explore the theory of dynamical instabilities and chaos in passive nonlinear resonators in both ring and Fabry-Perot configurations. The approach and content is essentially a tying together of results and ideas which we have developed since about 1981 in conjunction with a number of collaborators, whose contributions we gratefully acknowledge: those of E.M. Wright and E. Abraham deserve special mention.

The standard Maxwell-Bloch equations of semiclassical two-level systems theory are our starting point, and we immediately adiabatically eliminate the polarisation. The approach then undergoes a number of bifurcations, of which one of the most significant is into unidirectional and bidirectional propagation within the nonlinear medium, corresponding to ring and Fabry-Perot resonators, respectively. The former have been widely analysed, in this volume and elsewhere, so we restrict ourselves to a discussion of the physical basis of the instabilities on the one hand, and development of models adequate to describe the experimental results presented later, on the other.

Fabry-Perot resonators give effects qualitatively similar to those in ring resonators, but are more complicated to analyse (though simpler and better for experiments). Our first work in this field [9.12] established the existence of instabilities in Fabry-Perot resonators containing a Kerr medium of zero response time. We review here our subsequent elaboration of this model to include finite response times, in the course of which we were able to extract an accurate Feigenbaum cascade [9.13], and our more recent work in saturable media, spurred by the successful experiments described below. We will only occasionally consider the role of transverse effects in these phenomena: the reader is referred to Chap. 7. by J.V. Moloney for a much fuller discussion of transverse effects.

9.2.1 The Basic Model

Let us assume that we have a system of two-level atoms of transition frequency ω_0 and dipole moment μ irradiated by a field of frequency ω [9.14]. Using the density matrix formalism and splitting the field and polarisation into their positive and negative frequency parts

$$E = E^+ \exp(-i\omega t) + E^- \exp(i\omega t) \ ,$$
$$P = N_a \mu [P^+ \exp(-i\omega t) + P^- \exp(i\omega t)] \ , \tag{9.1}$$

with $E^+ = E^{-*}$, $P^+ = P^{-*}$, and N_a the number density of atoms, we obtain in the rotating-wave approximation, and assuming homogeneous broadening,

$$\frac{\partial P^\pm}{\partial t} = \left(\pm i\Delta - \frac{1}{T_2} \right) P^\pm \mp i\frac{\mu}{\hbar} E^\pm n \ , \tag{9.2}$$

$$\frac{\partial n}{\partial t} = -\frac{n+1}{T_1} - 2i\frac{\mu}{\hbar}(E^- P^+ - E^+ P^-) \ , \tag{9.3}$$

where n is the population inversion, T_1 is the longitudinal relaxation time, T_2 is the transverse relaxation time and $\Delta = \omega - \omega_0$. Equations (9.2 and 3) used in conjunction with Maxwell's wave equation (in the plane-wave approximation)

$$\frac{\partial^2 E}{\partial x^2} - \frac{1}{c^2}\frac{\partial^2 E}{\partial t^2} = \frac{1}{\varepsilon_0 c^2}\frac{\partial^2 P}{\partial t^2}$$

(9.4)

in the slowly varying envelope approximation (not yet made) constitute the Maxwell-Bloch equations. In the limit $T_2 \ll T_1$ we can eliminate $P^\pm$ adiabatically (i.e., $\partial P^\pm/\partial t \simeq 0$) from (9.2), so that

$$P^\pm = \mp i \frac{\mu T_2}{\hbar} E^\pm n \frac{(1 \pm i\tilde{\Delta})}{(1 + \tilde{\Delta}^2)} \ ,$$

(9.5)

where $\tilde{\Delta} = \Delta T_2$; substituting (9.5) into (9.3) and defining the "intensity" $I = E^+ E^-$, we obtain

$$T_1 \frac{\partial n}{\partial t} = -(n+1) - nI/I_\mathrm{s}$$

(9.6)

where I_s is the saturation intensity, given by

$$I_\mathrm{s} = \frac{\hbar^2}{4\mu^2}\frac{(1 + \tilde{\Delta}^2)}{T_1 T_2} \ .$$

Substituting (9.5) into (9.4) gives, for the positive frequency component,

$$\frac{\partial^2}{\partial x^2}[E^+ \exp(-i\omega t)] - \frac{n_b^2}{c^2}\frac{\partial^2}{\partial t^2}[E^+ \exp(-i\omega t)]$$

$$= \frac{-\omega^2}{\varepsilon_0 c^2}\frac{N_\mathrm{a}\mu^2 T_2}{\hbar(1 + \tilde{\Delta}^2)} n(\tilde{\Delta} - i)[E^+ \exp(-i\omega t)] \ ,$$

(9.7)

where n_b is the background refractive index of the medium. In the case where the sample is in a Fabry-Perot cavity, E^+ can be expressed as a sum of forward and backward fields with envelopes $E_\mathrm{F}(x,t)$ and $E_\mathrm{B}(x,t)$, respectively:

$$E^+(x,t) = E_\mathrm{F}(x,t)\exp(in_b k_0 x) + E_\mathrm{B}(x,t)\exp(-in_b k_0 x) \ ,$$

$$k_0 = \omega/c \ .$$

(9.8)

Upon substituting (9.8) into (9.7), and making the slowly-varying envelope approximation, we obtain,

$$\exp(ikx)\left(\frac{\partial E_\mathrm{F}}{\partial x} + \frac{n_b}{c}\frac{\partial E_\mathrm{F}}{\partial t}\right) + \exp(-ikx)\left(-\frac{\partial E_\mathrm{B}}{\partial x} + \frac{n_b}{c}\frac{\partial E_\mathrm{B}}{\partial t}\right)$$

$$= +\frac{\alpha_0}{2}\left(\frac{1 + i\tilde{\Delta}}{1 + \tilde{\Delta}^2}\right)nE^+(x,t) \ , \quad k = n_b k_0$$

(9.9)

where $\alpha_0 = \omega N_\mathrm{a}\mu^2 T_2/\varepsilon_0 n_b c\hbar$ is the on-resonance absorption coefficient.

For bidirectional propagation, n will be a rapidly varying function of z, due to the spatial modulation of I in (9.6). In order to obtain true slowly varying amplitude equations for the evolution of E_F and E_B, it is then necessary to solve (9.6), so as to find the Fourier components of the right-hand side of (9.9) which are phase-matched to one or other of the terms on the left. Before involving ourselves in such complexities, let us first consider ring cavities, and thus set $E_B \equiv 0$ in (9.9). Let us then define, for a medium length L,

$$D(t) = -\frac{1}{L} \int_0^L dx\, n\left(x, t + \frac{n_b x}{c}\right) , \tag{9.10}$$

which means that $D(t)$ is (minus) the inversion averaged along the light cone starting from $x = 0$ at time t. Some manipulation [9.4] then enables (9.6) to be written in averaged form as

$$T_1 \dot{D}(t) = [1 - D(t)] - |\varepsilon(t)|^2 \frac{(1 - e^{-\alpha L D(t)})}{\alpha L} \tag{9.11}$$

where $\varepsilon(t) = E_F(0, t)/I_s^{1/2}$ and $\alpha = \alpha_0/(1 + \tilde{\Delta}^2)$. This ansatz for the population enables an explicit integration of (9.9) along the light cone; the ring cavity boundary conditions can then be applied to obtain the following relation between ε at times separated by one cavity round-trip time t_R [9.9]

$$\varepsilon(t) = \varepsilon_i(t) + R\varepsilon(t - t_R)e^{i\theta} \exp\left[-\frac{\alpha L}{2}(1 - i\tilde{\Delta})D(t - t_R)\right] . \tag{9.12}$$

Here $\varepsilon_i(t)$ is the scaled incident field (transmitted into the cavity), R is the round-trip amplitude loss (by transmission and any absorption outside the nonlinear medium), and θ is the cavity tuning in the absence of the medium $(\alpha \to 0)$. In the latter case, we see that constant input field leads to a constant forward field

$$\varepsilon = \frac{\varepsilon_i}{1 - R \exp(i\theta)} \tag{9.13}$$

showing the usual resonance behaviour as θ is varied, e.g. by fine-tuning a cavity mirror. Equations (9.11 and 12), supplemented and slightly modified to account for reservoir effects, will be used to analyse experiments on ammonia gas in the following sections.

The effect of the nonlinear medium on the resonator response is best approached by considering first the *dispersive limit* $\alpha L \ll 1$; $|\tilde{\Delta}| \gg 1$; $I \ll I_s$. If we also assume $T_1 \ll t_R$, then (9.11) gives, approximately,

$$D(t) \approx \frac{1}{1 + |\varepsilon(t)|^2} \approx 1 - |\varepsilon(t)|^2$$

so that (9.12) becomes

$$\varepsilon(t) = \varepsilon_i(t) + R\varepsilon(t - t_R)e^{i\theta'}e^{iG|\varepsilon(t-t_R)|^2} \tag{9.14}$$

where $G = \alpha L\tilde{\Delta}/2$, and θ' includes the linear phase shift due to the gas. In this limit, the original system of differential equations and boundary conditions is reduced to a nonlinear *mapping* of the cavity field at intervals of t_R, the round trip time, as first obtained by *Ikeda* [9.4]. He showed that the fixed points of (9.14) are, in general, multiple-valued functions of the input field, giving optical multistability. More significant here, however, was his demonstration that the Ikeda map can show period-doubling cascades leading to chaos, in which $\varepsilon(t)$ wanders on a *strange attractor* (of fractional dimension) in its complex phase plane.

9.2.2 Physical Interpretation of the Ikeda Instability

The material nonlinearity responsible for nonlinear refraction, and thus bistability, can also generate side bands on the pump frequency. Physically, two photons with frequency ω are scattered to form a pair of photons at $\omega \pm \Delta\omega$, so the process is termed four-wave mixing. An additional input signal, the probe, detuned from the pump by $\Delta\omega$ will thus experience gain or loss according to the combined effect of the nonlinearity and the mode structure of the cavity. Clearly, it is most advantageous if *both* $\omega + \Delta\omega$ and $\omega - \Delta\omega$ are cavity resonant, i.e., if $2\Delta\omega$ is a multiple of the cavity's free spectral range. An *even* multiple (including zero) means that the pump is itself resonant, and this double resonance lies behind optical bistability and the side band instabilities analysed by *Lugiato* and co-workers [9.15]. An odd multiple, on the other hand, means that the pump is off resonance, lying exactly half-way between two modes: the beat note between the pump and the resonant sidebands (which will be self-excited if the nonlinearity is strong enough) then has period $2t_R$, identifying this as the Ikeda instability.

The nice thing about this system is that the above double resonance can be guaranteed: the refractive-index change induced by the strong pump beam actually moves the comb of longitudinal modes with respect to the pump frequency, and this "transphasing" of the mode spectrum alternately gives rise to bistable and Ikeda-type double resonances as the pump parameter ε_i is increased [9.16]. This is illustrated in Fig. 9.1.

The entire period-doubling cascade can be given a similar interpretation. As each new period $2^n t_R$ bursts into oscillation the effective free spectral range of the cavity is halved, because the generalised condition for resonance must be constructive interference after 2^n round trips (since only after $2^n t_R$ do the cavity's optical properties repeat themselves). At the $2^n t_R$ threshold these small signal modes are degenerate with the oscillating frequency spectrum, but as ε_i is increased, they "transphase" in frequency to halfway between the oscillating frequencies. If the gain there becomes large enough, these new modes

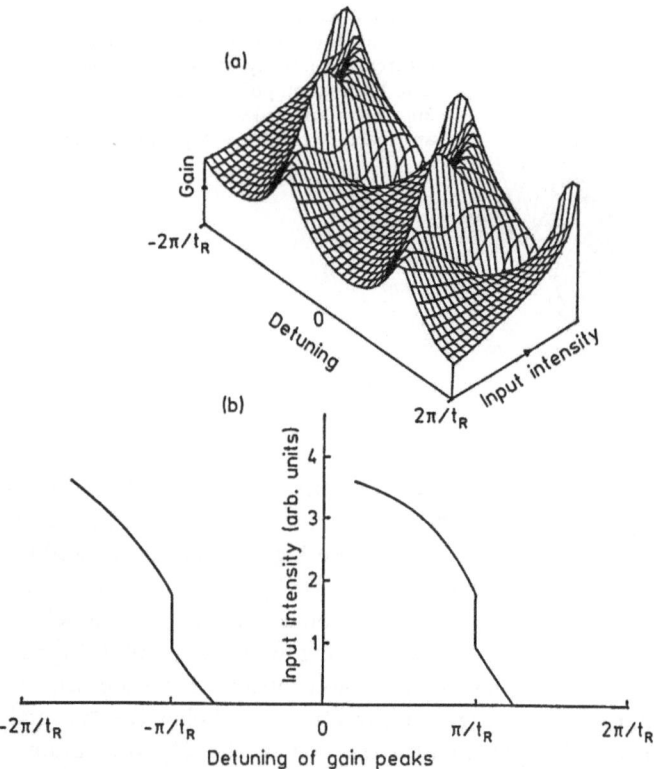

Fig. 9.1. Gain spectrum for a ring resonator ($B = 0.58$, $\theta_0 = 2.3$) showing "transphasing" of the cavity resonances as the input intensity is increased, together with locking at the Ikeda $2t_R$ resonance position (B is the amplitude feedback factor of the cavity, and θ_0 the mistuning)

burst into oscillation: if not, "transphasing" continues (Fig. 9.1) to a renewed coincidence with the oscillating frequencies, at which point an inverse period doubling cascade ensues, leading eventually to a steady-state response. Figure 9.2 illustrates this process, showing the small-signal gain spectrum as ε_i is increased through the $2t_R$ threshold, and the ensuing doubling of the spectrum followed by "transphasing" to the verge of $4t_R$ oscillation. It should be noted that the noise spectrum of the system should be very similar to Fig. 9.2, because of the filtering action of the cavity on any broadband noise source.

Considering now finite T_1, perturbation theory readily leads to the conclusion that the gain spectrum of four-wave mixing becomes Lorentzian, with halfwidth $\sim T_1^{-1}$, and the associated dispersion means that the free spectral range becomes a function of frequency. The former is usually the more important: clearly if $t_R \ll T_1$ then there will be no significant gain in the Ikeda situation, where the modes straddle the pump frequency: this is the physical origin of the requirement $t_R > T_1$ usually advanced for Ikeda instability.

Another effect of finite T_1 is to raise the degeneracy by which all symmetrically-placed pairs of side-band modes reach threshold simultaneously. On the

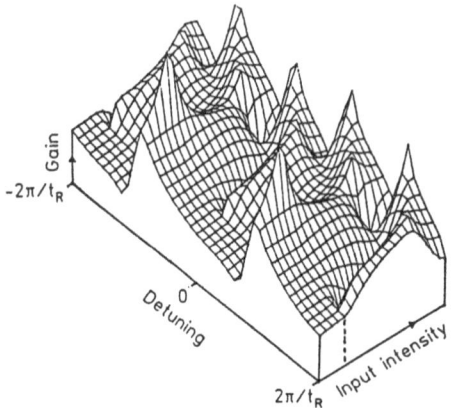

Fig. 9.2. Gain spectrum for a ring resonator showing the creation of new "modes" at a bifurcation (steady state →$2t_R$ oscillation). The bifurcation point is marked by the *dotted line*. The four peaks at the back indicate the approach to the $2t_R \rightarrow 4t_R$ bifurcation

one hand, the raising of the OB degeneracy permits self-pulsing at period t_R, as described by *Lugiato* et al. [9.15], while on the other, the Ikeda degeneracy splits to yield pulsing at $2t_R/3$, $2t_R/5$, etc. These high frequency instabilities have been extensively studied in hybrid systems [9.17]: we have observed the $2t_R/3$ oscillation and in all-optical system based on ammonia gas (see below).

As a final point in connection with the mapping, we have found [9.14] that in high-finesse resonators, viz $R \geq 0.8$, it possesses solutions of period $3t_R$ and higher periodicities, appearing by tangent bifurcation even below the "normal" Ikeda instability threshold. The basin of attraction of the $3t_R$ oscillation is intimately mixed with that of the fixed point, and strong $3t_R$ modulation is found in simulation of pulsed excitation under appropriate conditions. Related phenomena are discussed in detail in Chap. 7.

9.2.3 Fabry-Perot Resonators

We now resume discussion of the case where E_B is non-zero in (9.8), appropriate to Fabry-Perot resonators. For finite T_1, progress can only be made in the dispesive limit, where the spatial modulation of n in (9.6), induced by that implicit in I, is small, so that only Fourier components of low order need to be considered. This is not the case, however, for a highly saturated medium, such as we deal with in the experiments described below, so we first take the limit $T_1 \rightarrow 0$ in (9.6), so that

$$n\left(1 + \frac{I}{I_s}\right) = 1 \ . \tag{9.15}$$

Writing

$$\begin{aligned} I &= |E_F|^2 + |E_B|^2 + E_F E_B^* e^{2ikz} + \text{c.c.} \\ &= (I_0 + G + G^*)I_s \ , \end{aligned} \tag{9.16}$$

where G describes the spatial grating in the standing-wave field, we seek to solve (9.15) by the ansatz

$$n = n_0 + \sum_{j=1}^{\infty} (n_j G^j + \text{c.c.})$$

which, on insertion into (9.15), gives, for the spatially uniform (dc) term

$$(1 + I_0)n_0 + |G|^2 (n_1 + n_1^*) = 1 \tag{9.17}$$

and for the term $\sim e^{i2jkz}$:

$$(1 + I_0)n_j + n_{j-1} + |G|^2 n_{j+1} = 0 \ . \tag{9.18}$$

This equation has solutions in which

$$\frac{n_j}{n_{j+1}} = \frac{-(1 + I_0) \pm \sqrt{(1 + I_0)^2 - 4|G|^2}}{2} \qquad \text{for all} \quad j \geq 0 \ . \tag{9.19}$$

In order that the degree of spatial modulation decrease with j, the lower sign must be chosen. Furthermore, all n_j are real.

Use of (9.19) in (9.17) enables n_0 to be calculated

$$n_0 = ((1 + I_0)^2 - 4|G|^2)^{-1/2} \ , \tag{9.20}$$

which reduces to the previous expression in the case of unidirectionality ($G = 0$). The quantity n_1 can be obtained from (9.19):

$$n_1/n_0 = \frac{-(1 + I_0) + \sqrt{(1 + I_0)^2 - 4|G|^2}}{2|G|^2} \ . \tag{9.21}$$

The terms on the right-hand side of (9.9) phase-matched to the forward and backward fields can now be found

$$\langle nE^+ \rangle_{\text{F}} = n_0 E_{\text{F}} [1 + (n_1/n_0)|E_{\text{B}}|^2/I_s] \qquad \text{and} \tag{9.22a}$$

$$\langle nE^+ \rangle_{\text{B}} = n_0 E_{\text{B}} [1 + (n_1/n_0)|E_{\text{F}}|^2/I_s] \ . \tag{9.22b}$$

The population-grating term n_1 thus gives rise to *nonlinear nonreciprocity:* the absorption coefficient and refractive index experienced by the forward- and backward-travelling waves are *unequal.*

For weak fields, we find

$$n_0 \simeq 1 - I_0 \ ; \quad n_1 = -n_0$$

and thus we obtain from (9.22):

$$\langle nE^{+}\rangle_{\mathrm{F}} = E_{\mathrm{F}}[1 - (|E_{\mathrm{F}}|^{2} + 2|E_{\mathrm{B}}|^{2})/I_{\mathrm{s}}] \ ,$$

$$\langle nE^{+}\rangle_{\mathrm{B}} = E_{\mathrm{B}}[1 - (2|E_{\mathrm{F}}|^{2} + |E_{\mathrm{B}}|^{2})/I_{\mathrm{s}}] \ .$$

We see that the mutual nonlinear effect is double the self-effect.

Firth [9.12] showed that in the dispersive (Kerr) limit, the Maxwell equation can be integrated along characteristics. An equation similar to (9.14)is obtained, but now the nonlinear phase shift involves an integral over time because the field $\varepsilon(t - t_{\mathrm{R}})$ meets, and is phase-shifted by, fields $\varepsilon(t')$ over the whole range $t - 2t_{\mathrm{R}} \le t' \le t$.

Though more complicated than that of the Ikeda mapping, this model gives rise to qualitatively similar effects, with bands of chaos on the positive-slope branches of the input-output characteristics. We have not investigated routes to chaos in detail, but the transition to chaos seems rather abrupt, and only rarely have we found a period-doubling sequence.

One novel feature of the Fabry-Perot is the prediction of extra instabilities due to the nonlinear nonreciprocity: Firth and Wright found period-t_{R} oscillations due to this effect in a study of transverse effects in Fabry-Perot resonators (Fig. 9.3) [9.18].

Instabilities in Fabry-Perot resonators with saturable two-level media, which require the full equations (9.9 and 22), have received little, if any, study, but we present below results of computations based on these equations showing period-doubling and chaos.

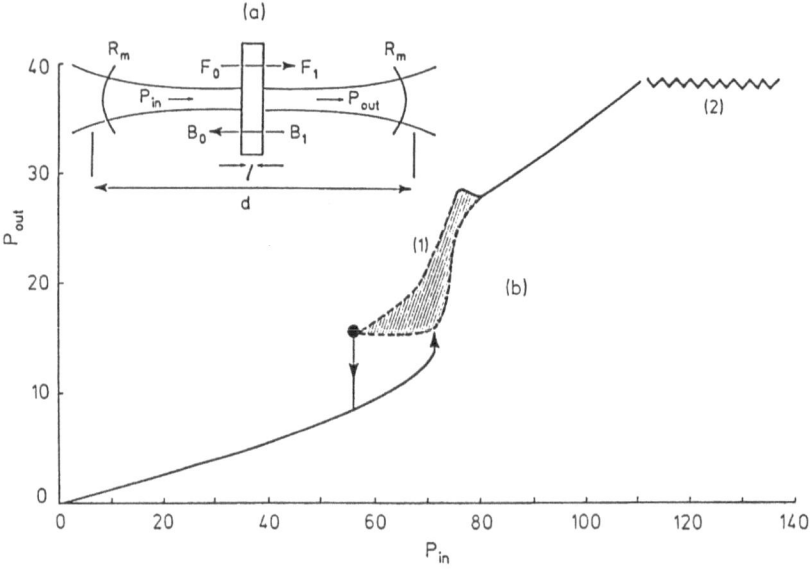

Fig. 9.3. (a) Resonator configuration (R_{m}: Radius curvature of the mirror, R : mirror reflectivity). (b) P_{out} versus P_{in}. Region (1) shows t_{R} oscillation; bifurcation and chaos appears in region (2)

As long as T_1 is zero, the grating terms do not induce any net energy transfer between the counter-propagating beams, but this becomes a problem for finite T_1. We conclude this theory section by presenting results [9.14] for finite T_1 in which atomic diffusion is assumed strong enough to "wash out" the grating, thus preventing back-scattering.

Taking also the dispersive limit, and setting

$$\varsigma = \frac{\omega n \mu^2 \tilde{\Delta} T_2}{2\epsilon_0 n_b \hbar (1 + \tilde{\Delta}^2)} \quad \text{and}$$

$$F \equiv \left(\frac{\varsigma L}{I_s}\right)^{1/2} E_F \ , \quad B = \left(\frac{\varsigma L}{I_s}\right)^{1/2} E_B \ ,$$

$$\chi^{NL} \equiv \varsigma(n+1) \ , \quad \tau \equiv T_1 \ , \quad \alpha = \frac{\alpha_0}{1 + \tilde{\Delta}^2}$$

the following equations:

$$\tau \frac{\partial \chi^{NL}}{\partial t} = -\chi^{NL} + |F|^2 + |B|^2 \ , \tag{9.23}$$

$$\frac{\partial F}{\partial x} + \frac{n_b \partial F}{c \partial t} = i \chi^{NL} F - \frac{\alpha}{2} F \ , \tag{9.24}$$

$$-\frac{\partial B}{\partial x} + \frac{n_b \partial B}{c \partial t} = i \chi^{NL} B - \frac{\alpha}{2} B \ , \tag{9.25}$$

are obtained from (9.6) and after neglecting rapidly varying terms $\exp(\pm 3 i n_b k x)$. Equations (9.23–25) are also known as the Maxwell-Debye equations; they apply to Kerr media in particular, but are equally suitable for atomic systems in the limits discussed above.

These equations, together with the boundary conditions for a Fabry-Perot resonator with input reflectivity R_1 and output reflectivity R_2

$$F(0,t) = (1 - R_1)^{1/2} E_i(t) + R_1^{1/2} B(0,t) \ ,$$

$$B(L,t) = R_2^{1/2} E_F(L,t) \exp(+i\theta) \tag{9.26}$$

where E_i is the amplitude of the incident field (up to a phase factor), and L is the length of the cavity, have been numerically integrated [9.3,4] in order to examine the influence of τ (i.e., T_1) and other parameters on the instabilities. Figure 9.4 shows the steady-state amplitude $|S| = |F(0,t)|$ as a function of input amplitude $A = (1 - R_1)^{1/2} E_i$, and displays the regions of positive slope instability as a function of τ/t_R. It will be observed that increasing τ has the effect of confining the instabilities to progessively-higher branches of the characteristic.

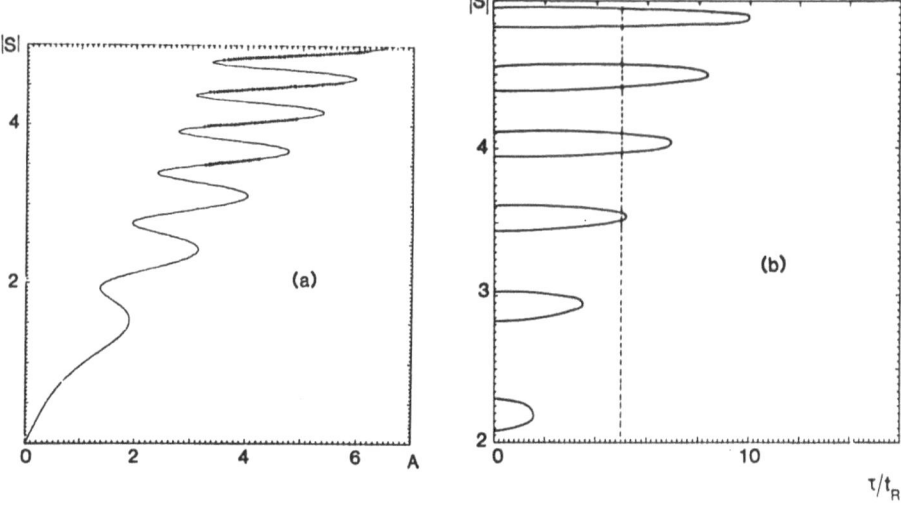

Fig. 9.4. (a) Plot of $|S|$ against A in the steady state (see the text) for $R_1 = R_2 = 0.8$, $\theta = 0$, and $\alpha L = 1$. The regions of negative slope are physically inaccessible (always unstable). (b) Positive-slope-instability edges as a function of t/t_R. The dashed vertical line corresponds to $t = 5t_R$, and its intersections with the curves give the unstable regions in the upper part of (a)

The numerical technique used to integrate (9.23–25) is the one applied previously to quantum-beat superfluorescence and other dynamic phenomena in optical bistability [9.19]. In Fig. 9.4 we show how oscillatory and chaotic behaviours develop for $\tau/t_R = 0.1$ and 5, respectively. The period of oscillation in Fig. 9.5 remains approximately the same as in the linear regime. For $|S| = 4.0$, $\alpha L = 1$, and $R_1 = R_2 = 0.81$, the calculated threshold value of the period is $\sim 2.8\,t_R$, whereas the actual one is $\sim 2.9\,t_R$, i.e., a "lethargic" $2t_R$ oscillation. We further extend the study of the dynamics to all branches for the same fixed input field ($A = 4.463$) of Fig. 9.5; the corresponding values of $|S|$ labelled, $1, 2 \ldots 8$ are shown in Fig. 9.6. When $\tau/t_R = 0.1$, the plot in the phase-plane

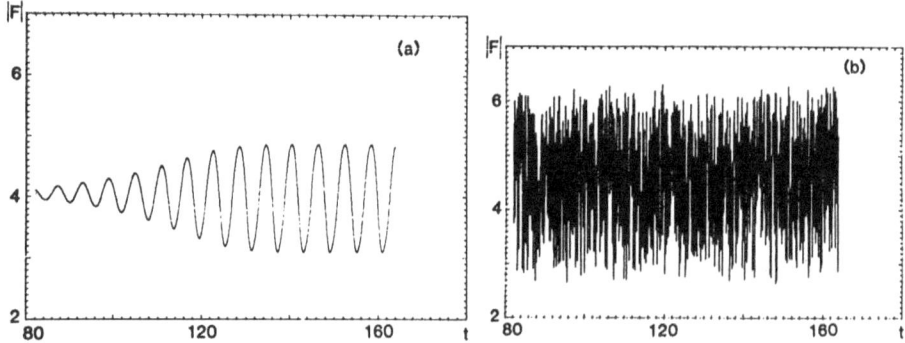

Fig. 9.5a,b. Development of an instability around $A = 4.463$, $|S| = 4.05$ for the same parameters and units as in Fig. 9.3a. (a) Periodic motion, $\tau = 5t_R$; (b) Chaos, $\tau = 0.1\,t_R$

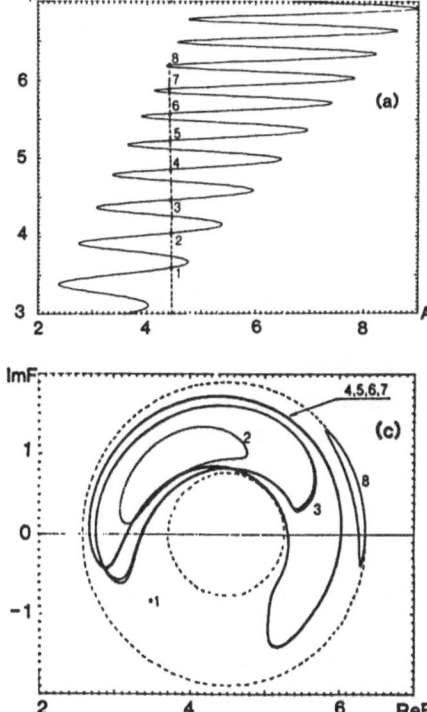

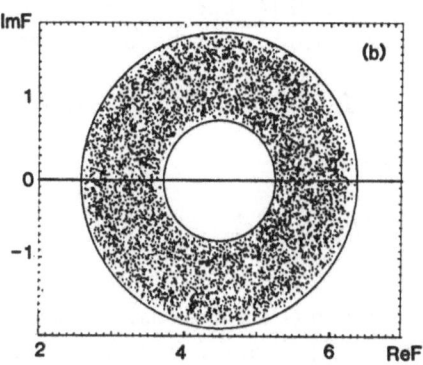

Fig. 9.6. (a) As Fig. 9.4a, the intersection points 1–8 standing for steady states corresponding to $A = 4.463$. (b) Slight perturbation of the steady states of (a) when $T = 0.1\,t_R$ leads to "strong turbulence" with a unique strange attractor. The *dots* correspond to the computed time series; as they wander erratically, the *dots* are not joined for the sake of clarity. (c) As (b), but for $t = 5t_R$ (self-oscillations). The numbers on the limit cycles correspond to the steady states of (a)

of $F(0,t)$ (Fig. 9.6) shows a *unique* strange attractor for all branches; here the "turbulence" is strong. For $\tau/t_R = 5$ we obtain periodic oscillations and there exist several limit cycles depending on the branch where the perturbation was made; note that the curves corresponding to 4, 5, 6 and 7 are, within the resolution of our graph plotter, identical; on the other hand 1 is a stable point.

9.2.4 Universality

In the analysis of problems with finite τ a major obstacle is that the period which doubles has no necessary relation to t_R. Time-domain analysis is this difficult and we examine instead the frequency domain by considering the evolution of the spectrum as a control parameter λ is varied.

At some value λ_n the period doubles from $2^n T$ to $2.2^n T$, where T is some base period of order t_R. Below λ_n, the spectrum containes all harmonics of $2\pi/2^n T$ while above λ_n a "new" component appears half-way between each pair of "old" components. It is thus convenient to represent the Fourier components of $F(t)$, as defined by

$$F(t) = \sum_m c_m^n \exp\left(it2\pi m/2^n T\right) \qquad \text{as a vector}$$

$$f = (c_1^n, c_2^n, \ldots, c_j^n, \ldots) \ .$$

Equation (9.14) and a broad class of its generalisations can then be Fourier-transformed to take the form (after Taylor expansion of the exponential)

$$Ef = G(\lambda)f + H(\lambda)ff + N(\lambda)fff + \ldots \tag{9.27}$$

where, e.g.,

$$ff = \sum_{jm} c_j^n c_m^n , \quad \text{etc.}$$

and E and G are diagonal matrices, the mth component of E being $\exp(-it_R 2\pi m/2^n T)$; H and N are connecting "tensors".

Above, but close to, the bifurcation point where $(\lambda - \lambda_n) \ll 1$, we have new components f_N of f which are very small compared with the old ones f_0. For our analysis we partition f as (f_0, f_N) in order to linearise in f_N, and write the relevant part of (9.27) as

$$E_n f_N = M_n(f_0, \lambda) f_N . \tag{9.28}$$

We can try to solve (9.28) if we know f_0 : there will be a non-trivial solution f_N^n at $\lambda = \lambda_n$.

Just above the bifurcation we take (9.27) to third order in f_N by assuming that $f_N \propto f_N^n$ where calculating the nonlinear effect of f_N or f_0 and itself. We then obtain a linear equation for f_N which matches smoothly onto (9.28) as $\lambda \to \lambda_n$ only if the dependence of M_n on λ is compensated by that due to f_N^n. As each term in (9.27) must have the same net frequency, it follows that f_N can affect f_0 only in second order and itself in third order. In both cases, M_n contains f_N^n at second order while it contains $(\lambda - \lambda_n)$ in first order. Hence,

$$f_N \sim (\lambda - \lambda_n)^{1/2} f_N^n . \tag{9.29}$$

This argument provides some justification for the extrapolation procedure explained below since the differences of corresponding maxima are functions of "new" components only.

At the next bifurcation point, $\lambda = \lambda_{n+1}$, a new component appears either side of each member of f_N. The former can be split into two sets f_N^+ and f_N^- according to whether the component has frequency above or below that of the corresponding member of f_N. Thus, working always to lowest order and in the limit of large n, we can show that $(\lambda_{n+1} - \lambda_n)$ can be expected to decrease *geometrically* with n as required for a Feigenbaum cascade. Noting the identity

$$E_{n+1} = wE_n + E_n/w \quad \text{with}$$

$$w = \exp\left(i\pi 2^{-n} t_R/T\right) \xrightarrow[n \to \infty]{} 1 + i\pi 2^{-n} t_R/T$$

we can write the next equivalent of (9.28) in block form

$$\begin{bmatrix} E_n w & 0 \\ 0 & E_n/w \end{bmatrix} \begin{bmatrix} f_N^+ \\ f_N^- \end{bmatrix} = \begin{bmatrix} M_{++} & M_{+-} \\ M_{-+} & M_{--} \end{bmatrix} \begin{bmatrix} f_N^+ \\ f_N^- \end{bmatrix} . \tag{9.30}$$

Physically, M provides the difference frequency which enables one new component to scatter and couple to another. For both M_{++} and M_{--}, this difference frequency belongs to f_0 and, provided $2^n T$ is large compared to all intrinsic response times of the system, *both* are equal to M_n of (9.28) (in lowest order). M_{+-} and M_{-+}, on the other hand, couple pairs of frequencies differing by a member of f_N so each element of them is proportional to $(\lambda - \lambda_n)^{1/2}$.

Finally, requiring a non-trivial solution of (9.30) while (9.28) remains true to lowest order, leads to

$$(w - 1)^2 \sim (\lambda_{n+1} - \lambda_n)$$

and hence using (9.29)

$$(\lambda_{n+1} - \lambda_n)/(\lambda_{n+2} - \lambda_{n+1}) \simeq (\mathrm{i}\pi 2^{-n})^2/(\mathrm{i}\pi 2^{n-1})^2 = 4 .$$

This demonstrates the required geometrical decrease in $\Delta\lambda$ and we also obtain an approximate value of Feigenbaum's $\delta (= 4.699\ldots)$. On the other hand, (9.27) is extremely general, covering ring as well as Fabry-Perot resonators.

9.2.5 Trapping Bifurcation Points

For a quantitative explanation of bifurcations in systems of differential equations, a considerable amount of computer time is needed to generate a single time sequence. In these circumstances it is difficult to pin down the most stable control value, while the bifurcation points themselves show poor convergence. We have found a technique, however, that has enabled us to find Feigenbaums's δ to $\leq 1\,\%$ (Table 9.1).

Table 9.1. Universal sequence as the control parameter λ approaches the limit point λ_∞ $[\delta_n = (\lambda_n - \lambda_{n+1})/(\lambda_{n+1} - \lambda_{n+2})]$

Control parameter λ	$\lambda_8 = P4 \rightarrow P8$	λ_∞	δ_8	δ_{16}	δ_{32}
A	4.476	4.463	3.902	4.786	4.612
θ_0	0.09046	0.000	4.411	4.601	4.658
τ/t_R	1.826	1.78636	4.506	4.643	4.618

The technique is based on the above hypothesis that at each bifurcation the new Fourier components grow as $(\lambda - \lambda_n)^{1/2}$ as any control parameter λ is increased through its nth bifurcation value λ_n. Consider any peak of $I(t) \equiv |F(t)|^2$ which first repeats a time T_n later. If λ is increased beyond λ_n, $I(t + T_n) - I(t) \neq 0$, and we assert that

$$|\Delta I|^2 = [I(t + T_n) - I(t)]^2 = \mathrm{const} \times (\lambda - \lambda_n) . \tag{9.31}$$

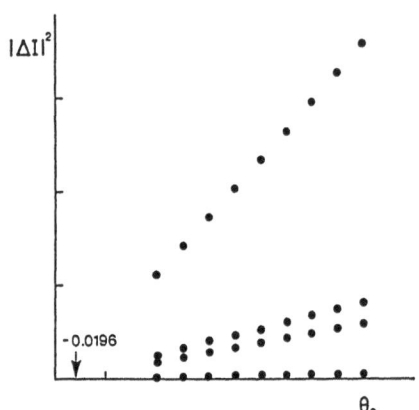

Fig. 9.7. Behaviour of $|\Delta I|^2$ as a function of detuning angle θ_0, in the neighbourhood of the 4P–8P threshold

Thus plotting (9.31) we get a straight line that intersects the λ axis at λ_n. As Fig. 9.7 shows, this rule is very well obeyed and the accuracy in fixing λ_n is enhanced because the number of straight lines doubles at each bifurcation, thus improving the statistics and partially offsetting the decline of $|\Delta I|^2$ for large n.

9.3 Nonlinear Absorption Characteristics of Molecules

Many molecular gases have near resonant vibrational-rotational transitions with the CO_2 laser in the 9–10 μm spectral region providing a range of frequency offsets well documented from photochemistry studies [9.20].

Of these ammonia is selected here as a particularly attractive candidate for investigation since it typifies a discrete level system in which excited state pumping is in general precluded by significant anharmonicity, inversion splitting and large rotational constant [9.21]. As such several of the $V - R$ transitions of the ν_2 fundamental band act as two-level systems when pumped by the CO_2 laser radiation. As fundamentally the simplest of nonlinear schemes, which can be furthermore fully quantised, the two-level medium merits special attention.

In contrast to NH_3, many polyatomic molecules have high density of states even at low energies. Although their spectroscopy is necessarily more complex the often low saturation intensity and high absorption cross sections of such molecules make them attractive with regard to eventual CW operation. We therefore also consider, although in more limited investigations, sulphur hexafluoride (SF_6), which is a representative and well documented example of these molecules.

Nonlinear refraction in near-resonantly excited media, which is the fundamental mechanism responsible for the instability phenomena considered here, arises from intensity dependent saturation of anomalous dispersion.

216

The laser-induced energy transfer processes which give rise to this may be understood from investigation of the accompanying effect of saturated absorption. Using this approach we find our data for both NH_3 and SF_6 are well modelled using a simplified energy-level scheme describing these molecules. This theory is subsequently applied to nonlinear optical resonators in the interpretation of the instabilities observed in these systems; viz. both ring and Fabry-Perot cavities.

Since the generation of resonant nonlinearity depends on partial saturation of the optically pumped transition, molecules with low saturation intensity are ideal. However, the counteracting requirement of fast medium response time for generation of instabilities implies high saturation intensities. To meet these conditions for the molecules considered here we are restricted to saturation effects arising from rotational relaxation rather than from vibrational-translational relaxation which is relatively slow. These points are further discussed below. Relatively high-power pump signals are therefore required necessitating the use of pulsed rather than CW laser systems.

In our experiments a transversely excited atmospheric (TEA) CO_2 laser operated on a single transverse and longitudinal mode was used, providing step-tunable (step interval $\sim 2\,\mathrm{cm}^{-1}$) smooth pulses full width at half maximum (FWHM) $\sim 100\,\mathrm{ns}$ and peak power $\sim 1\,\mathrm{MW}$. Mode control was achieved by injection locking [9.22] of the TEA laser using the signal from a frequency stabilised CW CO_2 laser. Alternatively a CW CO_2 gain section was incorporated within the cavity of the TEA system; the so-called hybrid system [9.23].

Absorption measurements were taken over an extended pressure range using short cell lengths, of 10–20 cm for NH_3 and 1 cm for SF_6 to minimise self-focussing effects so ensuring an essentially constant beam cross section throughout the medium. The TEA CO_2 laser input and transmitted signals was sampled by KBr beam splitters and monitored by photon drag detectors and a Tektronix 7104 oscilloscope, total response time $\leq 1\,\mathrm{ns}$.

9.3.1 Ammonia

Typical input and output pulse shapes are shown in Fig. 9.8 for a fixed input intensity of $\sim 3\,\mathrm{MW/cm}^{-2}$. Data here is for the $aR(1,1)$ transition which lies 1.23 GHz below the 10R(14) CO_2 laser pump line [9.24] and is representative of that obtained for other optically pumped transitions. Effects of saturation, clearly evident at low pressure, diminish with increased pressure resulting in a progressively weaker, narrower, and apparently delayed transmitted signal. Additional measurements of small signal absorption, using a low-power CW CO_2 input signal showed good pressure-squared (p^2) dependence for absorption coefficient confirming the system acts as an off-resonantly pumped, homogeneously broadened two-level system yielding a value for α of $0.025\,\mathrm{cm}^{-1}$ at 10 Torr.

9.3.2 Sulphur Hexafluoride

Pulsed transmission measurements for SF_6 are shown in Fig. 9.9. Results here are for the 10P(16) CO_2 laser line at $947.7\,\mathrm{cm}^{-1}$ coincident with several close

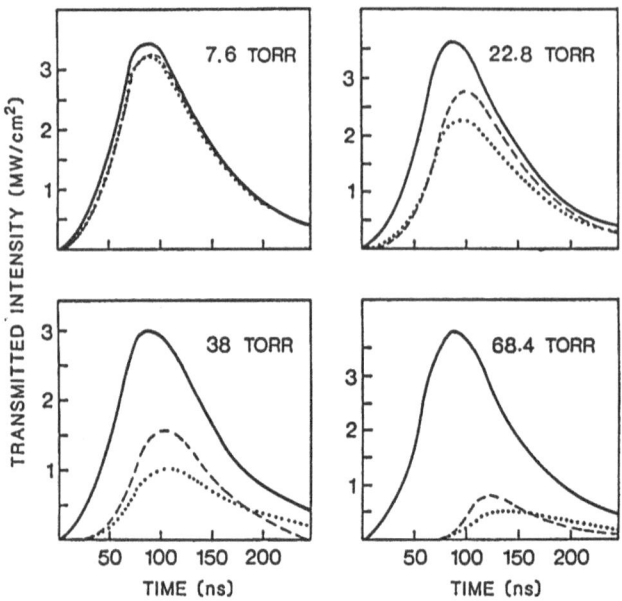

Fig. 9.8a,b. CO_2 laser pulse transmission through NH_3, for various pressures; input pulse: transmitted pulse, **(a)** experimental (---) and **(b)** computed ($\cdots$) lines

lying Q branch transitions near the ν_3 band centre. Similar results were obtained for other optically pumped transitions to the blue side of the absorption head. At low pressures the onset of saturation appears immediately indicating saturation intensities well below the limit of response of the detection system. Substantial absorption later in the pulse, in contrast to data for NH_3, is attributed to excited-state absorption as discussed below.

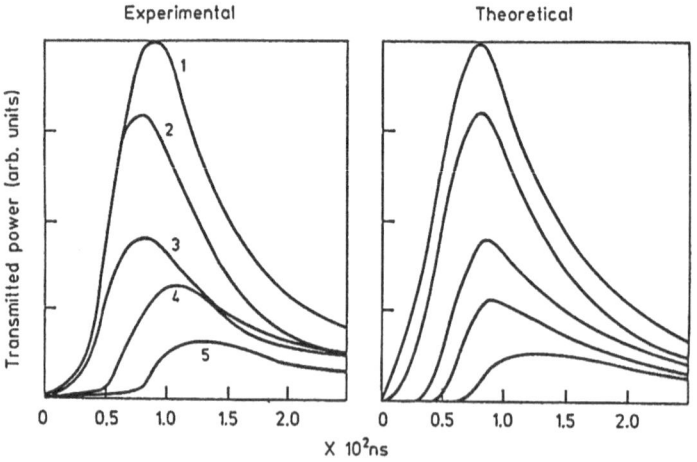

Fig. 9.9. CO_2 laser pulse transmission of 10P(16) line in SF_6. Curves *2–5* for pressures 0.75, 18.5, 26, 37 Torr, respectively. Cell length 1 cm, curve *1* for input pulse

9.3.3 Molecular Energy-Level Scheme

Population redistributions in optically pumped molecular systems is determined by the interplay between such factors as the pump intensity, pump-pulse duration, and the various molecular relaxation processes responsible for redistributing the excited population among the vibrational-rotational energy states of the molecule.

Of these, rotational inelastic collisions involve the smallest energy change [(rotational level spacing) $\sim 10^{-2} \times$ (vibrational level spacing)] and consequently rotational relaxation is usually the fastest of these processes ($\sim 10^{-9} - 10^{-10}$ atm s). This is therefore the dominant pressure-dependent contribution to line broadening. Following excitation of a particular vibrational-rotational transition, population is then rapidly thermalised among the manifold of rotational states of the ground and excited vibrational levels. Subsequent deexcitation of population to ground occurs at a considerably slower rate normally through vibrational-translational $(V - T)$ relaxation ($10^{-5} - 10^{-8}$ atm s). We note that, at least for low levels, vibrational-vibrational $(V - V)$ relaxation between like molecules, although often faster than $V - T$ relaxation, does not in general contribute appreciably to population redistribution. Similarly, spontaneous radiative emission, which is extremely slow in the infrared ($\gtrsim 10^{-2}$ s), has little effect.

a) Four State Model for NH_3

A typical energy-level scheme showing the relevant levels for the ν_2 fundamental band, of NH_3, is illustrated in Fig. 9.10a, showing the radiatively coupled vibrational-rotational transition and collisional relaxation coupled vibrational-rotational transition and collisional relaxation routes. For the purpose of analysis it is convenient to represent this scheme in the simplified form shown in Fig. 9.10b, where n_1 and n_2 are the number densities of the molecules in the vibrational-rotational level interacting with the CO_2 pump laser and N_1 and N_2 represent the number densities of all other molecules in the states of the same vibrational mode.

The radiative coupling between the pumped transitions is shown by continuous lines, and the corresponding rate coefficients are $W_{lm} = I(t)\sigma_{lm}/\hbar\omega$ [s^{-1}], where σ_{lm} is the absorption cross section, $I(t)$ the pump laser intensity, and $\hbar\omega$ the pump photon energy.

Collisional relaxations (both rotational and vibrational) are shown by wavy arrows and corresponding rate coefficients linking level l and m are denoted by k_{lm} [Torr^{-1}s^{-1}]. Vibrational cross relaxations are omitted for the sake of clarity in illustration.

The genesis of our model can be traced to the work of *Burak* et al. [9.25], in which they provided justification for treating the molecular densities N_1 and N_2 as single kinetic groups. The implication of this treatment is that rotational-level populations constituting the kinetic group N_1 (or N_2) relax together,

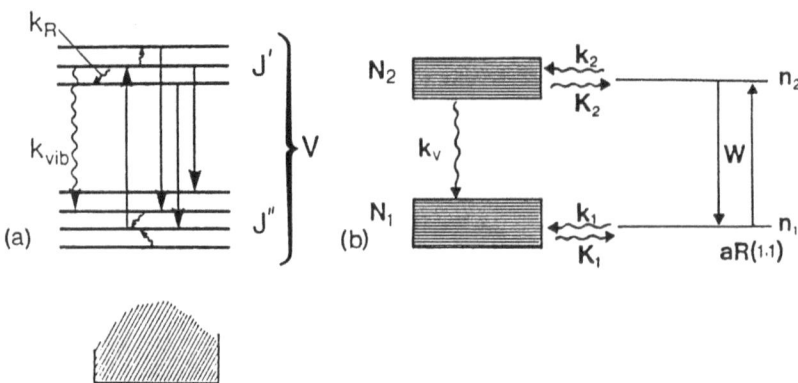

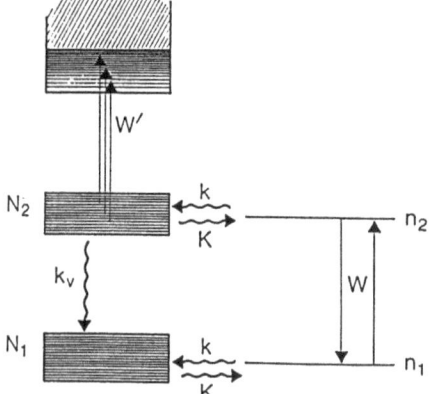

Fig. 9.10. (a) Energy levels and transitions for pumping and collisional processes. (b) Kinetic model scheme representative of the energy-level scheme in (a) describing NH_3, (c) equivalent scheme for SF_6 showing additional excited state absorption. Radiative coupling is represented by continuous lines and collisional relaxation (both rotational and vibrational) by wavy lines

maintaining their thermal equilibrium distribution. It was further shown that the equilibrium rate law between the rotational relaxation rates, namely,

$$k_{lm} = (f_m/f_l)k_{ml} ,\qquad(9.32)$$

where f_l is the thermal equilibrium fractional population of the level l, can reasonably be extended to the nonequilibrium case of pumping. The equations describing energy transfer are considerably simplified by considering only those collisional relaxation processes that make a dominant contribution to population redistribution among the levels. These are $V - V$ and $V - T$ or $V - R$ relaxation from the manifold of rotational states comprising N_2 to those comprising N_1 (rate constant k_v) and the rotational relaxation of population between the pumped levels and these manifolds. Specifying the rotational relaxation rates from levels 1 and 2 as k_1 and k_2, respectively, then the rates from the manifolds to these states, K_1 and K_2, respectively, are determined by detailed balance, see (9.32). The equations then can be written as

$$\frac{dn_1}{dt} = -W\left(n_1 - \frac{g_1}{g_2}n_2\right) - k_1n_1 + K_1N_1 ,\qquad(9.33)$$

$$\frac{dn_2}{dt} = W\left(n_1 - \frac{g_1}{g_2}n_2\right) - k_2n_2 + K_2N_2 ,\qquad(9.34)$$

$$\frac{dN_1}{dt} = -K_1 N_1 + k n_1 + k_v N_2 , \tag{9.35}$$

$$\frac{dN_2}{dt} = -K_2 N_2 + k_2 n_2 - k_v N_2 , \tag{9.36}$$

which imply the conservation of population condition

$$N = n_1 + n_2 + N_1 + N_2 . \tag{9.37}$$

General consideration of these equations has been given in [9.26]. Here we concentrate on the solution specific to our experimental conditions. For simplicity, we assume equal level manifold rate constants in the two levels $k_1 = k_2 = k$; $K_1 = K_2 = K$. The degeneracy factor g_1/g_2 is $3/5$ for the $aR(1,1)$ transition and for our conditions k_v, the $V - T$ rate is negligibly small.

These equations conserve $(n_1 + n_2) = n_e$ and $(N_1 + N_2) = N_e$ separately (e relating to thermal equilibrium); if k, K cannot be set equal in the two levels, generalisation is straightforward. Detailed balance requires that $k n_e = K N_e$; in the infinite reservoir limit K thus goes to zero and the system has effectively just two levels with appropriate dynamics. In NH_3 however, $n_e/N_e \sim 2\%$ and $K \sim 2 \, \mu s^{-1} \, Torr^{-1}$. The major effect of finite K is a leaching of ground-manifold population on a time scale K^{-1}, which is comparable to our pulse duration at the operational pressures, due to equilibration of population between the ground and excited manifolds.

In applying this model to the absorption data of Fig. 9.8 and subsequently the nonlinear resonator data, we have to consider pressure scaling and spatial effects. The rates k, K are assumed linear in pressure, and consequently the small-signal absorption coefficient is $\sim p^2$ at low pressure (off-resonance), as confirmed from small-signal absorption measurements. We adopt a value 13 MHz for the FWHM at 1 Torr [9.26], which leads to a value 80 MHz/Torr for $k : K$ is 2.1 % of k. These values lead to a saturation intensity of 2.31 MW cm^{-2} at 1 Torr (independent of pressure at low pressure). Provided diffraction is negligible, which is an excellent approximation for the single-pass experiments at least, we can account for both longitudinal and radial variation of the field in the gas cell. The former is achieved by a retarded-time spatial averaging as discussed in the previous section, giving

$$\dot{D} = k[(g+1)N - g - D - (g+1)|\varepsilon(t)|^2 (1 - e^{-\alpha LD})/2\alpha L] , \tag{9.38}$$

$$\dot{N} = K[(D+g)/(1+g) - N] , \tag{9.39}$$

where $g = g_1/g_2$, and [compare (9.10)]

$$D(t) = \langle n_1 - g n_2 \rangle / n_e$$

and $N(t) = \langle N_1 \rangle / N_e$, and where $n_{2e} = N_{2e} = 0$ is assumed, for simplicity.

We integrate these equations by a Runge-Kutta method, using a fit to the observed input pulse shape for $I(0,t)$. Transverse effects are included by using the identity for the transmitted power P_t,

$$P_t(t) = \int_0^\infty 2\pi r\, dr\, I_t(r,t)$$

$$= \tfrac{1}{2}\pi w^2 I_s \int_0^{I_0/I_s} dI\, e^{-\alpha L D(I,t)} , \qquad (9.40)$$

where I equals $|\varepsilon(t)|^2$. The transmitted power is thus calculated by summing the transmissions, calculated from (9.38–40), of pulses which all have the same time dependence, but whose peak intensities vary from zero in the wings to I_0 in the centre.

Using this model with the above parameter values, the observed pump powers and beam area $\pi w^2/2 = 0.35\,\mathrm{cm}^2$, we compute the dotted pulse shapes in Fig. 9.8, which represent a very satisfactory fit to experiment with no free parameters, and justify our model.

b) Five State Model for SF_6

In quantify the absorption characteristics of SF_6 allowance must be made not only for collisional population transfer within the rotational manifold as for NH_3 but also subsequent excited state absorption. The level scheme is illustrated in Fig. 9.10c. If we assume the excited state absorption does not have a significant effect on the population N_2 then (9.33–37) describe the population distribution among the appropriate levels. As for ammonia the k_v rate can be neglected as slow on the time scale of our pulse [9.27], and $n_1 + n_2 = n_e$ and $N_1 + N_2 = N_e$ where n_e and N_e are equilibrium values. The thermal population n_{1e} is that for the ground states of a number of transitions [9.28], notably $Q(38)$, $Q(43)$, $Q(45)$ unresolved by the pump radiation. Using detailed balance, $kn_e = KN_e$, where we have taken $n_e : N_e = 1 : 200$ (within the range of [9.29]). This gives a value of $K = 0.9\,\mu s^{-1}\,\mathrm{Torr}^{-1}$ using $k = 175\,\mu s^{-1}\,\mathrm{Torr}^{-1}$, taken as the rotational relaxation rate [9.30]. As before the major effect of K is a leaching of population on a time scale K^{-1} which is significant on the time scale of our pulse at the pressures used. k, K and also α (see above) scale linearly with pressure, with α, the small-signal absorption, measured experimentally as $0.55\,\mathrm{cm}^{-1}\,\mathrm{Torr}^{-1}$ for the 10P(16) CO_2 line. Using $\sigma = 3.4 \times 10^{-17}\,\mathrm{cm}^2$ [9.31] gives a saturation intensity I_s of $\sim 50\,\mathrm{kW\,cm}^{-2}\,\mathrm{Torr}^{-1}$, considerably lower than for NH_3.

The simplest way to incorporate excited state absorption into this model is to postulate an absorption cross-section proportional to the excited manifold population N_2. Strictly such an absorption should depopulate the manifold, but in the interest of simplicity, we ignore this – in a sense this involves a reinterpretation of N_2 as the manifold of all excited states other than those directly pumped. While a drastic over simplification the model seems adequate to interprete our results. In support of this the excited state absorption is likely to be weak and unsaturable due to smearing of absorption strength by anharmonic mixing of the terminal levels [9.32] arising from the high state denstiy $(10^3 \, \text{cm}^{-1})$. Hence the population N_2 of the upper rotational manifold will not be significantly altered. This has been confirmed from the analysis by comparing transmitted pulses with and without excited-state absorption. Restriction of this absorption to the ν_3 mode alone is justified for low levels of vibrational excitation where rotational relaxation dominates over intermolecular vibrational effects in transferring population. Furthermore anharmonicity of the ν_4 mode is adequately compensated for by various rotational states in the upper manifold and these will contribute most to excited state absorption.

The equations can be integrated through the medium exactly if excited state absorption is ignored. Including excited state absorption leads to the approximate set of equations:

$$\dot{D} = k\left[2N - 1 - D - |\varepsilon(t)|^2 \frac{(1 - e^{-\alpha LD})}{\alpha L} e^{-\alpha_2 L \hat{N}_2}\right] \qquad (9.41)$$

$$\alpha_2 = \alpha(\sigma_{12})^2 \qquad (9.42)$$

where $\hat{N}_2(t) = 1 - N(t)$, σ_{12} being the ratio of the excited state absorption cross-section to that for the ground state two-level transition; the equilibrium populations are $N_{1e} = N_e$, $n_{1e} = n_e$ and $n_{2e} = N_{2e} = 0$. Neglecting diffraction (an excellent approximation) means the transverse effects can be calculated in the same way as for NH_3.

Using this model and experimental values of $I_0 = 20 \, \text{MW cm}^{-2}$ and beam radius 3 mm the theoretical curves of Fig. 9.10c were calculated. The best fit occurred with the free parameter $\sigma_{12} = 0.14$, a value similar to that calculated in [9.33].

9.4 Optical Bistability and Instability Generation in Optical Resonators

9.4.1 Ring Resonator

The original optical scheme described by *Ikeda* [9.4] for the generation of oscila-tion and turbulence was a ring cavity containing a two-level nonlinear medium. In our initial cavity experiments we have therefore considered this scheme com-prising a unidirectional ring system containing NH_3 gas [9.9]. As discussed above, the $aR(1,1)$ transition of NH_3 is off-resonantly pumped at $10.3\,\mu m$ by pulsed emission from a TEA CO_2 laser. The scheme is illustrated in Fig. 9.11. The laser pulses are coupled with use of a single-surface Ge flat ($R = 36\,\%$) into a 3.5 m three-element ring cavity closed by $100\,\%$ gold mirrors, containing the gas cell. The input and cavity signals were sampled by KBr beam splitters, and monitored by photon-drag detectors and a Tektronix model 7104 oscillo-scope; total response time was $\leq 1\,ns$. For NH_3 pressures ~ 9–15 Torr, significant self-focussing was observed in single-pass experiments, confirming a nonlinear refractive-index contribution [9.34] substantial enough for dispersive optical bistability and Ikeda instability. Closing the ring caused a huge distortion of the pulse shapes (sampled after the NH_3 cell). In particular, a considerable pro-portion of these showed modulation at the 23.4 ns period expected for Ikeda oscillation in our system [9.9]. Figures 9.12b and c show representative exam-ples of this modulation, Fig. 9.12a shows the input pulse shape. To confirm the period, we have digitised and Fourier transformed the traces; the resulting spec-tra show pronounced peaks at $\sim 45\,MHz$, confirming our observation of Ikeda

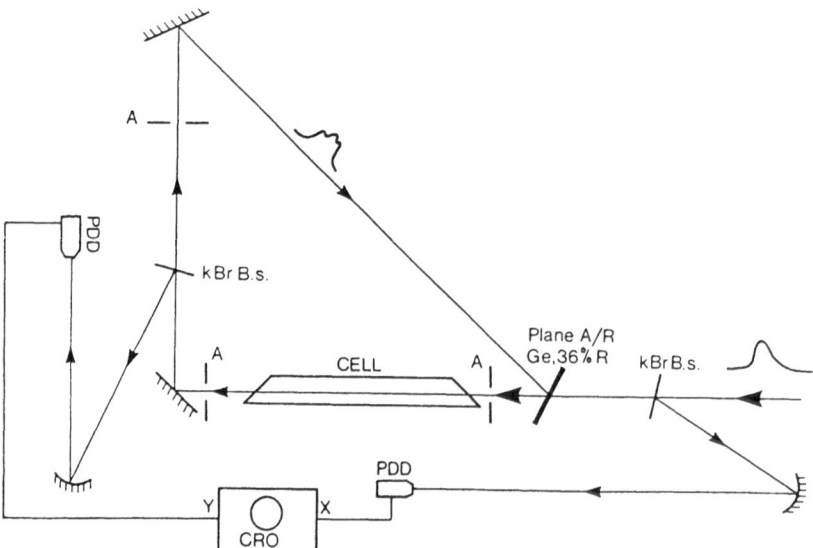

Fig. 9.11. Schematic diagram of ring-cavity system. (B.S.: beam splitter; PDD: photon-drag detector; A/R: anti-reflection coated)

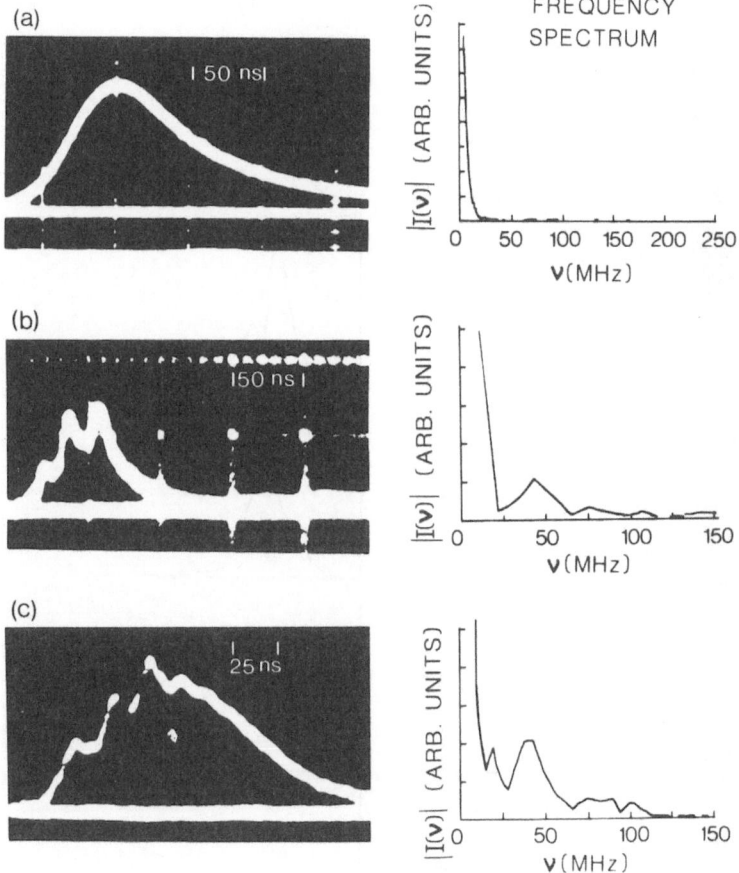

Fig. 9.12a–c. Sample oscilloscope traces of (a) the pump signal, and (b) and (c) the ring-cavity signal, together with their frequency spectra showing period $2t_R$ in (b) and (c) with indication of $4t_R$ in (c)

instability. Subsidiary peaks at $(4t_R)^{-1}$, possibly indicating a further bifurcation, have also been observed for some of the oscillatory signals. In contrast the input pulse (Fig. 9.12a), has an essentially featureless spectrum.

At other cavity settings strong pulse distortion indicatative of optical bistability was observed [9.35]. Examples are shown in Fig. 9.13a.

As discussed earlier, our molecular system is not quite a two-level system since allowance must be made for population transfer within the rotational manifolds. Applying the simplified molecular energy level scheme, developed earlier, which accounts for this, to the ring cavity system we obtain a considerable generalisation of Ikeda's model. The same assumptions previously used in modelling the nonlinear absorption data are used here.

We have numerically integrated (9.38 and 39) together with (9.12) using the pump pulse of Fig. 9.12a as input, and Fig. 9.13b shows the predicted intracavity pulses as a function of cavity tuning for representative parameter

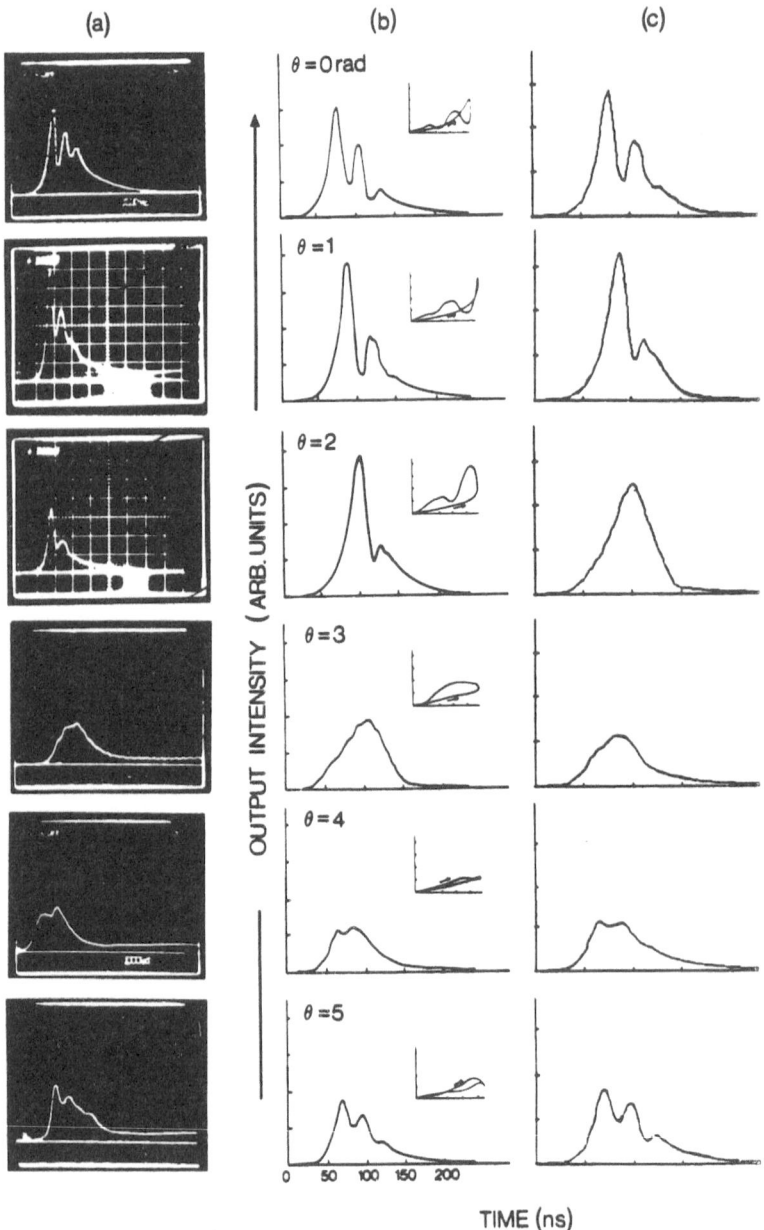

Fig. 9.13. Sample oscilloscope traces (a) the ring-cavity signal, together with corresponding theoretical pulse shapes for various detuning angles θ; for (b) the four-level scheme (also showing hysteresis curves of instantaneous cavity signal intensity against incident signal intensity and (c) the two-level scheme. ($\alpha L = 3$, $kt_R = 5$, $Kt_R = 0.1$, $|\varepsilon|^2_{\max} = 0.9$)

values. For comparison of our model with a pure two-level scheme we also show in Fig. 9.13c corresponding traces with K set equal to zero. Results are seen to be in good agreement. Considering data for our generalised four-level scheme (Fig. 9.13b) oscillation at $2t_R$, to be compared with traces of Fig. 9.13c, is manifest in the top and bottom traces which are for similar cavity detuning of $\theta = 0$ and 5 radiants while strong pulse distortion, associated with optical hysteresis [see inserts of instantaneous input (x axis) versus output (y axis) signal] occur at the opposite tuning as expected (Sect. 9.2). In view of the fact that our present model neglects self-focussing, this range of behaviours matches extremely well with the pulse shapes we observed (Fig. 9.13a).

9.4.2 Fabry-Perot Resonator

Compared to the ring cavity the Fabry-Perot resonator is a compact and versatile system particularly amenable to parametric studies of instability effects. Admittedly, interpretation of these phenomena is complicated by the effects of standing waves, as discussed earlier. The macroscopic behaviour of the system is nevertheless expected to be similar to that for the ring system.

To provide a basis for comparison with our data for the ring cavity we again concentrate on the $aR(1,1)$ NH_3 transition pumped by the 10.3 μm pulsed output from a single mode TEA CO_2 laser. Fabry-Perot resonators of various lengths 20–150 cm with intracavity NH_3 cells of 5–40 Torr were investigated (Fig. 9.14).

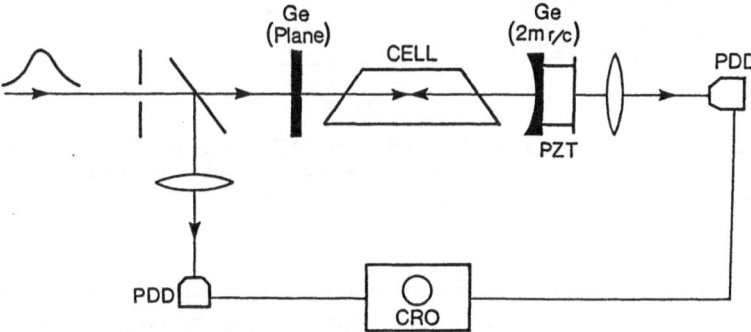

Fig. 9.14. Schematic diagram of Fabry-Perot cavity system

A 17-m optical delay line prevented significant feedback from the nonlinear resonator to the laser. The Fabry-Perot resonator comprised in most investigations a single-surface Ge flat input coupler of reflectivity $R_0 = 36$–85% and a single-surface Ge output coupler, of 2 m radius of curvature and reflectivity $R_L = 76\%$. The intracavity NH_3 gas cells, of length 10–100 cm, were terminated with KBr Brewster windows. As in the ring experiment, the input signal, sampled by a KBr beam splitter, and the output signal were monitored by photon-drag detectors and a Tektronix model 7104 oscilloscope. The output coupler was equipped with piezoelectric (PZT) tuning encompassing one free-spectral range of the Fabry-Perot resonator; for an empty cavity the temporal profile of the transmitted signal was identical with that of the input, and showed the expected variation in signal strength with PZT tuning (Fig. 9.15a).

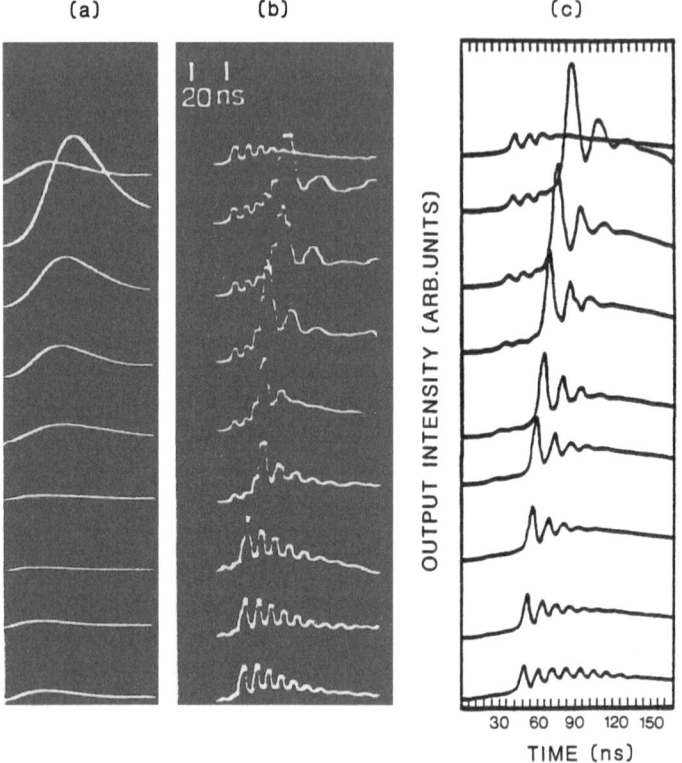

(a)   (b)   (c)

20 ns

OUTPUT INTENSITY (ARB. UNITS)

30   60   90   120 150
TIME (ns)

Fig. 9.15. PZT scan of (a) empty cavity; (b) cavity with 10 Torr NH_3 in 70 cm cell ($L =$ 86 cm); input and output reflectivities 67 % and 76 %, respectively; (c) computer traces for parameters corresponding to (b); $\alpha L = 1.5$, $\Delta \sim 10$; tick spacing t_R

The transverse intensity profile of the input signal was spatially monitored using conventional pinhole-sampling techniques and shown to be Gaussian with a 1/e spot diameter of ~ 3 mm.

The effects obtained with NH_3 were huge; the transmitted signal showed large pulse distortion and modulation, the structure of which was sensitive to NH_3 gas pressure, cavity tuning, and input-signal intensity. Of the various gas cells investigated optimum performance was obtained for lengths of 50–70 cm; optimum gas pressure decreased approximately linearly with increased cell length. The modulation period of the transmitted signal scaled linearly with cavity length, as expected.

Representative examples of the modulated output for PZT tuning are shown in Fig. 9.15b for cavity length 86 cm, cell length 70 cm and pressure 10 Torr with input coupler reflectivity 67 %. Strong Ikeda oscillation (period ≈ 13 ns, fairly close to $2t_R = 11.5$ ns), persistent throughout the pulse, is evident in the neighbourhood of minimum transmission, consistent with the four-wave mixing interpretation of this instability discussed earlier. (Note that in contrast to the ring resonator the Fabry-Perot geometry does not prescribe

$2t_R$ as the basic period for Ikeda oscillation). PZT tuning of the cavity leads progressively to "switching" behaviour with high peak transmission followed by damped oscillations of longer period. At lower pressures (4–8 Torr), where inhomogeneous broadening may be important, we also obtained strong and sustained $4t_R$ oscillation. A typical PZT scan showing these oscillations in the region of low transmission is shown in Fig. 9.16a. At higher pressures (20–30 Torr) much more complex pulse shapes were obtained. These features were enhanced for reduced input coupler reflectivity ($R_0{\simeq}36\%$), since large input coupling is needed to bleach the high absorption ($\alpha L = 6$) to achieve adequate cavity feedback. The PZT sequence shown in Fig. 9.16b for a pressure of 19 Torr shows $2t_R$ oscillation (top trace), developing to $2t_R/3$ oscillation on the higher branch, bifurcating to $4t_R/3$ before again evolving to lower-branch $2t_R$ modulation (bottom trace). Aperiodic pulse shapes, characteristic of chaos, are also evident here and in other data taken at similar pressures. An example for ammonia pressure 15 Torr is shown in Fig. 9.16c where aperiodic pulse shaping is clearly evident in the middle traces. We have undertaken spectral analysis of these pulse shapes, and find strong sharp lines for the oscillatory traces; the aperiodic traces give a broad spectrum with weak structure, as expected for chaos, though pulsed operation perforce reduces the utility of spectral evidence in the identification of chaos.

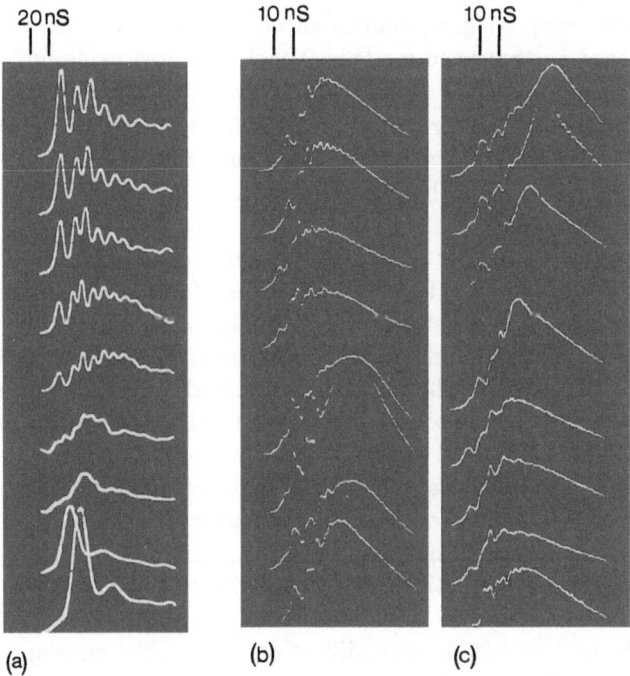

20 nS   10 nS   10 nS

(a)   (b)   (c)

Fig. 9.16. PZT scan showing (a) $4t_R$ modulation at 5.5 Torr; (b) $2t_R$, $2t_R/3$ and $4t_R/3$ modulation at 19 Torr and (c) aperiodicity in the modulation at 15 Torr. Cavity parameters as for Fig. 9.15 but with $R_0 = 36\%$

Our modelling of this system is based on the standing wave analysis of Sect. 9.2 for a fully saturable medium [9.10], but in the limit $T_1 \to 0$ necessary for the validity of the analysis. This is probably quite a good approximation at NH_3 pressures above 10 Torr, and at lower pressures inhomogeneous broadening ought to be considered in any case. The field equations are then similar to (9.23–25) except that (9.23) is dropped because $\tau \to 0$ and the field equations, using (9.6 and 22), become

$$\frac{\partial E_F}{\partial x} + \frac{n_b}{c} \frac{\partial E_F}{\partial t} = \frac{\alpha_0}{2} \left(\frac{1 + i\tilde{\Delta}}{1 + \tilde{\Delta}^2} \right) n_0 \left(1 + \frac{n_1}{n_0} \frac{|E_F|^2}{I_s} \right) E_F$$

$$= f(|E_F|^2 |E_B|^2) E_F - \frac{\partial F_B}{\partial x} + \frac{n_b}{c} \frac{\partial F_B}{\partial t} = f(|E_B|^2, |E_F|^2) E_B \qquad (9.43)$$

where n_0 and n_1 are given by (9.20 and 21). The simplest method of handling these equations is to replace the cell with a set of thin slices containing the same number of atoms [9.10,36].

Using this procedure, and an input pulse fitted to experiment, yields the transmitted pulse shapes in Fig. 9.15c, in pleasing agreement with the observed pulse shapes (Fig. 9.14b), especially since only measured parameters are used: α is $0.025\,\mathrm{cm}^{-1}$ at 10 Torr, and scales as p^2, $I_s \sim 2.3\,\mathrm{MW\,cm}^{-2}$ at 20 Torr ($2.5\,\mathrm{MW\,cm}^{-2}$ at 30 Torr). The value $I/I_s = 7$ is thus in line with the measured input intensities in the range $10–20\,\mathrm{MW\,cm}^{-2}$.

This good, if somewhat surprising, agreement encouraged the development of the numerical code based on that used for (Fig. 9.15c), but now using finite differences rather than slices. Reservoir effects are incorporated using (9.39), on the reasonable assumption that the reservoir grating is completely washed out. We also account crudely for transverse effects by summing the separate contributions from elemental slices of the transverse intensity distribution of the input signal.

The data for a pressure of 10 Torr is not significantly different to that predicted by plane wave analysis (Fig. 9.15c). In contrast for high pressure operation where chaotic emission is observed (Fig. 9.16b and c), predicted behaviour is sensitively dependent on the inclusion of both transverse effects and reservoir features. For comparison we show in Fig. 9.17 simulated PZT scans for the operating conditions applicable to the high pressure data of Fig. 9.16b where set (a) are for the simple two-level scheme under plane wave approximation, set (b) includes transverse effects, and set (c) also accounts for reservoir population transfer (the four-level scheme). The general features of scan (c) are seen to give quite good agreement with our experimental findings. Comparing scans (a) and (b), the significant reductions in the amplitude is attributable to different phasing of the oscillation at different radii. Inclusion of reservoir effects [compare traces (b) with (c)] leads to modification of the pulse envelope; to be expected since this manifests the dynamic population distribution in the radiatively coupled transition which in turn is affected by the finite relaxation time for population transfer from the reservoir levels (the rotational manifolds).

230

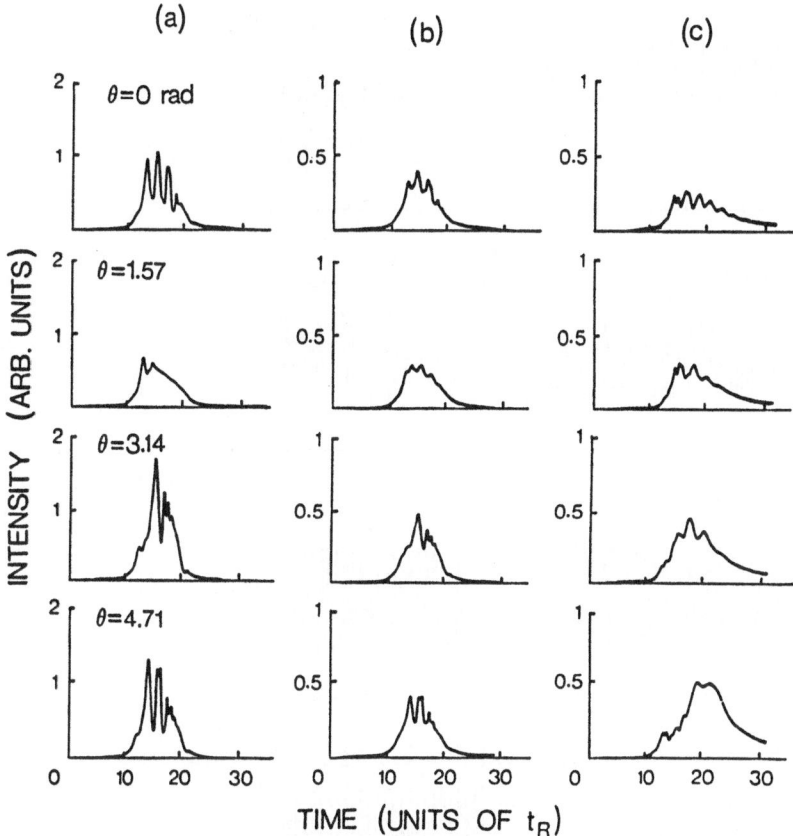

Fig. 9.17a–c. Predicted transmitted pulse shape at NH_3 pressure 20 Torr for **(a)** simple two-level scheme; **(b)** two-level scheme with transverse effects; **(c)** four-level scheme (reservoire population transfer). (Parameters values as for the experimental values of Fig. 9.15b)

As further corroboration of the instability phenomena predicted for our system and also towards eventual generation of these effects under CW conditions we have extended analysis to the CW regime. We consider the simple two-level scheme with parameter values similar to those for pulsed operation (Fig. 9.15b). Figure 9.18a shows a coarse cavity tuning scan for fixed CW input intensity of $7I_s$ as for data of Fig. 9.16b. Oscillation initially at period $2t_R$ bifurcates on cavity tuning to $4t_R$ and subsequently $8t_R$ oscillation at an optimum detuning setting of $\theta = 3.48$ rad. Further cavity tuning over its free spectral range results in reversal of the sequence back to $2t_R$ modulation as expected. Evidently the input intensity here is not quite sufficient to drive the cavity signal to a chaotic state. Selecting the detuning for period eight modulation the input intensity was varied over the small range $6.9I_s$ to $7.2I_s$ (Fig. 9.18b). Oscillation at period $4t_R$ for an input of $6.9I_s$ bifurcates to period $8t_R$ on increasing to $7I_s$; the original value. For further increase in signal strength to $7.2I_s$ the transmitted signal shows evidence of the onset of chaotic behaviour.

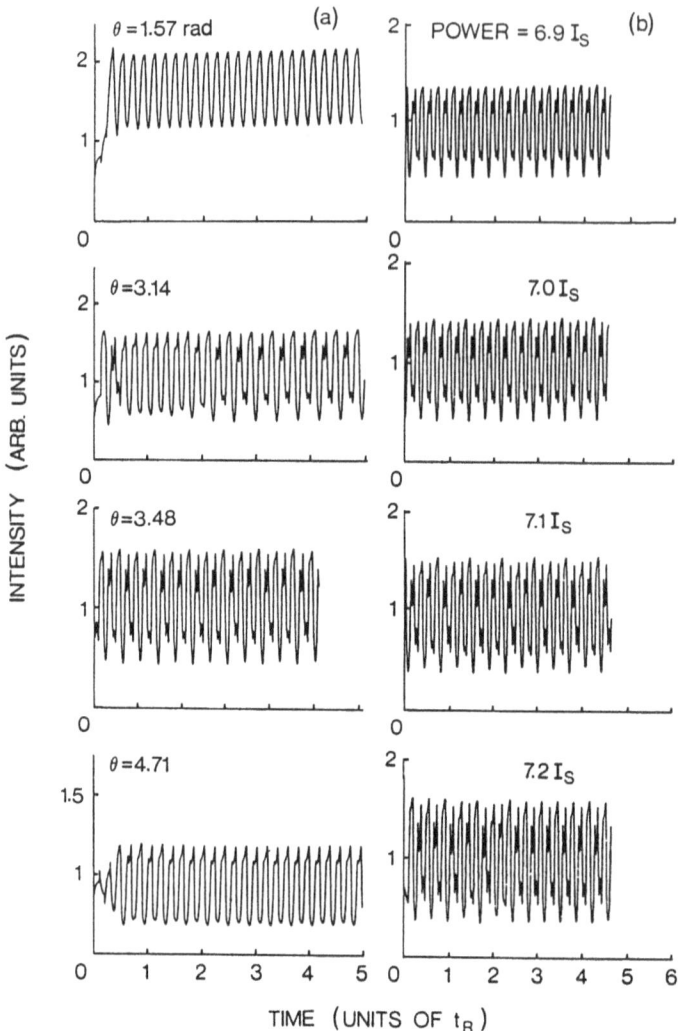

Fig. 9.18a,b. Predicted CW transmitted pulse shape at NH_3 pressure 20 Torr for simple two-level scheme (a) cavity tuning scan for fixed input intensity of $7\,I_s$; (b) input intensity scan for fixed cavity tuning ($\theta = 3.48\,\text{rad}$). (Other parameter values as for Fig. 9.17)

The practical realisation of these effects under CW conditions is, of course, precluded by the relatively high saturation intensity of this transition requiring input signals of $\sim 10\,\text{MW\,cm}^{-2}$. Molecules with transitions in closer coincidence with the pump wavelengths and with larger transition moments such as sulphur hexafluoride (see below) are clearly favoured here.

For ammonia we finally note the generality of our results with the observations of bistability and $2t_R$ oscillation using other TEA laser lines again close to resonance and exhibiting self-focussing in ammonia.

9.4.3 Fabry-Perot Resonator with Sulphur Hexafluoride

Various lines of the 10P band of the CO_2 laser were used to excite the dense and broad spectral featurs of the ν_3 vibrational mode of SF_6. Effects of switching, power limiting and overshoot, with nanosecond response time have been routinely obtained [9.11] although generation of instability was precluded in our system since operation was in the bad cavity limit.

Significantly, for SF_6 which exhibits both low saturation intensity and high absorption cross section, switching has been readily obtained in cavities with gas cells as thin as 1 mm.

Here we concentrate on results obtained using the 10P(16) CO_2 line at 947.7 cm^{-1} coincident with several close lying Q branch transitions near the ν_3 band centre [9.28], as discussed in Sect. 9.1. The Fabry-Perot resonators ranging in length from 2–14 cm, with intracavity gas cells of 1 mm–10 cm were operated at SF_6 gas pressures ranging from ~1 to 200 Torr, dependent on cell length. As for the NH_3 system the resonators comprised single surface Ge output coupler, of 2 m radius of curvature and reflectivity $R_L = 76\%$; the output coupler was equipped with PZT tuning encompassing one free-spectral range of the Fabry-Perot resonator.

Sample traces of the transmitted signal for PZT tuning at equal increments through half a spectral range are shown in Fig. 9.19b; cavity length 14 cm, cell length 10 cm with an SF_6 pressure of ~5 Torr pumped at 947.7 cm^{-1} (10P16 CO_2 line) with an input power ~1.5 MW. Corresponding recordings of instantaneous input (x axis) and output (y axis) intensity (Fig. 9.19c) show pronounced

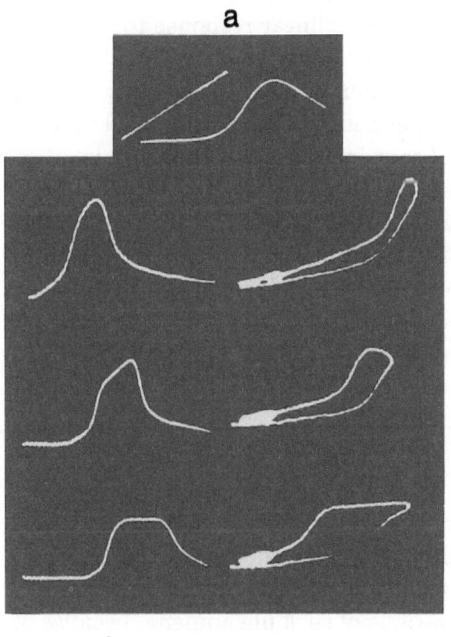

Fig. 9.19. (a) Typical single mode input pulse together with $x - y$ trace of instantaneous input and output intensity for empty Fabry-Perot. (b) PZT scan (over 1/2 free-spectral range) of transmitted signals; SF_6 pressure 5.5 Torr, cavity length 14 cm, cell length 10 cm. (c) Corresponding $x - y$ traces of input and transmitted signal intensity

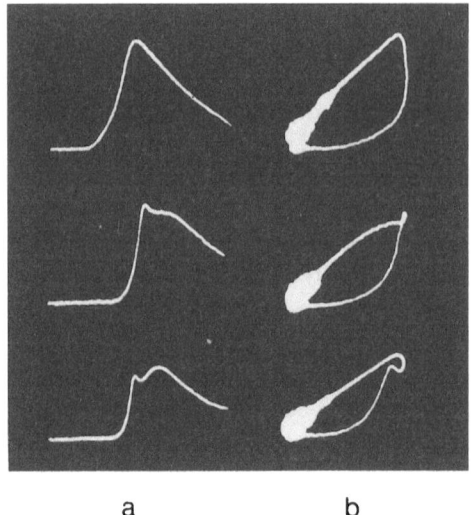

Fig. 9.20. PZT scan (over 1/2 free-spectral range) of transmitted signal (SF_6 pressure 90 Torr, cavity length 2 cm, cell length 1 mm)

a   b

hysteresis effects (to ensure proper synchronisation, the system was adjusted with the cavity empty to obtain the linear trace in Fig. 9.19a). Pronounced and reproducible pulse distortion effects and bistable action are clearly evident. These were not strongly dependent on cavity finesse and were maintained over a pressure range from ~2 to 10 Torr.

Figure 9.20 shows similar recordings taken for a 2 cm cavity containing a 1 mm cell at an increased SF_6 pressure of ~90 Torr. In contrast to the data of Fig. 9.19, the hysteresis curves here, which show the transmitted signal intensity to fall linearly with the input intensity, suggest linear response to the latter part of the input signal. Such behaviour is consistent with the absorption characteristics considered earlier. Saturation behaviour evident in the traces results from eventual total equalisation of population between the ground and first excited ν_3 levels through pumping and subsequent rapid rotational relaxation which is enhanced at these high pressures.

In spite of the masking effects of excited state absorption the low saturation intensities (at low pressures) support future prospects of pulse switching and limiting in optimised systems at considerably reduced input intensities ~50 kW cm^{-2} Torr. Our observations also suggest SF_6 as a promising gas for generation of instabilities at low input intensities.

9.5 Conclusion

In summary, we have observed optical bistability, period doubling, higher-harmonic oscillations and chaos in all-optical, ring and Fabry-Perot resonators containing NH_3 gas as the nonlinear medium. All these effects have been shown to be in qualitative agreement with a simple mathematical model. These results establish the suitability of gases for observations of such phenomena, because of flexibility in the degree of resonant enhancement through choice of transition, of response time through pressure variation and of linear optical properties (finesse, etc.) through simple independent control of cavity length, cell length and, again, pressure. The optical systems discussed in this chapter are of particular interest because of the possibility of using such systems as a bridge between turbulence and quantum systems. The oscillatory instabilities may lead to the development of passive all-optical modulators [9.37], which could find wide application in optoelectronics and laser spectroscopy.

In the experiments reported here we have argued that rotational relaxation is the significant energy-transfer mechanism responsible for the instability phenomena we observe. For our pulsed experiments the considerably slower processes of $V-V$ and $V-T$ relaxation have little effect within the timescale of the input signal. However, operation under truly CW conditions will be necessarily controlled by the $V-T$ relaxation rate since this is the limiting process for recycling of population from the rotational level manifold of the excited vibrational level state back to ground. Relaxation times τ_v for $V-T$ processes are typically microseconds; for NH_3 intensities associated with this process is therefore advantageously very low (for $SF_6 \sim 6\,W\,cm^{-2}$ [9.31]) though at the expense of the long medium-response time which will normally preclude generation of instability effects. However, a significant virtue of the gas phase is in the use of buffer gas to control response time. This versatility should readily enable optimisation of a system for CW conditions.

Acknowledgements. This work was supported by the Science and Engineering Research Council.

References

9.1   C.M. Bowden, H.M. Gibbs, S.L. McCall (eds.): *Optical Bistability* 2 (Plenum, New York 1983);
H.M. Gibbs, P. Mandel, N. Peyghambarian, S.D. Smith (eds.): *Optical Bistability III*, Springer Proc. Phys., Vol. 8 (Springer, Berlin, Heidelberg 1986)
9.2   L.P. Kadanoff: Phys. Today, **36**, 46 (December 1983);
H. Haken (ed.): *Chaos and Order in Nature*, Springer Ser. Syn., Vol. 11 (Springer, Berlin, Heidelberg 1981)
9.3   K. Ikeda, H. Daido, O. Akimoto: Phys. Rev. Lett. **45**, 709 (1980)
9.4   K. Ikeda: Opt. Commun. **30**, 257 (1979)
9.5   H.M. Gibbs, F.A. Hopf, D.L. Kaplan, R.L. Shoemaker: Phys. Rev. Lett. **46**, 474 (1981)
9.6   H. Nakatsuka, S. Asaka, H. Itoh, K. Ikeda, M. Matsuoka: Phys. Rev. Lett. **50**, 109 (1983)
9.7   A.C. Walker, F.A. P. Tooley, M.E. Prise, J.G.H. Mathew, A.K.Kar, M.R. Taghizadeh, S.D. Smith: Phil.Trans. R. Soc. Lond. A**313**, 249 (1984)
9.8   F. Shimizu: J. Chem. Phys. **52**, 3572 (1970)
9.9   R.G. Harrison, W.J. Firth, C.A. Emshary, I.A. Al-Saidi: Phys. Rev. Lett. **51**, 562 (1983)
9.10   R.G. Harrison, W.J. Firth, I.A. Al-Saidi: Phys. Rev. Lett. **53**, 258 (1984)
9.11   R.G. Harrison, I.A. Al-Saidi, E.D. Cummins, W.J. Firth: Appl. Phys. Lett. **46**, 532 (1985)

9.12   W.J. Firth: Opt. Commun. **39**, 343 81983);
W.J. Firth, E. Abraham, E.M. Wright: Appl. Phys. B**28**, 170 (1982)

9.13   M.J. Feigenbaum: J. Stat. Phys. **19**, 25 (1978)

9.14   E. Abraham, W.J. Firth: Opt. Acta **30**, 1541 (1983)

9.15   L.A. Lugiato, M. Gronchi, R. Bonifacio: Opt. Commun. **30**, 129 (1979)

9.16   W.J. Firth, E.M. Wright, E.H.J. Cummins: In *Optical Bistability* 2, ed. by C.M. Bowden, H.M. Gibbs and S.L. McCall (Plenum, New York 1983) p. 111

9.17   F.A. Hopf, M.W.Derstine, H.M. Gibbs, M.C. Rushford: In *Optical Bistability* 2, ed. by C.M. Bowden, H.M. Gibbs and S.L. McCall (Plenum, New York 1983) p. 67

9.18   W.J. Firth, E.M. Wright: Opt. Commun. **40**, 233 (1982)

9.19   E. Abraham: Ph.D. Thesis, University of Manchester (1979)

9.20   W. Fuss, K.L. Kompa: Prog. Quantum Electron. **7**, 117 (1981)

9.21   C.H. Townes, A.L. Schawlow: *Microwave Spetroscopy* (McGraw-Hill, New York 1955)

9.22   J.L. Lachambre, P. Lavigne, G. Otis, M. Noel: IEEE J. QE-**12**, 756 (1976)

9.23   A. Gondhalekar, N.R. Heckenberg, E. Holzhauer: Phys. Lett. A**46**, 229 (1973)

9.24   J.S. Garing, H.H. Nielsen, K. Narahari Rao: J. Mol. Spect., **3**, 496 (1959);
T.Y. Chang: Opty. Eng. **20**, 220 (1981)

9.25   I. Burak, P.L. Houston, D.G. Soutton, J.I. Steinfeld: IEEE J. QE-**7**, 73 (1971)

9.26   P.K. Gupta, R.G. Harrison: IEEE J. QE-**17**, 2238 (1981)

9.27   R.V. Ambartzumian: In *Tunable Lasers in Applications*, ed. by A. Mooradian, T. Jaeger, and P. Stoksett Springer Ser. Opt. Sci., Vol. 3 (Springer, Berlin, Heidelberg 1976) p. 150

9.28   P. Aubourg, J.P. Beltini, G.P. Agrawal, P. Coltin, D. Guerin, O. Meunier, J.OL. Boulnois: Opt. Lett. **6**, 383 (1981)

9.29   J.I. Steinfeld, I. Burak, D.G. Sutton, A.V. Nowak: J. Chem. Phys. **52**, 5421 (1970)

9.30   P.F. Moulton, D.M. Larsen, J.N. Walpole, A. Mooradian: Opt. Lett. **1**, 51 (1977)

9.31   H. Burnet: IEEE J. QE-**6**, 678 (1970)

9.32   J.L. Lyman, G.P. Quigley, O.P. Judd: In *Multiple Photon Excitation and Dissociation of Polyatomic Molecules*, ed. by C. Cantrell, Topics Current Phys., Vol. 35 (Springer, Berlin, Heidelberg 1986), p. 25

9.33   J.J. Armstrong, O.L. Gaddy: IEEE J. QE-**8**, 797 (1972)

9.34   I.A. Al-Saidi, D. Biswas, C.A. Emshary, R.G. Harrison: Opt. Commun. **52**, 336 (1985)

9.35   R.G. Harrison, W.J. Firth, C.A. Emshary, I.A. Al-Saidi: Appl. Phys. Lett. **44**, 716 (1984)

9.36   H.J. Carmichael, J.A. Hermann: Z. Phys. B**38**, 365 (1980)

9.37   R.G. Harrison, W.J. Firth, I.A. Al-Saidi, E. Cummins: In Proc. Topical Meeting on Digital Optical Circuit Technology, Schiersee, Germany (1984) Advisory Group for Aerospace Research and Development (AGARD) Conf. Proc. No. 362, ed. by B.L. Dove (AGARD, Neuilly Sur Seine, France 1985) p. 4-1

10. Transmission Properties of a Sodium-Filled Fabry-Perot Resonator

G. Giusfredi, S. Cecchi, E. Petriella, P. Salieri, and F.T. Arecchi

With 8 Figures

We report on systematic studies of the static and dynamic behavior of intensity and polarization of the light transmitted by a sodium-filled Fabry-Perot (FP) cavity. The device displays a rich variety of phenomena, including optical bistability, tristablity, pitchfork bifurcation, as well as intrinsic oscillations. The behavior can be traced back to the simultaneous presence of Zeeman pumping and hyperfine pumping in the sodium ground state, combined with velocity selective hole-burning effects.

10.1 Background

After the first observations of optical bistability (OB) and a number of theoretical predictions, optical resonators filled with nonlinear passive media began to be considered important instruments in nonlinear quantum optics as well as useful all optical nonlinear devices [10.1]. Stimulated by all these possiblities, we performed several experiments, briefly discussed here, by using a Fabry-Perot cavity filled with sodium (Na) vapor and by tuning the laser frequency across the $D1$ line. Our idea was to use the mirror substrates as windows for the sodium cell to reduce the non-saturable losses of the cavity to a minor value. More important, by avoiding Brewster windows we could achieve complete cylindrical symmetry around the laser beam axis. With such an arrangement any light polarization was allowed in the cavity. This disclosed a new class of phenomena induced by the Zeeman pumping over the Na ground state sublevels. First we observed optical bistability [10.2] and tristability (OT) [10.3], with circularly and linearly polarized input light respectively. In these experiments a buffer gas was added to allow for Zeeman pumping over the whole line. The behavior becomes increasingly complex when the buffer gas pressure is reduced. More recently we explored the completely inhomogeneously broadened case (no buffer gas), where single homogeneous velocity groups interact with the optical field [10.4,5]. This gave us the opportunity to study systems where Zeeman and hyperfine pumping are both effective. Even though the theoretical analysis tends to be very complicated, this system offers an interesting variety of phenomena, such as multistability in the form of superposition of OB by hyperfine pumping and OT. It also shows oscillations in the polarization of the output light even in the absence of stationary magnetic fields, while period doubling and intermittency behaviors appear in the presence of a magnetic field.

10.2 Experimental Set-Up

The device (Fig. 10.1) used in these experiments was a vacuum-sealed stainless-steel cell, with internal mirrors that formed the FP interferometer. Two AR-coated windows were at the ends. The mirrors themselves, ~8 cm apart, confined the sodium vapor in the FP cavity and were mounted with the reflecting layers outward to prevent any damage. The Na reservoir was contained in a glass tube at the bottom of the FP chamber. Stable alignment and finetuning of the cavity were provided by deformable bellows, three quartz spacers, and three piezoelectric transducers. The input power was locked to the signal of a waveform generator by means of an electro-optic modulator and a photodiode both enclosed in "noise eater" loop. The input polarization was controlled by a retardation plate or by a Pockels cell. The power and polarization of the transmitted light were measured depending on an experimental parameter, typically the input power (P_{in}) or the cavity mistuning with respect to the laser frequency (related to the cavity length). A triangular waveform at low frequency (10 to 20 Hz) was generally used to scan P_{in}. The operating density (N) was $\sim 9 \times 10^{11}$ cm^{-3} in the OB and OT experiments and ranged between 0.5×10^{11} and 9×10 cm^{-3} in the others. The laser frequency was tuned across the $D1$ sodium line.

→ TO VACUUM LINE

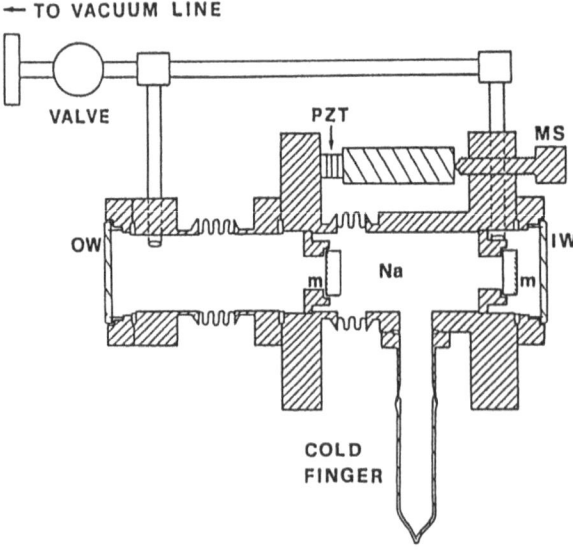

Fig. 10.1. Schematic of the sodium-filled FP interferometer (m: mirror, IW and OW: input and output window, PZT: piezoelectric, MS: micrometer screw)

10.3 Discussion

10.3.1 Optical Bistability by Zeeman Pumping

It is well known that in a λ-shaped 3-level system in which a ground-state sublevel is coupled to the excited level by a resonant field, the atoms are pumped to the other ground-state sublevel. If the pumping rate P is weak enough with respect to the spontaneous decay rate of the excited state, the dynamics of the system can be described by the equation

$$\dot{n} = PN - \left(P + \frac{1}{\tau}\right)n \tag{10.1}$$

where n is the population difference between the ground state sublevels and N is the total population. τ is the ground state relaxation time, generally attributable to some collision mechanism, while the total relaxation term is the sum $P + 1$.

In the OB experiment [10.2] circularly polarized input light was used as the pumping field, while the argon pressure was in the range of 20 Torr. In these conditions the atoms are pumped in the $F = 2$, $m_F = 2$ (or -2) ground state, i.e., the atomic spins are oriented along the beam propagation axis. The homogeneous broadening of the line and the velocity changes (due to collisions with the buffer gas) make the pumping process effective over the whole line, even with argon pressure of few torr only. In our device the atoms were pumped in a small fration of the cell, and were diffused all around (cell and beam diameter were 40 mm and ~2 to 3 mm, respectively). The spin orientation is only destroyed by collision with the cell walls, so that in equilibrium, in our almost cylindrical geometry, the beam is surrounded by clouds of oriented atoms of ~10 mm in FWHM diameter. Therefore the relaxation time in near-equilibrium conditions is close to the diffusion time spent by the pumped atoms to leave this cloud diameter. The atomic polarization, calculated by an approximate 3-level model (Fig. 10.2), is [10.2]

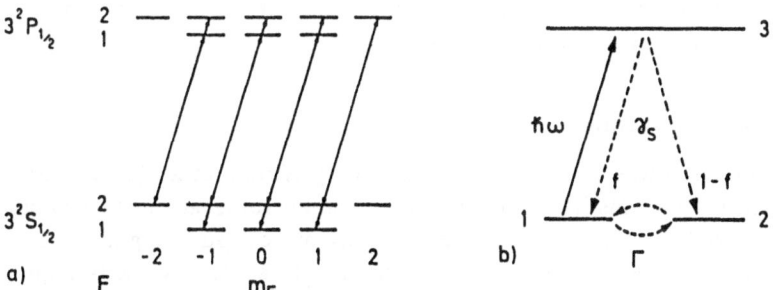

Fig. 10.2. (a) Zeeman structure and allowed transitions of Na under illumination with circularly polarized light tuned at the $D1$ line. (b) Three-level simplified model, $f = 7/8$

$$S = \frac{7\alpha N}{8\gamma_+} \frac{1 - i\Delta}{1 + \Delta^2 + 4|\alpha|^2/\Gamma\gamma_\perp} \tag{10.2}$$

where α is the Rabi frequency, $\gamma_\perp$ the decay time of the optical coherence, Δ the detuning normalized to $\gamma_\perp$, and $1/\Gamma = 7\tau/128$. With respect to the laser power S has the same trend as the polarization of 2-level atoms, but the required saturation power is reduced by a factor $\Gamma/\gamma_{||}$, where $\gamma_{||} = 1/(16\,\text{ns})$ for the sodium $D1$ line.

Bistability was observed at input powers of few milliwatt with frequency detuning ranging from -5 to $+8\,\text{GHz}$ with respect to the $F = 2$ resonance peak (Fig. 10.3). It was dispersive in character and absorptive OB was not seen at zero laser detuning. This means that the inhomogeneous line plays an important role in the obseved phenomenology. Indeed, with the used buffer gas pressure, the homogeneous width was about $1/8$ of the Doppler linewidth $\gamma_\perp = \gamma_{||}/2 + 2\pi p \times 10^7\,\text{s}^{-1}$, with p the argon pressure measured in Torr [10.6]. Moreover, the cell was not optically thin. Since the mean-field analysis cannot be applied in these conditions, we followed that of *Gronchi* and *Lugiato* [10.7], accounting for the field propagation and Doppler broadening. By incorporating the above expression for S into their equations, we finally reached a good agreement with the experimental results. In our computation, τ was used as a free parameter, ranging around some millisecond, to fit the experimental data. However, for the analysis of the dynamical behavior, in particular the jump between the two transmission states, to assign a single time constant to the medium is an oversimplification, and more careful treatment is necessary, as discussed in [10.8]. The experiments reported there were performed in a similar device (a confocal FP with beam waist of $120\,\mu\text{m}$ and $100\,\text{Torr}$ of argon) by using transverse optical pumping; critical slowing down and transient bimodality [10.9] were observed.

As can be seen in Fig. 10.3, a considerable degree of asymmetry is present between the hysteresis loops when scanning in frequency above and below resonance. This behavior can be attributed to a gradient in the atomic susceptibility due to the beam intensity profile and to the diffusive motion of the atoms. This gradient forms a lens yielding self-focusing in the high frequency side and self-defocussing in the low frequency side of the resonance [10.10]. This effect, even though not being the main source of OB, can enhance the hystereses at higher frequencies or depress them at lower ones.

10.3.2 Optical Tristability

The OT phenomenon was theoretically predicted by *Kitano* et al. [10.11] for a FP cavity filled with atoms having degenerate Zeeman sublevels in the ground state, and is caused by spin orientation induced by optical pumping in the frequency region of anomalous dispersion. The used atomic scheme was a λ-shaped three-level model where the two circular components of the laser field separately couple the two ground-state sublevels to the excited state, while ground state

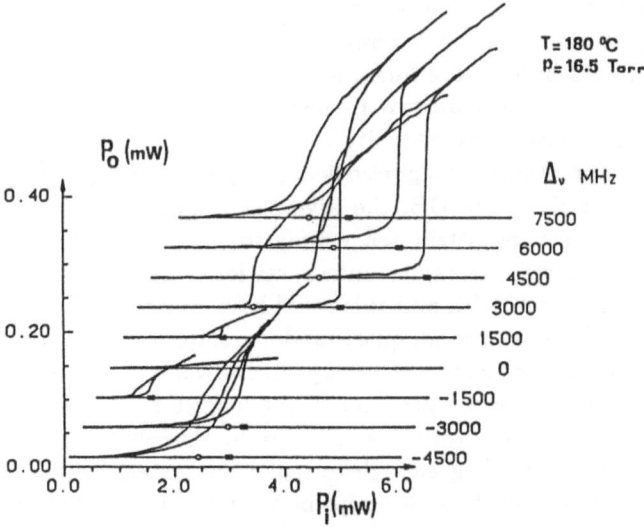

Fig. 10.3. Power out (P_o) vs. power in (P_{in}) observed at different detuning measured from the peak of the $F = 2$ hyperfine transition argon pressure 16.5 Torr

coherence effects are not present. The basic equation for the atomic system becomes [10.11,12]

$$\dot{n} = (P_+ - P_-)N - \left(P_+ + P_- + \frac{1}{\tau}\right)n \qquad (10.3)$$

where P_+ and P_- are the pumping rates for the left and right light polarization components respectively. This result can be well applied (far from the line saturation) to a $J = 1/2\langle - - - \rangle J' = 1/2$ atomic transition, which is a good approximation for the Na $D1$ line at argon pressure around 100 Torr or more and/or large detuning ($\sim 10\,\text{GHz}$) [10.12,13].

Let us also recall some polarization features of OT. For linearly polarized incident light, the transmitted light can be in one of three stable states of polarization: namely linearly polarized; left (L) dominant (almost left-circularly polarized), and right (R) dominant (almost right-circularly polarized). In the linearly polarized state, low power enters the cavity and the atomic spins are almost randomly oriented, while the field is out of resonance with the cavity. When the input power is increased above a threshold level, this linear state becomes unstable and a symmetry breaking transitions to the L or R state occurs. The nonlinear coupling of the two circularly polarized light components, due to their competitive pumping and the feedback provided by the cavity, regeneratively amplifies fluctuations of their competitive pumping and feedback provided by the cavity, regeneratively amplifies fluctuation of their power ratio. Therefore the atomic spins are forced to orient themselves parallel (or antiparallel) to the propagation axis, while the left (right) component is pushed toward resonance within the FP cavity and the other one is pushed away. On decreasing the input power, it is only at a lower power level that spin orientation is lost and the output switches back to the linear state.

We obtained OT in a relatively narrow frequency range around 1.5 GHz above the $D1$ line center at low input intensity ($\sim 3\,\mathrm{mW/mm^2}$) over a 3-mm-wide laser beam [10.3]. A longitudinal magnetic field (~ 3 Gauss) was generated by a couple of Helmholtz coils. As in OB, the asymmetric behavior with respect to the laser detuning can be attributed mainly to transverse effects in the beam propagation. Moreover, the cavity finesse was lower than in the OB experiment, namely >6 instead of ~ 30, and this could have contributed to kill OT in the low-frequency side. As expected, quasi-random jumps from the linear state to the L or R states were observed at a very precise value of the input polarization (not exactly linear for the presence of the longitudinal magnetic field). A variation of 1 V or less in the voltage applied to the Pockels cell, whose quarter-wave voltage is 600 V, was big enough to make the system switch in one direction only. An hysteresis cycle with jumps between the L and R states was observed by scanning the input polarization at given P_{in} value. This gave further evidence of OT, showing that $R(L)$ state was possible even with left (right) dominant input polarization. Indeed, in the above unbalanced scanning of P_{in}, the jump toward the other state is prevented by the crossing of an unstable (and therefore repulsive) state. Figure 10.4 sketches the steady state of the system when P_{in} is scanned for different values of input polarization around the symmetric condition, and also when input polarization is scanned with P_{in} being constant. OT has been successively observed by several authors. *Giacobino* [10.12] has introduced a useful graphical way to compute the OT steady states, while the effect of the line saturation has been reported in [10.13].

10.3.3 Multistability

In the experiment without buffer gas we observed a variety of phenomena of multistability and self-pulsing [10.4,5], whose complexity arises from the $D1$ line structure and the presence of two counter-propagating waves in the FP cavity. However some simplifications arise from the absence of buffer gas, since,

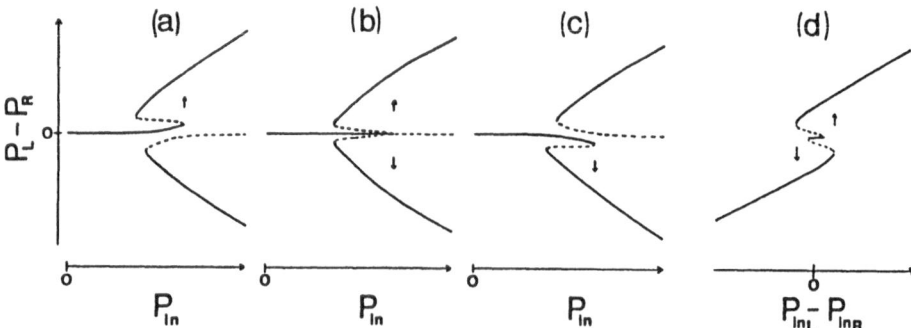

Fig. 10.4. Difference in transmission between the two circular components vs. P_{in} (a,b,c) or vs. input polarization (d). *Solid line, stable branch; dashed line, unstable branch*

in that condition, the atomic motion is free inside the cavity. Indeed, at our vapor density the mean-free path of an atom, calculated from elastic and spin exchange cross sections [10.14] is much longer than the cavity length. Thus the ground state relaxation time is close to the mean transit time across the beam ($\sim 9\,\mu s$). It is much slower than $1/\gamma_\parallel$ or the field decay time ($\sim 10\,ns$) of our cavity and the field adiabatically follows the evolution of the ground state population. Therefore the observation of the output polarization and intensity allows for the determination of the state of the vapor. The Doppler broadening is much wider than the natural linewidth (~ 1600 vs. $10\,MHz$) and is comparable to the hyperfine separation ($1.77\,GHz$). Hyperfine pumping leads to peaks and dips in the velocity distributions of both $F = 1,2$ atomic populations. The width of these peaks and dips grows with intensity and, already at moderate values ($\sim 10\,mW/cm^2$) [10.15], is of the order of the excited state hyperfine separation. By considering this broadening for the excited state, there are four atomic groups in the velocity distribution which are resonant with the forward (F) or backward (B) wave in the FP cavity. These waves act as pump and probe, one with respect to the other: each wave is essentially absorbed by its resonant groups and dispersively affected by the pumping of the other in its own groups.

In our set-up three mutually orthogonal Helmholtz coils were added to compensate the background magnetic field within $10\,mGauss$, while a Faraday isolator was used to decouple the laser from the FP cavity. The cavity finesse was ~ 18, while at the cavity input the available power was $\sim 50\,mW$ over a 3-mm beam diameter.

We systematically examined the system by varying the laser frequency (ν_L) and the atomic density. Together with OB and OT, pitchfork bifurcations (PB) and other peculiar statical phenomena, designated for brevity with OT+OB, PB+OB, and OB+PB, were observed in various zones (Fig. 10.5). The OB cycles, as expected to arise from hyperfine pumping [10.16], are characterized by low and high symmetrical transmission states (LS and HS), in which the two circular polarized components of the laser beam are equally transmitted. Instead, left and right asymmetrical transmission states (LA and RA), with elliptically polarized output light, coexist in OT. The PB phenomenon does not appear as a hysteresis cycle upon P_{in} scanning, but it appears just as a bifurcation of the LS branch into the LA and RA branches. It can be considered an "incomplete OT" in which the LS state no longer coexists with the asymmetrical states. Furthermore, it is possible to transform OT in PB in a continuous way by changing the cavity mistuning. The mixed hystereses OT+OB and PB+OB, observed when ν_L is below ν_2 ($F = 2$ transition), can be seen as an OT or a PB taking place in the low transmission branch of an OB cycle. The opposite happens between ν_0 (center of $D1$ line) and ν_1 ($F = 1$ transition) at the higher sodium densities (OB+PB).

Below ν_2, the resonant population is mainly provided, for both waves, by the $F = 2$ level, while the $F = 1$ resonant groups are wide apart. As ν_L gets closer to ν_2, strong reciprocal dispersive effects are produced by the f and B

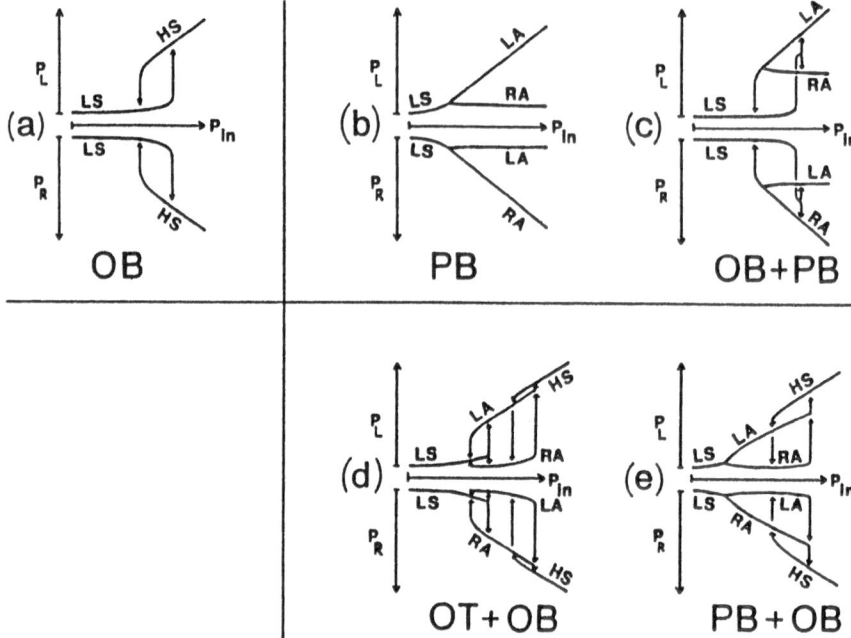

Fig. 10.5a–e. Static transmitted signals vs. input power (P_{in}) at constant cavity length (L) as they appear in the various frequency regions. **(a)** regions between ν_2 and ν_0 and above γ_1; **(b)** region between ν_0 and ν_1 at low Na densities (---) PB is not a hysteresis cycle and appears with precise linear polarization, however, a hysteresis cycle appears when scanning the input polarization instead of P_{in}; **(c)** OB+PB, as in (b) but at high densities; **(d)** PB+OB, region below ν_2 (---) the LS branch still bifurcates into the LA, RA branches, but at higher P_{in} an inverse OT cycle brings the system to an HS state; **(e)** OT+PB, as in (d) but for a different L (---) an OT cycle replace the PB.

waves that cooperatively pump the groups symmetrically displaced around the zero velocity. Thus, as in OT, symmetry breaking in the polarization is generated. However, at higher intensity, the cavity rejection is no longer able to prevent the lower transmission component from depleting this resonant population toward level $F = 1$, leading to high transmission for both components. Similar switching toward a symmetrical state, induced by saturation of the optical transition, has been reported in [10.13]. With ν_L above ν_1, the resonant population is mainly provided by the $F = 1$ level. In this case, since an effective orientation is not feasible, the observed symmetrical OB can be attributed to hyperfine pumping.

Between ν_0 and ν_1, in particular at frequencies close to ν_0, the pumping over the level $F = 1(2)$ by the F wave is not well separated, in the velocity distributions of atoms, from that over the level $F = 2(1)$ by the B wave. The atoms for which the interaction with the laser field is stronger are pumped from one hyperfine level into the other, so that they contribute to the symmetric nonlinearity. However, for those atomic groups whose velocity lies in between those of the resonant atoms, the interaction competitively acts with the $F = 1$ and $F = 2$ levels. So these atoms can be oriented and can unbalance the light polarization. The dynamics of Fig. 10.4c can be understood accordingly: first hyperfine pumping on resonant atoms partially bleaches the cavity, then orientation of the vapor takes place. In the presence of a longitudinal magnetic field the symmetry between the two circular polarizations is broken and, under proper conditions, polarization switchings between LA and RA branches appear in the P_{in} scanning [10.4]. A similar phenomenon has been reported for samarium [10.7].

In Fig. 10.6 we report the recording of dispersion and absorption profiles. In Fig. 10.7, a phase diagram summarizes the relevant phenomenology; the most interesting features are found in a region that closely follows the transmission curve, somehow indicating that a proper number of atoms effectively interacting with the field is required.

10.3.4 Self-Pulsing

In addition to the static phenomena reported above, instabilities can be observed when the cavity mistuning is changed [10.5]. The instabilities appear mainly is polarization oscillations, suggesting that they are provided by the same atomic groups giving the static polarization asymmetries. Indeed, they

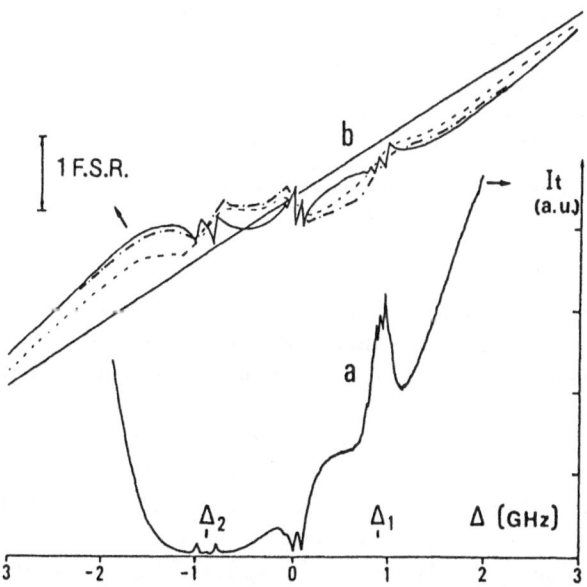

Fig. 10.6. (a) Peak intensity in transmission as the laser frequency ν_L is varied ($\Delta = \nu_L$, $\Delta_i = \nu_i - \nu_c$, $i = 1, 2$). Input intensity $I_{in} = 12$ mW, zero magnetic field ($B = 0$), Na density $n = (3\pm0.6) \times 10^{11}$ cm^{-3}. **(b)** PZT voltage required to keep the laser frequency tuned to the same cavity axial mode as Δ is varied. Dispersion is proportional to the offset from the central straight line (empty cavity signal): (——) linearly polarized input light (low power $\leq$12 mW), (-·-) 50 mW linear polarization, (- - -) 50 mW circular polarization ($B = 0$). Off-resonance, Zeeman pumping is intrinsically more effective than hyperfine pumping, hence saturation for (- - -) is higher than (-·-), especially for high Δ values

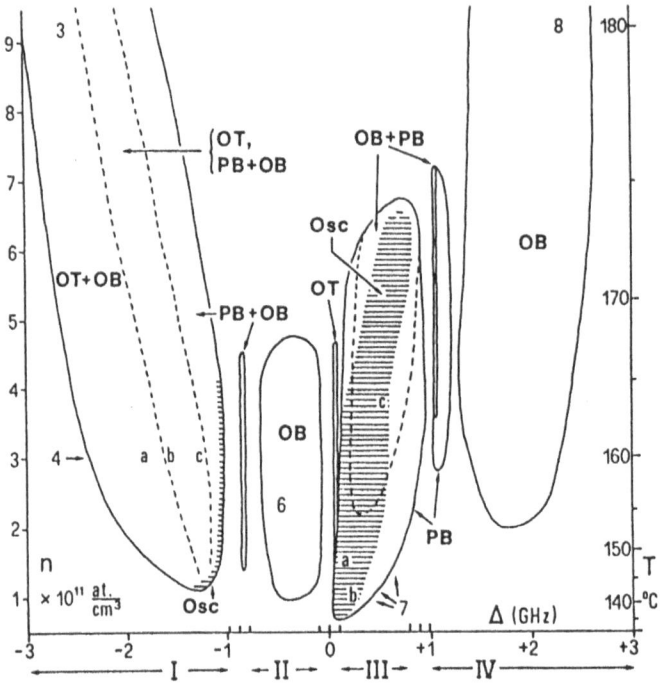

Fig. 10.7. Phase diagram summarizing the observed behavior with $B = 0$ in linearly polarized input light (power up to $\sim 6\,\mathrm{mW}$). The width of the various zones depends on power and cavity tuning (always optimized). The Na density is known within $\pm 20\,\%$ of its value. The OT, OB, PB terms are described in the text, "+" is used for brevity and without commutative meaning (Figs. 10.4,7). Narrow featurs near $\nu_L = \nu_1, \nu_2$ are related to strong nonlinearities in two transmission peaks and are localized to the right of these peaks. The strong induced variations in dispersion enhance all these features (also near ν_c). In the hatched regions we observe (at $B = 0$) wide oscillations in polarization between the two asymmetrical branches. Numbers and small letters refer to the following figures

were found in a narrow range ($\sim 50\,\mathrm{MHz}$ wide) just below ν_2, and in a broader band ($\sim 400\,\mathrm{MHz}$ wide) between ν_0 and ν_1, where PB also appears for proper cavity mistuning. When changing either input power or cavity length, they arise through a Hopf bifurcation and then often die as the system falls on one of the two stationary asymmetrical states. The opposite occurs by reversing the scanning: the oscillations, after starting with an abrupt destabilization of the LA and RA states, smoothly end on the symmetrical state. Usually amplitude and period of the oscillations progressively are increased by proceeding away from the HB point, and frequency tends to zero by approaching the falling point. It is possible to change widely the oscillation frequency by acting on the mistuning and we obtained values up to $\sim 500\,\mathrm{Hz}$ even though the most common values are between 50 and 200 Hz.

A main difference exists between the oscillations in the two frequency ranges. Below ν_2 the oscillations start smoothly from the HS state and their frequencies decrease upon decreasing P_{in}. Instead the opposite happens above

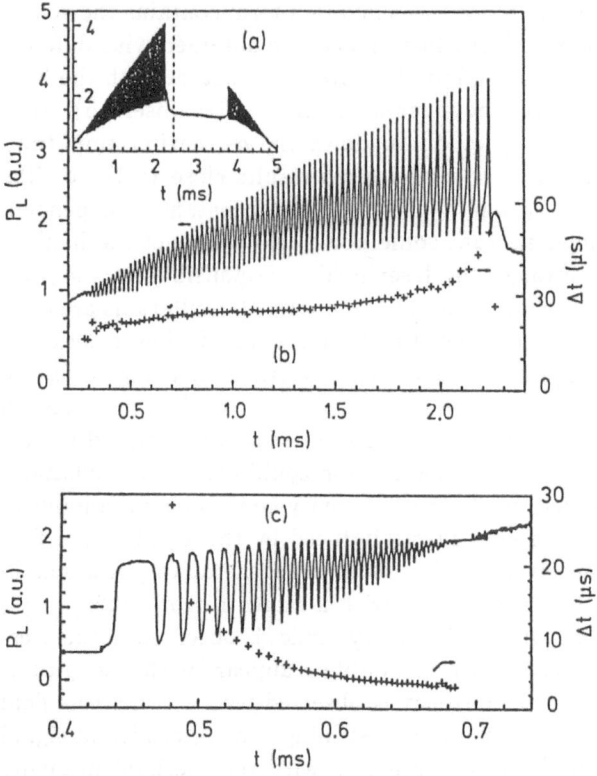

Fig. 10.8. P_{in} with triangular waveform at 200 Hz. *(a)* region above ν_0, $\nu_L - \nu_0 = 0.08\,\text{GHz}$. $P_{in} = 0$ to $\sim 50\,\text{mW}$, $N = 1.2 \times 10^{11}\,\text{cm}^{-3}$. *(b)* magnification of *(a)*. *(c)* region below ν_2, $\nu_L - \nu_0 = 1.08\,\text{GHz}$. $P_{in} \sim 15$ to $\sim 25\,\text{mW}$, $N = 2 \times 10^{11}\,\text{cm}^{-3}$. Each cross gives the Δt interval between the corresponding peak and the previous one

ν_0 (Fig. 10.8), where the oscillations start smoothly from the LS state and their frequency decreases upon increasing P_{in}.

In principle, a reliable model explaining the above experimental results requires a large number of degrees of freedom to account for the atomic structure and the Doppler broadening. However, the regularity of the oscillations suggests that a sensible description can be made in terms of small number of independent variables. Indeed, the birth of a limit cycle via a Hopf bifurcation suggests behavior as a Van der Pol oscillation, that requires two dynamical variables. As proposed by *McCall* [10.17], in bistable system instabilities can be provided by two contribution of opposite signs and different time constants. This has also been observed in FP etalons filled with various materials, such as GaAs [10.18] or liquid crystals [10.19]. Two contributions of opposite signs in the relative dephasing of the two circular components can be provided by two atomic groups in the velocity distribution, and this would explain the oscillations observed below ν_2. One of us [10.21] obtained results close to the experimental behavior with a simple model of optical pumping (as for 10.3) including the hyperfine saturation. Since the relaxation term contains the sum of the pumping rates, two different time constants can be obtained with different detunings for the two atomic groups, but they are not constant with the laser power. As a consequence the frequency increases with P_{in} (as observed), while the effect of hyperfine pumping is to dump down the oscillation amplitude. Surprisingly enough an even simpler model gives results close to the oscillation behavior observed above ν_0 but it uses two time constants independent from the laser field. This suggests that some non-field-dependent mechanism, such as the atomic motion through the laser field is responsible for those oscillations. Moreover, the frequency values of the observed oscillations are close to the inverse of the transit time across the beam. Thus, during a period of oscillation the atomic motion can effectively change (without being so fast to destroy it) a gradient of spin orientation generated by the optical pumping. In particular, the atoms pumped in the center of the beam are displaced toward the periphery. As a possibility, the atomic motion could drive the oscillations with sort of a delayed feedback, while the FP cavity could give an amplification of the focussing and defocussing effects induced by the gradient itself. A similar explanation has been given for the intrinsic regenerative pulsations in self-trapping optical bistability in Na vapor with a single mirror [10.22].

As for the static behavior, more complex dynamics is observed in the presence of a magnetic field. Above ν_0, period doubling appears with a longitudinal field ($\sim$40 mGauss), while intermittency is observed with a transverse field ($\sim$0.7 Gauss) at very precise value of cavity mistuning. The intermittent signal could arise from coupling of the spin precession with the mechanism giving the oscillations at zero magnetic field. Oscillation driven by the spin precession induced by transverse magnetic field was reported for FP cavity filled with Na vapor and buffer gas [10.23].

In summary, we have reported on phenomena as OB, OT, PB, multistability and self-pulsing, and discussed several underlying mechanisms. The fact that the period and the amplitude of the Hopf bifurcations vary as we scan the input intensity (Fig. 10.8) suggests that we are in the presence of a line of Hopf bifurcations, eventually ending into a codimension-two bifurcation point [10.24]. Nevertheless, further investigations are necessary to provide a more complete picture.

Acknowledgements. This work is partially supported by EJOB (European Joint Project on Optical Bistability). We thank the INO technical staff, namely S. Acciai, L. Albavetti, S. Euzzor, S. Mascalchi, P. Poggi, and A. Tenani for their assistence. We also thank G. Assanto for his assistence in writing the manuscript and W. Lange for useful discussions and helpful comments.

References

10.1   A recent review of all the work on OB is H.M. Gibbs: *Optical Bistability: Controlling Light with Light* (Academic, New York 1985)
10.2   F.T. Arecchi, G. Giusfredi, E. Petriella, P. Salieri: Appl. Phys. B29, 70 (1982)
10.3   S. Cecchi, G. Giusfredi, E. Petriella, P. Salieri: Phys. Rev. Lett. 49, 1928 (1982)
10.4   G. Giusfredi, P. Salieri, S. Cecchi, F.T. Arecchi: Opt. Commun. 54, 39 (1985)
10.5   P. Salieri, G. Giusfredi, S. Cecchi, F.T. Arecchi: Phys. Rev. A
10.6   D.G. McCartan, J.M. Farr: J. Phys. B9, 985 (1976);
       R.H. Chartham, A. Gallangher, E.L. Lewis: J. Phys. B13, L7 (1980)
10.7   M. Gronchi, L.A. Lugiato: Opt. Lett. 5, 108 (1980)
10.8   F. Mitschke, R. Deserno, J. Mlinek, W. Lange: Opt. Commun. 46, 135 (1983);
       F. Mitschke, R. Deserno, J. Mlinek, W. Lange: IEEE J. QE-21, 1435 (1985);
       W. Lange, F. Mitschke, R. Deserno, J. Mlinek: Phys. Rev. A32, 1271 (1985)
10.9   G. Broggi, L.A. Lugiato: Phys. Rev. A29, 2949 (1984)
10.10 A. Javan, P.L. Kelley: IEEE J. QE-2, 470 (1966)
10.11 M. Kitano, T. Ybuzaki, T. Ogawa: Phys. Rev. Lett. 46, 926 (1981)
10.12 E. Giacobino: Opt. Commun. 56, 249 (1985)
10.13 M.W. Hamilton, W.J. Sandle, J.T. Chilwell, J.S. Satchell, D.M. Warrington: Opt. Commun. 48, 190 (1983)
10.14 A. Moretti, F. Strumia: Phys. Rev. A3, 349 (1971)
10.15 P.G. Pappas, R.A. Forber, W.W. Quivers, Jr., R.R. Dasari, M.S. Feld, D.E. Murnick: Phys. Rev. Lett. 47, 236 (1981)
10.16 H.M. Gibbs, S.L. McCall, T.N.C. Venkatesan: Phys. Rev. Lett. 36, 1135 (1976)
10.17 C. Parigger, P. Hannaford, W.J. Sandle, R.J. Ballagh: Phys. Rev. A31, 4043 (1985)
10.18 S.L. McCall: Appl. Phys. Lett. 32, 284 (1978)
10.19 H.M. Gibbs, J.L. Jewell, S.S. Tarng, A.C. Gossard, W. Wiegmann: IEEE J. QE-17, 42 (1981)
10.20 M.M. Cheung, S.D. Durbin, Y.R. Shen: Opt. Lett. 8, 39 (1983)
10.21 G. Giusfredi: unpublished work
10.22 H.M. Gibbs, M.W. Derstine, F.A. Hopf, D.L. Kaplan, M.C. Rushford, R.L. Shoemaker, D.A. Weinberger, W.H. Wing: Talk FL2 in *Conference on Laer and Electro-Optics Technical Digest*, (IEEE, New York 1982). See also [Ref. 10.1, pp. 246–247]
10.23 M. Kitano, T. Yabuzaki, T. Ogawa: Phys. Rev. A24, 3156 (1981);
       F. Mitschke, J. Mlinek, W. Lange: Phys. Rev. Lett. 50, 1660 (1983);
       W. Lange, F. Mitschke, R. Deserno, J. Mlinek: J. Opt. Soc. Am. B1, 468 (1984)
10.24 J. Guckenheimer, P. Holmes: *Nonlinear Oscillations, Dynamical Systems, and Bifurcations of Vector Fields* (Springer, Berlin, Heidelberg 1983)

Subject Index